Motion-Based Recognition

Computational Imaging and Vision

Motion-Based Recognition

Edited by

Mubarak Shah
Computer Vision Laboratory,
Computer Science Department,
University of Central Florida,
Orlando, Florida, U.S.A.

and

Ramesh Jain
Electrical and Computer Engineering,
University of California,
San Diego, California, U.S.A.

KLUWER ACADEMIC PUBLISHERS
DORDRECHT / BOSTON / LONDON

A C.I.P. Catalogue record for this book is available from the Library of Congress.

ISBN 978-90-481-4870-7

Published by Kluwer Academic Publishers,
P.O. Box 17, 3300 AA Dordrecht, The Netherlands.

Sold and distributed in the U.S.A. and Canada
by Kluwer Academic Publishers,
101 Philip Drive, Norwell, MA 02061, U.S.A.

In all other countries, sold and distributed
by Kluwer Academic Publishers,
P.O. Box 322, 3300 AH Dordrecht, The Netherlands.

Printed on acid-free paper

To

My parents, who *taught* me to be tough, dedicated, confident, and simple.

The residents of Lakhi Ghulam Shah, a small village named after my grandfather, where I went to *School* up to the eighth grade.

——Mubarak Shah

To

All my *students*; Who facilitated my learning.

——Ramesh Jain

Contents

Contributors

- Adam Baumberg
  School of Computer Studies
  University of Leeds
  Leeds LS2 9JT, UK

- Michael J. Black
  Xerox Palo Alto Research Center
  Palo Alto, CA 94304, USA

- Aaron F. Bobick
  MIT Media Laboratory
  20 Ames St.
  Cambridge, MA 02139, USA

- Christoph Bregler
  Computer Science Division
  U.C. Berkeley
  Berkeley, CA 94720, USA

- James W. Davis
  MIT Media Laboratory
  20 Ames St.
  Cambridge, MA 02139, USA

- Larry Davis
  Computer Vision Laboratory
  Center for Automation Research
  University of Maryland
  College Park, MD 20742, USA

- Shawn Dettmer
  Computer Vision Lab
  Computer Science Department
  University of Central Florida
  Orlando, FL 32816, USA

- Charles R. Dyer
  Department of Computer Sciences
  University of Wisconsin
  Madison, WI 53706, USA

- Irfan Essa
  College of Computing
  Georgia Institute of Technology
  Atlanta, GA 30332-0280, USA

- Oscar N. Garcia
  Wright State University
  Dayton, OH 45435, USA

- Nigel H. Goddard
  Pittsburgh Supercomputing Center
  4400 Fifth Avenue
  Pittsburgh, PA 15213, USA

- Alan J. Goldschen
  The Mitre Corporation
  McLean, VA 22101, USA

- David Hogg
  School of Computer Studies
  University of Leeds
  Leeds LS2 9JT, UK

- Ramesh Jain
  Electrical and Computer Engineering
  University of California, San Diego
  San Diego, CA 92137, USA

- Shanon X. Ju
  Department of Computer Science
  University of Toronto
  Toronto, Ontario M5S 1A4, Canada

- Li Nan
  Computer Vision Lab
  Computer Science Department
  University of Central Florida
  Orlando, FL 32816, USA

- Randal Nelson
  Department of Computer Science
  University of Rochester
  Rochester, New York 14627, USA

- Stephen M. Omohundro
  NEC Research Institute, Inc.
  4 Independence Way
  Princeton, NJ 08540, USA

- Alex Pentland
  Room E15-383
  The Media Laboratory
  Massachusetts Institute of Technology
  20 Ames Street
  Cambridge MA 02139, USA

- Eric D. Petajan
  Bell Laboratories - Lucent Technologies
  Murray Hill, NJ 07974, USA

- Ramprasad Polana
  MicroStrategy Inc.
  8000 Towers Crescent Drive
  Vienna, VA 22182, USA

- K. Rohr
  Arbeitsbereich Kognitive Systeme
  Fachbereich Informatik, Universität Hamburg
  Vogt-Kölln-Str. 30
  D-22527 Hamburg, Germany

- Steven M. Seitz
  Department of Computer Sciences
  University of Wisconsin
  Madison, WI 53706, USA

- Mubarak Shah
  Computer Vision Lab
  Computer Science Department
  University of Central Florida
  Orlando, FL 32816, USA

- Thad Starner
  Room E15-383
  The Media Laboratory
  Massachusetts Institute of Technology
  20 Ames Street
  Cambridge MA 02139, USA

- Andrew D. Wilson
  MIT Media Laboratory
  20 Ames St.
  Cambridge, MA 02139, USA

- Yaser Yacoob
  Computer Vision Laboratory
  Center for Automation Research
  University of Maryland
  College Park, MD 20742, USA

Preface

The last few years have been a period of immense activities in multimedia and the Web. Suddenly video has become a central data in computing. Most of the activity has been, however, in communication, storage, and networking aspects of video. To make video achieve its true potential and become a central source of information, techniques for understanding video are essential. Many researchers in computer vision have been active in the area of Dynamic Scene Understanding. Techniques from dynamic scene understanding are likely to be central to the next generation human-machine interfaces and in video asset management. This book was motivated by our desire to bring several important techniques together in one place and make it available to researchers in relevant fields.

This book is suited for active researchers in Human-Computer Interfaces, Video Asset Management, and Computer Vision. Researchers in Psychology, Computer Graphics, and Multimedia may also find this book interesting. The book assumes the knowledge of basic concepts in computer vision, pattern recognition, computer graphics, and mathematics. An introductory course in computer vision should provide a reasonable background to understand the material in this book. The key concepts used in various chapters of this book include: Kalman filter, K-L transform, robust statistics, splines, moments, clustering, Hidden Markov Models, etc. The complexity of the material presented is at the advanced graduate level.

One of us (Mubarak Shah) wrote a survey paper on motion-based recognition with a M.S. student, Claudette Cédras in 1995. The paper summarized the work in this emerging area. Soon after the paper appeared, we started getting invitations from publishers to edit a collection of papers on this topic. During ICCV-95 in Boston, and IWFGR-95 in Zurich, we discussed the idea about the book with our colleagues in the field. The response was very enthusiastic. As a result, we submitted a short proposal about the book to Max Viergever, editor for Kluwer Academic Publisher's Computational Imaging and Vision series. His response was very encouraging. Consequently, we sent out the formal invitations, in January of 1996 to fifteen active researchers, for contributing chapters in the proposed book. All invited people accepted the invitation, and committed to contribute to the book. All chapters were received by June 1996. Each chapter was sent out to one referee for review. The authors were sent the reviewers' comments, and some instructions regarding the style and format in November of 1996. The final and revised version of chapters were received by December 15, 1996. We organized the chapters into three parts, and wrote an introduction chapter. The final product is in your hands.

We want to thank the contributors for meeting the deadlines, and making this book possible. We want to thank Max Viergever, and Paul Ross and Monique van der Wa at Kluwer Academic Publishers for their help. Mubarak Shah wants to thank his students Shawn Dettmer and Chris Efford, for providing help in latex; Don Harper, system programmer, for making ample disk space available for the book; Kari Mitchell, Susan Brooks, and Angel Robinson, for the secretarial help. Finally, we want to thank our families for their constant support throughout the course of the book.

Mubarak Shah

Ramesh Jain

March 4, 1997

VISUAL RECOGNITION OF ACTIVITIES, GESTURES, FACIAL EXPRESSIONS AND SPEECH: AN INTRODUCTION AND A PERSPECTIVE

MUBARAK SHAH
Computer Vision Lab
Computer Science Department
University of Central Florida
Orlando, FL 32816

AND

RAMESH JAIN
Electrical and Computer Engineering
University of California, San Diego
La Jolla, CA 92093-0407

1. Introduction

Computer vision has started migrating from the peripheral area to the core of computer science and engineering. Multimedia computing and natural human-machine interfaces are providing adequate challenges and motivation to develop techniques that will play key role in the next generation of computing systems. Recognition of objects and events is very important in multimedia systems as well as interfaces. We consider an object a *spatial* entity and an event a *temporal* entity. Visual recognition of objects and activities is one of the fastest developing area of computer vision.

Objects and events must be recognized by analyzing images. An *image* is an array of numbers representing the brightness of a scene, which depends on the camera, light source, and objects in the scene. Images look different from each other mainly due to the fact that they contain different objects.

[1]The first author acknowledges the support of DoD STRICOM under Contract No. N61339-96-K-0004. The content of the information herein does not necessarily reflect the position or the policy of the government, and no official endorsement should be inferred.

M. Shah and R. Jain (eds.), Motion-Based Recognition, 1–14.
© 1997 *Kluwer Academic Publishers.*

The most important visual attribute which distinguishes one object from the other is its *shape*. The shape represents the geometry of the object, which can be 2-D or 3-D. Edges, lines, curves, junctions, blobs, regions, etc. can be used to represent 2-D shape. Similarly, planes, surface patches, surface normals, cylinders, super-quadrics, etc. can be used to represent 3-D shape.

The shape plays the most dominant role in model-based *recognition*. In the most simple case of recognition, 2-D models and 2-D input are used. In this case, the computations are simple, but 2-D shape can be ambiguous; for several 3-D shapes may project to the same 2-D shape. In the the most general case of recognition, 3-D models and 3-D input are used, however, this approach is computationally expensive. In between is the approach in which 3-D models and 2-D input are used. In this case, the object's pose (3-D rotation and translation) of the model needs to be computed such that when it is projected on the image plane it exactly matches with the input.

A *sequence* of images represents how images change due to *motion* of objects, camera, or light source; or due to any of the two, or due to all three. Among all sequence of images, the most common sequences are the sequences which depict the motion of objects. The motion can be represented using difference pictures, optical flow, trajectories, spatiotemporal curvature, joint curves, muscle actuation, etc. Motion can also be represented by the deformation of shape with respect to time.

This book is about *motion-based recognition*. Motion-based recognition deals with the identification of objects or motion based on object's motion in a sequence [6]. In motion-based recognition, the motion is directly used in recognition, in contrast to the standard structure from motion (sfm) approach, where recognition follows reconstruction. Consequently, in some cases, it is not necessary to recover the structure from motion. Another important point here is that it is crucial to use a large number of frames; for it is almost impossible to extract meaningful motion characteristics using just two or three frames. There exists a distinction between *motion-based recognition* and *motion recognition*: motion-based recognition is a general approach that favors the use of motion information for the purpose of recognition, while motion recognition is one goal that can be attained with that approach.

This is an exciting research direction, which will have ever lasting effects on Computer Vision research. In the last few years, many exciting ideas have started appearing, but at disparate places in literature. Therefore, to provide the state of art in motion-based recognition at one place, we have collected key papers in this book. It consists of a collection of invited chapters by leading researchers in the world who are actively involved in this area.

The book is divided into three main parts: human activity recognition, gesture recognition and facial expression recognition, and lipreading. The next three sections introduce each part of the book, and summarize the chapters included.

2. Human Activity Recognition

Automatically detecting and recognizing human activities from video sequences is a very important problem in motion-based recognition. There are several possible applications of activity recognition. One possible application is in automated video surveillance and monitoring, where human visual monitoring is very expensive, and not practical. One human operator at a remote host workstation may supervise many automated video surveillance systems. This may include monitoring of sensitive sites for unusual activity, unauthorized intrusions, and triggering of significant events. Another area is detection and recognition of animal motion, with the primary purpose of discriminating it from the human motion in surveillance applications. Video games could be made more realistic using an activity recognition system, where the players control navigation by using their own body movements. A virtual dance or aerobics instructor could be developed that watches different dance movements or exercises and offers feedback on the performance. Other applications include athlete training, clinical gait analysis, military simulation, traffic monitoring, video annotations (most videos are about people) and human-computer interface.

Given a sequence of images, usually the first step in activity recognition is to detect a motion in a sequence; if the sequence represents a stationary scene, there is no point analyzing such a sequence for activity recognition. Difference pictures have been widely used for motion detection since the original paper of Jain et al [18]. The simple difference picture has some limitations, for instance, with covering and uncovering of image regions. Several ways to deal with this limitations has been reported in the literature. An example is to use the difference between the stationary background image and the current image, resulting only in the moving components. Here, the difficulty is how to obtain the background image. One possibility is to use the first image taken before the objects start moving in front of the camera as a background image. The other possibility is to reconstruct the background image for each pixel, by applying the median filter to all pixels gray values at a given location in a sequence; this is more general, but it is time consuming. In fact, in numerous application, the stationary background image is very easily available and can be used effectively.

The difference picture identifies the changed pixels in an image. The changed pixels need to be grouped into regions corresponding to the human body using a connected component algorithm. Also, due to the non-rigid nature of human body, there may be several small adjacent regions, which need to be merged.

Some variations of this change detection method include: computing the difference in a small neighborhood (e.g. 5 × 5) around a pixel instead of a pixel by pixel difference; or computing the difference between the current and the previous two and the next two frames (accumulated difference picture), as compared to the difference between just two frames (current and previous frames).

The optical flow has been widely used in motion analysis work to extract motion information from a sequence of images. Optical flow computes displacement vectors for each pixel between frames. One problem with optical flow, in general, is that it is susceptible to the aperture problem, which, in some conditions, only allows the precise computation of the *normal flow*, i.e. the component parallel to the gradient. Therefore, several researchers in motion-based recognition have employed normal flow, instead of full flow.

Human motion occurs at a variety of scales from fine to coarse (e.g., motion of lips to motion of legs). In the second chapter of this book, Yacoob and Davis present a scale-space based optical flow method for computing motion from sequences of humans. Their method first uses a simple model of optical flow, which assumes the optical flow to be constant, and employs robust estimation. However, human motion contains an acceleration component as well. Therefore, they extend their model to include acceleration. They also extend their approach to compute optical flow using parameterized models: affine and planar models.

Both change detection and optical flow are bottom-up approaches. An alternate approach is to use some *a priori* knowledge about the object being tracked. Snakes provide means to introduce a priori knowledge [20, 32]. The user-imposed constraint forces can guide the snakes near features of interest. Recently, another approach, using active shape models, was proposed, which use the Point Distribution Model (PDM) to build models by learning patterns of variability from a training set [8]. The major difference between snakes and PDM is that in PDM the shape can deform only when it is consistent to the training sets.

In chapter three of this book, Baumberg and Hogg present a method for automatically generating deformable 2-D contour models for tracking human body using the PDM approach. The conventional PDM approach requires a hand generated set of labeled points from the training images. The PDM approach is extended to automate the process of extracting a training set and building the model automatically. The results on tracking sequences of humans with scaling, change of view, translation and rotation are shown. In this approach, a B-spline is used to represent contour, and real time tracking is performed using a Kalman filter framework .

Many motions in nature are cyclic, like walking, a heartbeat, birds flying, a pendulum swinging, etc. The presence of cyclic motion in a sequence of images can reveal a lot about the object showing that type of motion. The cyclic motion detection problem was first introduced in Computer Vision by Allmen and Dyer in 1990. Based on studies of the human visual system, Allmen and Dyer [4] and Allmen [3] argue that cyclic motion detection: (1) does not depend on prior recognition of the moving object, i.e. cycles can be detected even if the object is unknown; (2) does not depend on the absolute position of the object; (3) needs long sequences (at least two complete cycles); (4) is sensitive to different scales, i.e. cycles at different levels of a moving object can be detected. They detected cyclic motion by identifying cycles in the curvature of a spatiotemporal curve using some form of A* algorithm. Polana and Nelson, and Tsai *et al* [30] proposed methods for cyclic motion detection using the Fourier transform. cyclic detection using Fourier transform. Polana and Nelson (see chapter five) first compute what they call the reference curve, which essentially is a linear trajectory of a centroid, the frames are aligned with respect to this trajectory. If the object presented some periodic motion, it will create some periodic gray level signals. They use the Fourier transform to compute periodicity of those gray level signals. The periodicity measure of several gray level signals are combined using some form of non maxima suppression.

The problem with these approaches to cyclic motion is that they only deal with strictly cyclic motions; the motions that repeat but are not regular are not dealt with. More important, these methods are not view invariant, they do not allow the camera to move. In chapter four, Seitz and Dyer describe view-independent analysis of cyclic motion using affine invariance. They introduce *period trace*, which gives a set of compact descriptions of near periodic signals.

There are two classes of approaches for human activity recognition: 3-D and 2-D. In a 3-D approach, some 3-D model of the human body and human motion is used. A projection of the model from a particular pose and particular posture in a cycle of activity is compared with each input image to recognize an activity. The advantage of this approach is that since a 3-D model is used it is not ambiguous. However, it is computationally quite expensive. Hogg [15] was the first to use a 3-D model-based approach for tracking humans. Instead of using a 3-D model and 2-D input, another approach is to use a 3-D input and 3-D model. Bobick *et al* [5] use 3-D point data to recognize ballet movements.

In 2-D approaches, no model of a 3-D body is used, only 2-D motion, e.g. optical flow, is employed to compute features in a sequence of frames to recognize activities. The advantage of this approach is that it is quite simple. In this book (chapters five to seven), the approaches of Polana and Nelson, Bobick and Davis, and Goddard are basically 2-D approaches.

Besides recognizing activities from motion, the motion can also be used to recognize people by their gaits. From our own experience, it is relatively easy to recognize a friend from the way he or she walks, even though this person is at a distance so that the face features are not recognizable. Rangarajan *et al* [25] describe a method for recognizing people by their gaits. They use trajectories of joints of a human body performing walking motions. Niyogi and Adelson [1] has proposed a method based on XT-trace to discriminate people. Boyd and Little [21] has described a method for recognizing people based on the phase of the weighted centroid of the human body.

Most approaches employ only one camera to capture the activities of a person. It may happen that due to self-occlusion, some parts of a person are not visible in the image, which may result in not having enough information for recognition. Another possible problem is that due to the limited field of view of one camera, the person may move out of the field of view of the camera. In order to deal with these difficulties some researchers have advocated the use of multiple cameras [22, 12, 27, 2]. However, with the introduction of additional cameras there is additional overhead, and we need to answer the following questions: How many views should be employed? Should information from all cameras be used or from only one? How to associate the image primitives among images obtained by multiple cameras? etc.

In chapter five of this book, Polana and Nelson classify motion into three categories: *events*, *temporal textures*, and *activities*. The events consists of isolated simple motions that do not exhibit any temporal or spatial repetition. Examples of motion events are opening a door, starting a car, throwing a ball, etc. The temporal textures exhibit statistical regularity but are of indeterminate spatial and temporal extent. Examples include ripples on water, the wind in the leaves of trees, a cloth waving in the wind. Activities consists of motion patterns that are temporally periodic and possess compact spatial structure. Four features of normal flow: the mean flow magnitude divided by its standard deviation, the positive and negative curl and divergence, the non-uniformity of flow direction, and the directional difference statistics in four directions are used to recognize temporal textures. Their activity recognition approach also uses normal flow. First, cycles in sequences of frame are detected using Fourier transform of reference curves. Each image is divided into a spatial grid of $X \times Y$ divisions. Each activity cycle is divided into T time divisions, and motion is totaled in each temporal division corresponding to each spatial cell separately. The feature vector is formed from these spatiotemporal cells, and used in a nearest centroid algorithm to recognize activities.

Bobick and Davis, in chapter six, first apply change detection to identify moving pixels in each image of sequence. Then MHI (Motion History Images) and MRI (Motion Recency Images) are generated. MRI basically is the union of all changed detected images, which will represent all the pixels which have changed in a whole sequence. MHI is a scalar-valued image where more recent moving pixels are brighter. In their system, MHI and MRI templates are used to recognize motion actions (18 aerobic exercises). Several moments of a region in these templates are employed in the recognition process. The templates for each exercise are generated using multiple views of a person performing the exercises. However, it is shown that during recognition only one or two views are sufficient to get reasonable results.

Next, in chapter eight, Goddard argues that the significant advances in vision algorithms come from studying extant biological systems. He presents a structured connectionist approach for recognizing activities. A scenario is used to represent movement, which is not based on 3-D, it is purely 2-D. The scenario represents a movement as a sequence of motion *events*, linked by the intervals. Input to the system is a set of trajectories of the joints of an actor performing an action. A hierarchy of models starting with the segment level is used, which include thigh, upper arm, fore arm; these segments are combined to get components like legs, arms. The components are combined into assemblies, and assemblies into objects. The system is triggered by the motion's events, which are defined as change in angular velocity of a segment, or a change in the orientation of a segment.

Finally, in the last chapter of this part of the book, chapter eight, Rohr presents a model-based approach for analyzing human movements. He uses cylinders to model the human body, and joint curves to model the motion. The joint curves were generated from the data of sixty normal people of different ages. This method is comprised of two phases. The first phase, called the initialization phase, provides an estimate for the posture and three-dimensional position of the body using a linear regression method; the second phase, starting with the estimate from the first phase, uses a Kalman filter approach to incrementally estimate the model parameters.

3. Gesture Recognition and Facial Expression Recognition

In our daily life, we often use gestures and facial expressions to communicate with each other. Gesture recognition is a very active area of research [16]. Some earlier work includes the work of Baudel *et al* [29] who used a mechanical glove to control the computer presentation; Fukomoto *et al* [14] also designed a method to guide a computer presentation, but without using any glove. Cipolla *et al* [7] used a rigid motion of a triangular region on a glove to control the rotation and scaling of an image of a model. Darrell and Pentland [9] used model views, which are automatically learned from a sequence of images representing all possible hand positions using correlation. Gesture models are then created, for each view, correlation is performed with each image of sequence, and the correlation score is plotted. Matching is done by comparing correlation scores. They were able to recognize hello and good bye gestures, they needed time warping, and special hardware for correlation.

There are two important issues in gesture recognition. One, *what* information about the hands is used, and *how* it is extracted from images? Two, how the variable length sequences are dealt with? Some approaches use gloves or markers on hands, consequently the extraction of information from images is very easy. In other approaches, the point based features (e.g., fingertips) are extracted, which carry the motion information, but do not convey any shape information. In some other approaches, however, a blob or region corresponding to a hand is identified in each image, and some shape properties of the region are extracted, and used in recognition. In addition, some approaches also use global features using the whole image (e.g., eigen vectors).

Most approaches to gesture recognition are 2-D. In these approaches, only 2-D image motion and 2-D region properties are used. Some 3-D gesture recognition approaches have also been proposed in which the 3-D motion and the 3-D shape of the whole hand, or fingers are computed and used in recognizing gestures. For example, Regh and Kanade [26] describe a model-based hand tracking system called *DigitEyes*. This system uses stereo cameras and special real-time image processing hardware to recover the state of a hand model with 27 spatial degrees of freedom. Kang and Ikeuchi [19] describe a framework for determining 3-D hand grasps. An intensity image is used for the identification and localization of the fingers using curvature analysis, and a range image is used for 3-D cylindrical fitting of the fingers. Davis and Shah [10] first identify the fingers of the hand and fit a 3-D generalized cylinder to the third phalangeal segment of each finger. Then six 3-D motion parameters (translation and rotation) are calculated for each model corresponding to the 2-D movement of the fingers in the image plane.

A gesture can be considered as a trajectory in a feature space. For example, a motion trajectory is a sequence of locations (x_i, y_i), for $i = 1 \dots n$, where n is the number of frames in a sequence. A motion trajectory can thus be considered as a vector valued function, that is, at each time we have two values (x, y). However, a single valued function is better suited for computations, and therefore parameterization of trajectories is necessary. A trajectory can be parameterized in several ways; for instance $\phi - S$

curve, speed and direction, velocities v_x and v_y, and spatiotemporal curvature. The first parameterization completely ignores time; two very different trajectories might have the same $\phi - S$ curves. The remaining parameterization are time dependent. Trajectories representing the same gesture or action may be of different length. The trajectories can be temporally aligned by non-linear time-warping. The difficulty with time warping methodology is that it is computationally extensive.

An alternate approach to time warping is to model a gesture as a Finite State Machine (FSM). Davis and Shah [11] identify four main phases in a generic gesture, and use a FSM to model these phases. The user is constrained to the following four *phases* for making a gesture. (1) Keep hand still (fixed) in start position until motion to gesture begins. (2) Move fingers smoothly as hand moves to gesture position. (3) Keep hand in gesture position for desired duration of gesture command. (4) Move fingers smoothly as hand moves back to start position.

Hidden Markov Models (HMM) have been known in the literature for a long time [24]. HMMs can be employed to build a stochastic model of a time-varying observation sequence by removing the time dependency. A HMM consists of a set of states, a set of output symbols, state transition probabilities, output symbol probabilities, and initial state probabilities [6]. The model works as follows. Sequences are used to train HMMs. Matching of an unknown sequence with a model is done through the calculation of the probability that an HMM could generate the particular unknown sequence. The HMM giving the highest probability is the one that most likely generated that sequence [17].

The FSM essentially is a simplified version of HMM with state transition probabilities equal to zero or one. The important difference is that FSM was generated by the user using the conceptual four phases of a generic gesture. However, a large number of training sequences are used to automatically generate HMMs.

In chapter nine, Bobick and Wilson use a time collapsing technique to achieve time invariance. They start with trajectories in a time-augmented configuration space and compute the principal curve using least squares fit of near by points. Trajectories and principal curve are slightly compressed in time, and a new principal curve is computed. The process is repeated until time is reduced to zero. Next, the sample points of the prototype curve are clustered. Each cluster is assigned a state. Bobick and Wilson define a gesture as a sequence of states. A dynamic programming algorithm is used for recognizing gestures.

Their approach is very similar to the HMM approach. One important difference is their use of the time-collapsing technique for converting trajectories. HMMs need a large set of training samples. However, Bobick and

Wilson claim rapid training with very few samples.

Starner and Pentland, in chapter ten, present a method for recognizing American Sign Language consisting of a 40 word vocabulary involving 500 sentences. They use tracking based on color (in one case the user had to wear colored gloves, in other case the color of skin is used). A blob corresponding to each hand is identified, and eight features (centroid, angle of orientation and eccentricity of bounding ellipse) are used in the HMM based approach.

The problem of recognizing facial expressions from video sequences is a challenging one. Since Ekman and Frisen's work [13] on *Facial Action Coding System* or FACS, there has been a lot of interest in facial expression recognition in Psychology and Computer Vision. For facial expression recognition, there are also two classes of approaches: 2-D and 3-D.

In chapter eleven, Black *et al* present a method for recognizing facial expression using 2-D motion. In their method, the rectangular windows corresponding to different parts of the face (eyes, brows, mouth) are identified, and optical flow is computed. The relative motion of parts of the face is used to recognize the expressions. Therefore, absolute motion of the face is first estimated to stabilize the motion of parts of the face in a warped sequence. The authors employ a parameterized model of optical flow using eight parameters for each patch. The eight parameters have the qualitative interpretation of the image motion in terms of translation, curl, deformation, divergence, and curvature. The approach then is extended to compute articulated motion. In articulated motion, each patch is connected to only one preceding patch and one following patch, for example a thigh patch may be connected to a preceding torso patch and following calf patch.

Essa and Pentland, in chapter twelve, present a 3-D approach for facial expression recognition. They employ a 3-D dynamic muscle based model of a face. Simoncelli's multi-scale, coarse-to-fine Kalman filter based method is used to compute optical flow. Using this optical flow, the velocities of each node of the face model are computed. Next, using a physically-based modeling technique, the forces that caused the motion are computed. Finally, a control theoretic approach is employed to obtain the muscle actuation.

The authors present two method for facial expressions. In the first one, they use peak actuation of each of 34 muscles between the application and release phases of the expression as the feature vectors. In another approach, they use spatio-temporal motion energy templates. In both methods they get a 98% recognition rate for five expressions: smile, surprise, rise eyebrows, anger, and disgust.

4. Lipreading

Automatic Speech Recognition (ASR) has been a hot research topic for a long time. Currently, there are a variety of ASR systems which are speaker independent, which recognize continuous speech, and which perform quite well. However, the recognition rate is not 100% yet. On the other hand, speech recognition by humans is extremely accurate. For, in addition to speech, we use lip movement, facial expression and sometimes gestures to supplement our speech. Starting from the original work of Petajan [23] on visual lipreading, there has been growing interest in visual lipreading (see [28], which provides the state of art in lipreading.). To some extent, it has been demonstrated that combined audio and video speech recognition outperforms any single modality (audio or video), particularly in noisy environments like offices, bars, parties, etc.

The problem of visual lipreading have proved to be more complex than speech recognition using sound. There are several reasons for this. First, the visual image contains more data than a sound sample. An image is two dimensional, and typically a 100 × 100 image captures the mouth region, which is 10, 000 bytes of data compared to one byte of data per sample for the speech signal. With a reasonable frame rate, at least 15 to 20 frames are needed to capture a single utterance, making this data even larger. Second, images are very sensitive to the motion of the speaker and his or her head movement, as compared to sound signal. Third, images are sensitive to lighting conditions, and there is not much contrast between the lips and the rest of the face.

In lipreading, there are two important issues: feature extraction and feature matching for recognition. The most simple approach is to use gray levels as a feature vector. As it has been shown in the context of face recognition using static images [31], the gray levels, surprisingly, perform quite well for the lipreading problem (see chapter fifteen). Another possible approach is to use the lip contour as a feature vector. The difficulty in this connection is the reliable extraction of a contour from a mouth image. Finally, some region properties of the mouth region can also be used as a feature vector.

Recognition and training strategies have smoothly migrated from speech recognition to the lipreading domain. Two popular methodologies are: Hidden Markov Models (HMM) and Neural Networks (NN). Both HMM and NN need a large training set, which in some cases is quite difficult to obtain. The advantage of HMM is that it is able to deal with variable length sequences. It is common to have sequences of the same utterance which are not of the same length.

In chapter thirteen, Bregler and Omohundro present a method for tracking lip contour, for the lipreading problem. One obvious way to do this is to use snakes. In the original snakes paper [20], lip tracking was demonstrated. However, the problem was that one of the authors had to put on lipstick in order to create an artificial contrast between lips and rest of the face to be useful in snakes. Otherwise, snakes get caught into local minima. In Bregler and Omohundro's method, the training lip images are initially labeled by the snakes algorithm; sometimes snakes select the boundary of an incorrect neighboring object, like the nose instead of the mouth, which are removed by hand. Next, using these lip contours a nonlinear manifold of all possible lip configurations is learned. During lip tracking, the contour from the learned manifold is found which maximizes the image gradient along the contour in the image. They report experiments with a four word vocabulary using only visual information with a 95% recognition rate. They also report the results on the database of six speakers on sequences of 3-8 letter names using audio only, and combined audio and video. The best recognition rate of about 55% is obtained with combined audio and video input in the presence of crosstalk.

Goldschen *et al*, in chapter fourteen, present an ambitious project for recognizing continuous speech using lipreading. Even though, their system gets only 25% recognition rate, it highlights several important issues in lipreading. They use fifteen features from the oral-cavity shadow images of the speaker. These images were taken by specialized apparatus, consequently detection of the oral-cavity images was not difficult. Their system uses HMMs for recognition, as does Bregler and Omohumdro's. The HMM's were trained to recognize a set of sentences using visemes, trisemes (triplets of visemes), and generalized trisemes (clustered trisemes).

Finally, in the last chapter of this book, chapter fifteen, Nan *et al* present a non-HMM method for lipreading. In their method, the feature vector consists of gray levels of all pixels in all images in a sequence. Several such vectors corresponding to the training sequences are used to compute eigenvectors (eigensequence), for each spoken letter. The recognition of an unknown sequence representing a spoken letter is performed by computing the ratio of energy of projection of the sequence on the model eigenspace and the energy of the sequence. The region around mouth area in each image is extracted by using a simple correlation method. A mouth template from the previous image is run through the current image, and correlation computed. The window around the location with the highest correlation is taken as the mouth region. They employ time warping to get equal length sequences. The recognition rate (85% for ten letter vocabulary) is good, even though their method is computationally expensive due to warping.

5. Conclusion

The papers presented in this book are representative of the directions being explored in dynamic vision that are very relevant to many applications in video analysis, video databases, and different advanced human-computer interfaces. Progress in these areas is essential to design computing environments needed in the next generation systems. On the other hand, techniques developed in these areas are in turn forcing computer vision researchers to look at computer vision as a dynamic system in which information exists at different abstraction levels. Clearly, we are at the beginning of an exciting research field. Many difficult problems have not even been articulated. We hope that this book will help researchers find relevant material at one place and encourage new researchers to explore some of the exciting and challenging directions presented in this book.

References

1. Adelson, E.H. and Niyogi, S. A. Analyzing and recognizing walking figures in XYT. In *IEEE CVPR-94*, pages 469–474, 1994.
2. A. Katkere, S. Moezzi, D. Kuramura, P. Kelly, and R. Jain. Towards video-based immersive environments. *Multimedia Systems Journal*, Spring 1997.
3. M. C. Allmen. *Image Sequence Description Using Spatiotemporal Flow Curves: Toward Motion-Based Recognition*. PhD thesis, University of Wisconsin–Madison, 1991.
4. M. C. Allmen and C. R. Dyer. Cyclic Motion Detection Using Spatiotemporal Surfaces and Curves. In *Proc. 10th Int. Conf. Pattern Recognition*, pages 365–370, 1990.
5. Bobick, A. F. and Campbell, L.W. Recognition of human body motion using phase space constraints. In *IEEE ICCV-95*, pages 624–630, 1995.
6. Cédras, C., and Shah, M. Motion-based recognition: A survey. *Image and Vision Computing*, 13(2):129–155, March 1995.
7. Cipolla, R.,Okamoto, Y. and Kuno, Y. Robust structure from motion using motion parallax. In *IEEE Proceedings on International Conference on Computer Vision*, 1993.
8. Cootes, T. J., taylor, C.J., and Graham, J. Active shape models-their training and application. *Computer Vision, and Image Understanding*, 61:38–59, 1995.
9. Darrell, T., and Pentland, A. Space-time gestures. In *CVPR*, pages 335–340. IEEE, 1993.
10. Davis, J., and Shah, M. Three-dimensional gesture recognition. In *Asilomar Conference on Signals, Systems, And Computers*, 1994.
11. Davis, J., and Shah, M. Visual gesture recognition. *IEE Proceedings Vision, Image and Signal Processing*, 141(2):101–106, 1994.
12. Davis, L. S. and Gavrila, D. M. 3-d model-based tracking of human upper body movement: A multi-view approach. In *IEEE CVPR-96*, pages 73–80, 1996.
13. Ekman, P. and Friesen, W. A.. *Facial Action Coding System*. Consulting Psychologist Press, 1978.
14. Fukumoto, M., Mase, K., and Suenaga, Y. Real-time detection of pointing actions for a glove-free interface. In *IAPR Workshop on Machine Vision Applications*, pages 473–476, December 1992.
15. D. C. Hogg. *Interpreting Images of a Known Moving Object*. PhD thesis, University of Sussex, 1984.

16. Huang, T., Pavlovic, V. Hand gesture modeling, analysis, and synthesis. In *Proc. International Workshop on Automatic Face and Gesture Recognition*, pages 73–79, 1995.

17. J. Schlenzig, E. Hunter and R. Jain. Recursive identification of gesture inputs using hidden markov models. In *Proc. IEEE Workshop on Applications of Computer Vision*, pages 187–194, 1994.

18. Jain, R.C., Militzer, D., and H.-H. Separating non-stationary from stationary scene components in a sequence of real world tv-images. In *IJCAI-77*, pages 612–618, 1977.

19. Kang, S.B., and Ikeuchi, K. Toward automatic robot instruction from perception – recognizing a grasp from observation. *IEEE Transactions of Robotics and Automation*, 9:432–443, August 1993.

20. Kass, M., Witkin, A., and Terzopoulos, D. Snakes: Active contour models. In *Proceedings of First International Conference on Computer Vision*, pages 259–269, London, 1987.

21. Little, J. and Boyd, J. Describing motion for recognition. *Int. Symposium on Computer Vision-95*, pages 235–240, November 1995.

22. Metaxas, D. and Kakadiaris, I.A. Model-based estimation of 3d human motion with occlusion based on active multi-viewpoint selection. In *IEEE CVPR-96*, pages 81–87, 1996.

23. Petajan, E. *Automatic Liprading to Enhance Speech Recognition*. PhD thesis, University of Illinois, 1984.

24. L.R. Rabiner and B.H. Juang. An Introduction to Hidden Markov Models. *IEEE ASSP Magazine*, pages 4–16, January 1986.

25. Rangarajan, K., Allen, Bill, and Shah, M. . Matching motion trajectories. *Pattern Recognition*, 26:595–610, July, 1993.

26. Rehg, J., and Kanade, T. Visual tracking of high dof articulated structures: an application to human hand tracking. In *ECCV*, pages 35–46, May 1994.

27. R. Jain and K. Wakimoto. Multiple perspective interactive video. In *Proceedings of the International Conference on Multimedia Computing and Systems*, pages 202–211. Computer Society Press, May 15-18 1995.

28. Stork, D. and Hennecke, M. *Speechreading by humans and machines*. Springer, 1996.

29. Baudel T. and Beaudouin-Lafon M. Charade: Remote control of objects using free-hand gestures. *CACM*, pages 28–35, July 1993.

30. Tsai, Ping-Sing, Keiter, K., Kasparis, T., and Shah, M. Cyclic motion detection. *Pattern Recognition*, 27(12), 1994.

31. Turk, M., and Pentland, A. Eigenfaces for recognition. *Journal of Cognitive Neuroscience*, pages 71–86, 1991.

32. Williams, D. and Shah, M. Greedy algorithm for active contour and curvature estimation. *Computer Vision, Graphics, and Image Processing*, pages 14–26, January, 1992.

Part I

Human Activity Recognition

ESTIMATING IMAGE MOTION USING TEMPORAL MULTI-SCALE MODELS OF FLOW AND ACCELERATION

YASER YACOOB AND LARRY S. DAVIS

Computer Vision Laboratory
Center for Automation Research
University of Maryland, College Park, MD 20742, USA

1. Introduction

Image motion estimation involves relating spatial and temporal changes in image intensity to estimates of image flow. Articulated and deformable motions such as those encountered in images of humans in motion give rise to image sequences having, instantaneously, a wide range of flow magnitudes ranging from very small sub-pixel motions, whose recovery is inhibited by typical signal to noise constraints, to very large multiple pixel motions that can be recovered using expensive correlation methods or multi-resolution approaches. Here, we focus on the problem of estimating dense image flow for image sequences in which instantaneous flows range from 2-4 pixels/frame down to $1/16 - 1/32$ pixel/frame. The practical problem, of course, is that we do not know a priori which parts of the image are moving with which speed. Our solution is a scale-space like solution [11] in which we estimate image flow over a wide range of temporal scales, and combine these estimates (using both spatial and temporal constraints) using a combination of robust estimation and parametric modeling as in [5].

To motivate both the problem and our proposed solution consider a pendulum arm moving in front of a camera. The image flow will vary depending upon the distance of the measured point from the hanging point (see Figure 1). As we move towards the pendulum hanging point the instantaneous flow becomes very small and will fall in the noise range of the imaging system. As a result, two frame estimation and subsequent integration of these flow measurements over time will be highly noisy. In the

[1]The support of the Defense Advanced Research Projects Agency (DARPA Order No. C635) under Contract N00014-95-1-0521 is gratefully acknowledged.

M. Shah and R. Jain (eds.), Motion-Based Recognition, 17–37.

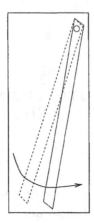

Figure 1.   Pendulum movement illustrating varying velocities along its motion path

context of human motion, the coincidence of lip motion with body and head motion, or the calf rotation around the knee create similar scale variations in the flow field.

The majority of published algorithms for estimation of image flow are based on two images (for a recent survey see [2]). Several approaches, however, consider the incremental estimation of flow [4, 13]; then, temporal continuity of the flow applied over a few images (for example, assuming constant acceleration) can improve the accuracy of the flow estimate. These approaches, are based on computations between consecutive images. Other approaches use velocity-tuned filters (i.e., frequency-based methods) [8, 10] to compute the flow, and can be extended to flow estimation from several frames. The use of scale-space theory to compute optical flow was recently proposed by Lindeberg [12]. The proposed algorithm focused on scale selection in the spatial dimension so that different size image structures lead to different selection of scales for flow computation. The algorithm estimates flow from two images and addresses spatial multiscales.

The approach presented in this chapter simultaneously estimates

- small and large flows (spatially and temporally).
- dense flow and acceleration.
- parameterized flow and acceleration.

The chapter is organized as follows. Section 2 illustrates, using an image sequence, the inadequacy of single scale flow estimation. In Section 3 we describe the motion model employed for estimating image flow from multiple scales. Section 4 describes the implementation issues and is followed by experimental results in section 5. Section 6 provides the extension of the model to compute image acceleration. In Section 7 we describe the estimation process when parameterized motion models are used. Finally, in section 8 discussion, summary and future applications of our approach are provided.

2. A Motivating Example

We will use *scale* = 1 to denote flow estimation between two consecutive images (i.e., the finest temporal resolution available), *scale* = 2 to denote flow estimation between images that are two frames apart, etc. To illustrate the limitation of image flow estimates from any single scale we employ an image sequence of an arm moving in front of a camera. The sequence was taken with a high-frame-rate camera (500 frames per second) which allows us to capture the natural rapid motion of the arm. The arm (see Figure 2) is moving in a pendulum-like motion (with the hand rotating around the arm during the motion) in front of a lightly textured background [2]. Notice that there is a shadow created by the hand, leading to non-zero flow estimates of the shadow as well as the arm. The arm's intensity pattern consists of two parts: the arm itself is highly textured (allowing better flow estimation) while the hand is somewhat uniform in brightness. Figure 2 shows eight images from the sequence (chosen two frames apart). While the motion of the arm between two frames is very small, it will become apparent when the flow estimates are shown.

Figure 3 shows the image flow magnitudes for six scales (falling on a geometric scale 1,2,4,8,16, and 32 frames apart). The finest scale provides detailed estimates of the flow magnitude at the hand but quite noisy estimates along the arm, while the coarsest scale results in accurate estimates along the arm but considerably blurred and inaccurate estimates on the hand.

Figure 4 is a rescaled version of Figure 3 in which the small flow values along the arm can be more easily observed. The flow estimation along the arm at fine scales is dominated by the noise of the imaging system. As scale increases, better estimates are computed along the arm at the cost of blurring the flow of the hand. As a consequence, if motion segmentation into parts is sought, the finest scale would result in highly fragmented components, while the coarsest scale would lead to highly inaccurate boundaries for the hand.

---

[2]The quadrants' boundary intensity variation of the background is because the video-camera consists of four separate A/D banks. As a result, flow estimation at the quadrant boundaries is inaccurate. The problem could be overcome by local gain compensation.

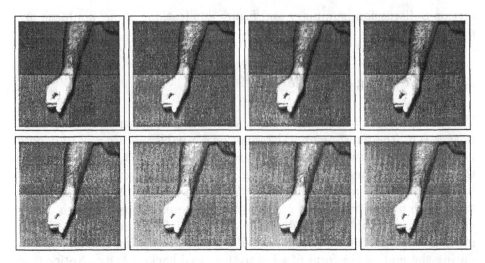

Figure 2.   Eight images (each two frames apart) from a long sequence of a moving arm

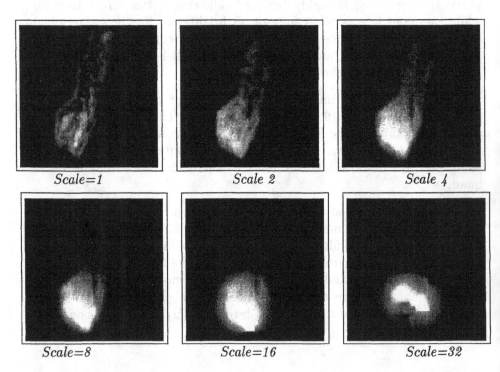

Figure 3.   Flow magnitudes at 1,2,4,8,16 and 32 scales (top left to bottom right respectively)

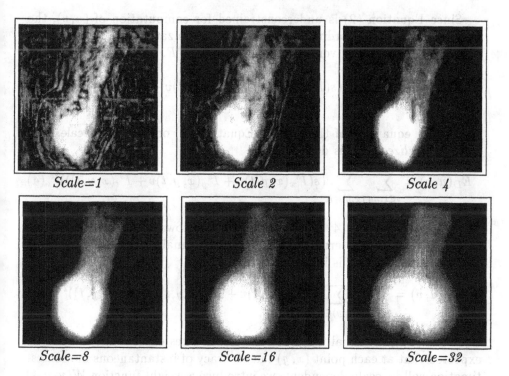

Figure 4. Enhanced flow display to show arm estimation at 1,2,4,8,16 and 32 scales (top left to bottom right respectively)

3. A Multi-scale Model for Image Flow Estimation

Let $I(x, y, t)$ be the image brightness at a point (x, y) at time t. The brightness constancy assumption at scale s is given by

$$I(x, y, t) = I(x + su\delta t, y + sv\delta t, t + s\delta t) \tag{1}$$

where (u, v) is the horizontal and vertical image velocity at (x, y), δt is small. We assume, for now, that the instantaneous velocity (u, v) remains constant during the time span $s\delta t$ (leading to a displacement $(su\delta t, sv\delta t)$). This assumption is less likely to hold with the increase of scale and can lead to violations of brightness constancy. Let the range of scales over which flow is to be estimated be $1, .., n$. Expanding Equation (1) using a Taylor Series approximation (assuming locally constant flow) and dropping terms results in

$$0 = s(I^s_x(x, y, t)u + I^s_y(x, y, t)v + I^s_t(x, y, t)) \tag{2}$$

where I^s is the s-th frame (forward in time relative to I) of the sequence, and I^s_x, I^s_y and I^s_t are the spatial and temporal derivatives of image I^s relative to I.

Since Equation (2) is underconstrained for computation of (u, v), the estimation of (u, v) is ordinarily posed as a minimization of a least squares error of the flow over a very small neighborhood, R, of (x, y), leading to

$$E(u, v, s) = \sum_{(x,y) \in R} (s(I^s_x(x, y, t)u + I^s_y(x, y, t)v + I^s_t(x, y, t)))^2 \quad (3)$$

We have n equations of the form of Equation (3) one for each scale. The *scale-generalized* error is defined as

$$E_D(u, v) = \sum_{s \in 1...n} \sum_{(x,y) \in R} (s(I^s_x(x, y, t)u + I^s_y(x, y, t)v + I^s_t(x, y, t)))^2 \quad (4)$$

Notice that Equation (4) biases the error term towards coarser scales due to the multiplication term s. Therefore, we normalize the error terms so that the minimization is in the form[3]

$$E_D(u, v) = \sum_{s \in 1...n} \sum_{(x,y) \in R} (I^s_x(x, y, t)u + I^s_y(x, y, t)v + I^s_t(x, y, t))^2 \quad (5)$$

Equation (5) gives equal weight to the error values of all scales. Since it is expected that at each point (x, y) the accuracy of instantaneous motion estimation will be scale-dependent, we introduce a weight function $\dot{W}(u, v, s)$ designed (see below) to minimize the influence of residuals of the relatively inaccurate scales. Equation 5 now becomes

$$E_D(u, v) = \sum_{s \in 1...n} \sum_{(x,y) \in R} (W(u, v, s)(I^s_x u + I^s_y v + I^s_t))^2 \quad (6)$$

Instead of the least squares minimization in Equation 6 we choose a robust estimation approach as proposed in [4], resulting in

$$E_D(u, v) = \sum_{s \in 1...n} \sum_{(x,y) \in R} \rho(W(u, v, s)(I^s_x u + I^s_y v + I^s_t), \sigma_e) \quad (7)$$

where ρ is a robust error norm that is a function of a scale parameter σ_e. Since the weight function $W(u, v, s)$ should also reflect the degree of accuracy of the flow estimation we redefine it to include a scaling parameter σ_w, $W(u, v, s, \sigma_w)$. The choice of the weighting function W should satisfy the following constraints:

– It should take on values in the range $[0..c]$, c typically chosen as 1.0 for computational convenience.

---

[3]The same effect could have been achieved by dividing the right side of Equation (2) by s for all scales prior to error summation.

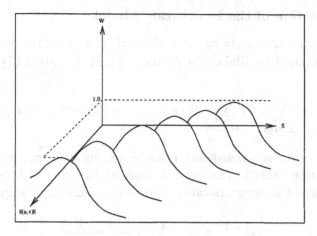

Figure 5. The weighting function as a function of s and flow magnitude $||(u, v)||$

– For a large σ_w, W should approach 1.0 regardless of (u, v) and s.
– Given σ_w, larger estimated flow (u, v) at point (x, y) should lead to higher weights for the lower scales of the error term $I^s_x u + I^s_y v + I^s_t$, while a small flow should lead to higher weights of the highest scales.

Figure 5 reflects qualitatively the desired shape of the weighting function for a fixed σ_w. It illustrates the weighting as a function of scale s and flow magnitude $||(u, v)||$ at (x, y).

The following Gaussian function satisfies the above requirements

$$W(u, v, s, \sigma_w) = e^{-(s - \frac{n}{(\alpha||(u,v)||^2 + 1.0)})^2 / 2\sigma_w^2} \qquad (8)$$

where $||(u, v)||^2$ is the squared magnitude of the current flow estimate at (x, y), and α is a constant. Notice that when $||(u, v)||^2 << 1.0$ the maximal weight occurs at the highest scale n, while higher values of $||(u, v)||^2$ lead to a maximal weight at lower scales; specifically the Gaussian is centered at $\frac{n}{\alpha||(u,v)||^2 + 1.0}$. The scale parameter σ_w determines the width of the Gaussian, and the constants α and 1.0 can be changed to further shift the maximal weight scale location. The application of the weighting function in the estimation is as follows: in the first iteration, all scales are given equal weight (1.0) by selecting a large σ_w. Afterwords, iteratively, the estimates are refined by decreasing σ_w.

This temporal multi-scale procedure is accompanied by a spatial coarse-to-fine strategy [3] that constructs a pyramid of the spatially filtered and sub-sampled images (for more information see [4]) and computes the flow initially at the coarsest level and then propagates the results to finer levels.

4. Computation of the Multi-scale Model

The computational aspects of the multi-scale model follow, generally, the approach proposed by Black and Anandan [4, 5]. Equation (7) can be rearranged as

$$E_D(u,v) = \sum_{(x,y)\in R} \rho \left(\sum_{s\in 1...n} W(u,v,s,\sigma_w)(I^s_x u + I^s_y v + I^s_t), \sigma_e \right) \quad (9)$$

The minimization is carried out using a descent method, Simultaneous Over-Relaxation (SOR). The minimization of $E_D(u,v)$ with respect to u and v is achieved using an iterative update equation, so at step $n+1$

$$u_{x,y}^{(n+1)} = u_{x,y}^{(n)} - \omega \frac{1}{T(u_{x,y})} \frac{\partial E_D}{\partial u_{x,y}} \quad (10)$$

where $0 < \omega < 2$ is an overrelaxation parameter which is used to overcorrect the estimate of $u_{x,y}^{(n+1)}$ at stage $n+1$ (a similar treatment is used for v). The value of ω determines the rate of convergence. The term $T(u_{x,y})$ is an upper bound on the second partial derivative of E_D

$$T(u_{x,y}) \geq \frac{\partial^2 E_D}{\partial^2 u_{x,y}} \quad (11)$$

To achieve a globally optimal solution the robust error norm ρ is started with a large enough scale parameter σ_e to find a solution using the SOR technique. Then this process is iteratively repeated while decreasing σ_e and starting with the last estimate. The choice of a large enough σ_e guarantees convexity of the error function at the beginning of the process, which is followed by the use of the Graduated Non-Convexity method developed in [7]. The iterated decrease in σ_e reduces the influence of the outlier measurements and thereby refines the estimates.

The robust error norm chosen is the one proposed by Geman-McClure [9] (see Figure 6)

$$\rho(x,\sigma_e) = \frac{x^2}{\sigma_e + x^2} \quad (12)$$

and its derivative

$$\psi(x,\sigma_e) = \frac{2x\sigma_e}{(\sigma_e + x^2)^2} \quad (13)$$

The weighting function W is controlled by a scale parameter σ_w. In each SOR cycle the value of σ_w is started high enough so that all scales equally contribute to the solution; it is then decreased and the SOR is repeated using the last estimate of flow to determine the value of W.

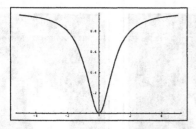 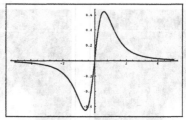

Figure 6. The robust error norm ρ (left) and its derivative (right) taken from Ge-man-McClure

5. Experimental Results

In the following figures we show the results of image flow computation when $\sigma_w = 20.0$ and is decreased at a rate of 0.85 for five iterations, and $\sigma_e = 100.0$ and is decreased also at a rate of 0.85 for 40 iterations. The computation is performed over 16 scales.

Figure 7 illustrates the weights at several scales during the computation of image flow (the brighter the intensity the higher the weight; weights across scales were normalized in these images to allow for comparisons). At $scale = 1$ only the hand area is given high weights while the arm and the background are given very low weights. As the scale increases the weights are increased along the arm and the background while a decrease on the hand gradually takes place. At the highest scale ($scale = 16$) the hand's weight is very low while the arm and the background receive a high weight. Figure 8 shows the effect of the iterative refinement of the weighting function W for $scale = 1$ (the finest scale) on the relative weights for different regions. The values are normalized across the five images to allow comparison. Notice that the first iteration gives high weights to the hand, and the weights given to the arm and the background are somewhat significant. The fifth iteration also gives high weights to the hand while the arm and the background have the lowest weight, and they are much lower than after the first iteration. This behavior is reversed when we consider the coarsest scale, $scale = 16$ (see Figure 9).

Figure 10 (top and middle rows) shows graphs of the individual scale flow magnitudes computed along a line drawn down the center of the arm (bottom right). These graphs correspond to the scale computations shown in Figure 3. Since the arm is *approximately* moving like a pendulum with the hand simultaneously rotating around the wrist (see Figure 15), the flow should increase slowly along the arm then jump considerably on the hand. This is clearly visible in these graphs. The dip in these graphs (occurring between 125-145) is a result of the intensity discontinuity of the four quadrants of the camera. Figure 10 also shows the multi-scale flow magnitude

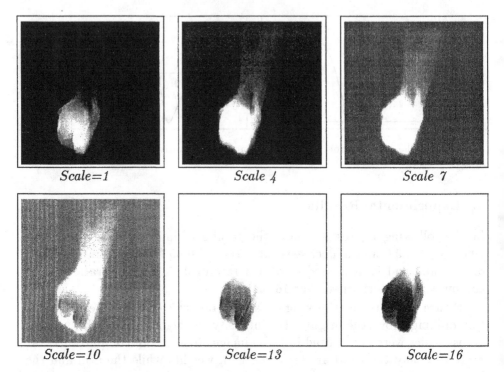

Figure 7. The weighting function W as computed at the scales 1,4,7,10,13 and 16 scales (top left to bottom right respectively) expressed as an intensity image.

results (bottom left). The flow boundary is quite sharp and the corresponding flow magnitude along the line is shown (bottom center), it measures a very smooth change in the flow along the arm and significant increase at the hand (with maximal flow at the finger).

In order to compare the performance of single scale ($scale = 1$) and multi-scale flow estimation, we generated a sequence of images using a synthetic flow model where we have ground-truth data. Figure 11 (top) shows an image of a person during a walking activity. The synthetic sequence is generated by warping the image patch of the "calf" foreward according to a multi-scale parameterized motion model for several frames (assuming constant velocity). The estimated multi-scale (12 scales) flow magnitudes are shown (top right). A quantitative comparison along a line on the "calf" between the original flow (bottom, solid line) the single scale flow (dotted line) and the multi-scale (dashed line). The multi-scale estimate is closer to the synthetic flow than the single scale estimation. Accurate recovery of the flow is actually limited by interpolation side effects in generating the synthetic motion.

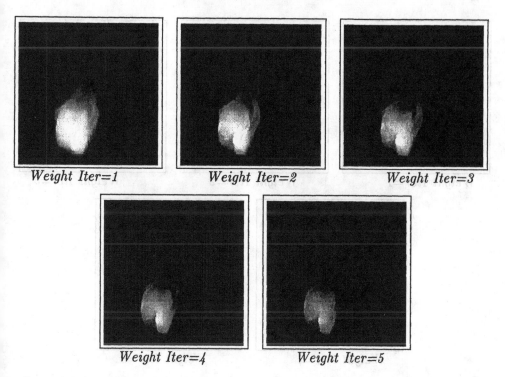

Figure 8. The weighting function W at scale 1 (finest scale) as evolved in five iterations

6. Estimation of Image Acceleration

The scale-generalized brightness constancy assumption given in Equation (1) assumes constant flow at all scales. This can be extended to include acceleration models. Let the image flow as a function of scale s be $(u(s), v(s))$. Then the brightness constancy assumption at scale s becomes

$$I(x, y, t) = I(x + \sum_s u(s)ds, y + \sum_s v(s)ds, t + s) \qquad (14)$$

As a special case, if image motion is assumed to be subject to a constant acceleration, the flow can be given by

$$u(s) = x_0 + x_1 s \qquad (15)$$

$$v(s) = x_2 + x_3 s \qquad (16)$$

where x_1 and x_3 are the horizontal and vertical acceleration terms. Note that in the context of a long sequence this model supports a piecewise constant acceleration assumption. If acceleration fluctuations within the scales involved in the estimation are small or fall within the performance

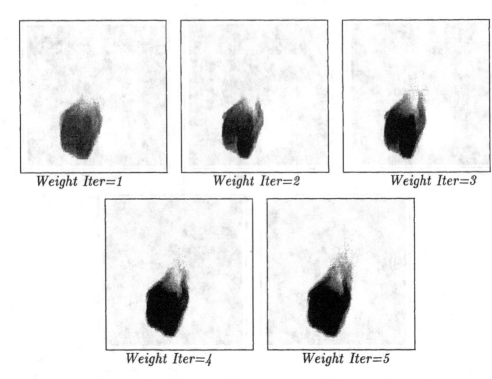

Weight Iter=1   *Weight Iter=2*   *Weight Iter=3*

Weight Iter=4   *Weight Iter=5*

Figure 9.   The weighting function W at scale 16 (coarsest scale) as evolved in five iterations

range of the robust estimator (about 35%-40% outliers) this model holds. This flow model leads to a brightness constancy assumption of the form

$$I(x,y,t) = I(x + \sum_{i \in 1...s}(x_0 + x_1 i), y + \sum_{i \in 1...s}(x_2 + x_3 i), t + s) \qquad (17)$$

Using a Taylor Series expansion and dropping terms (including scale normalization) we arrive at

$$0 = I^s{}_x(x_0 + x_1\frac{s+1}{2}) + I^s{}_y(x_2 + x_3\frac{s+1}{2}) + I^s{}_t \qquad (18)$$

The new scale-generalized error function is given by

$$E_D(u,v) = \sum_{s \in 1...n} \sum_{(x,y) \in R} \rho(W(u,v,s,\sigma_w)(I^s{}_x(x_0 + x_1\frac{s+1}{2}) + \qquad (19)$$

$$I^s{}_y(x_2 + x_3\frac{s+1}{2})) + I^s{}_t), \sigma_e)$$

A minimization procedure similar to that discussed in Section 4 can be employed to estimate these parameters.

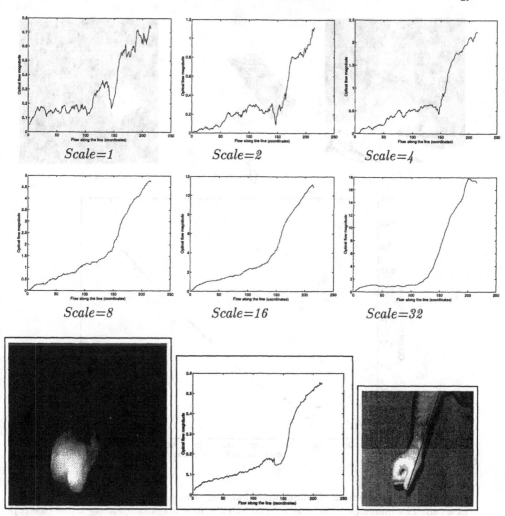

Figure 10. The flow magnitude along a line (bottom right) computed using a single scale (*s* is chosen from 1,2,4,8,16 and 32 scales; top and middle rows), the multi-scale flow magnitudes (bottom left), and the multi-scale flow magnitudes along the line (bottom center)

Figure 12 shows the dense flow and acceleration estimated for a book-falling sequence (see also Figure 16). The top row shows the the weighting function's values assigned for each scale (normalized to enhance the contrast). At low scales the book's region is assigned high weight while the background is assigned very low weight. This is reversed as scale is increased, so at the top scale the motion of the book is so large that little weight is given to the book area. The bottom row shows the dense veloc-

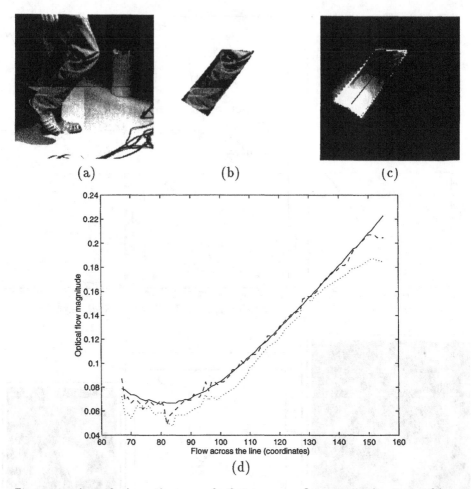

Figure 11. A synthetic motion example that compares flow magnitudes on a real image
of a calf. The image (see (b)) was warped and the flow magnitudes along a line (see (c))
are shown as a solid line (see (d)). The estimates of flow magnitudes using 1 and 12 scales
over the same line are shown ((d), dotted and dashed lines, respectively).

ity magnitude (left) and the vertical and horizontal accelerations (center
and right, respectively). Notice that the estimated horizontal acceleration
is almost uniformly zero.

7. Parameterized Image Motion Models

Dense flow computation generates large data sets that may not be easily
used in higher level vision tasks. Recently, it has been demonstrated that
parameterized flow models can provide a powerful tool for reasoning about
image motion between successive images (see the chapter by Black et al. in

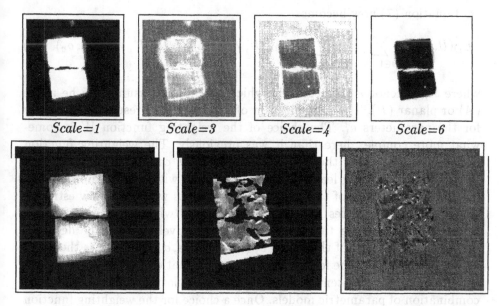

Figure 12. The weights (upper row) at scales 1, 3, 4 and 6, respectively (out of 6 scales), and the flow magnitude and vertical and horizontal accelerations (bottom row) for a falling book.

this volume). The multi-scale flow estimation algorithm can be extended in a straightforward way to parameterized models of image flow. In this section we describe the extension of the muti-scale framework to affine and planar parameterized image motion models.

Recall that the flow constraint given in Equation (2) assumes constant flow over a small neighborhood around the point (x, y). Over larger neighborhoods, a more accurate model of the image flow is given by low-order polynomials [1]. For example, affine motion is given by

$$U(x, y) = a_0 + a_1 x + a_2 y \tag{20}$$
$$V(x, y) = a_3 + a_4 x + a_5 y \tag{21}$$

where a_i's are constants and (U, V) is the instantaneous velocity vector. The affine model generally holds when the region modeled is not too large or subject to significant perspective effects. A more general model is the planar motion model [1] which is an approximation to the flow generated by a plane moving in 3-D under perspective projection. The planar motion model is given by

$$U(x, y) = a_0 + a_1 x + a_2 y + a_6 x^2 + a_7 xy \tag{22}$$
$$V(x, y) = a_3 + a_4 x + a_5 y + a_6 xy + a_7 y^2 \tag{23}$$

Equation (7) now becomes

$$E_D(U,V) = \sum_{s \in 1...n} \sum_{(x,y) \in A/P} \rho(W(U,V,s,\sigma_w)(I^s_x U + I^s_y V + I^s_t), \sigma_e) \quad (24)$$

where A/P denotes the region in which the flow is assumed to be affine (A) or planar (P). The minimization of Equation (24) results in estimates for the parameters a_i. The choice of the weighting function W is somewhat more complex here than it was previously. The weighting function can be designed using the current flow estimates computed by the model (U,V). This weighting leads to different weights within the region according to the magnitude of the flow so that at points where the flow estimate is low the coarser scales will be more dominant while the larger flow estimates will determine the finer scales. Alternatively, W can be designed using the parameters of the model a_i (i.e., $W(\bar{a}, s, \sigma_w)$ where $\bar{a}$ is the set of model parameters). The former leads to a computation based on a weighted combination of spatio-temporal derivatives while latter leads to a weighted combination of parametric models. Once a choice for the weighting function has been made the computation of the parameters of the model follows the approach proposed in [4].

In the examples in this chapter we adopt the weighted combination of parametric models. Recall that the parameters of the affine and planar models capture several aspects of the region's motion. The translation in the horizontal and vertical directions are captured by a_0 and a_3, respectively, while the *divergence, deformation* and *curl* are captured by the following equations (see Figures 13 and 14)

$$\text{divergence} \overset{\triangle}{=} a_1 + a_5 = (U_x + V_y), \quad (25)$$

$$\text{curl} \overset{\triangle}{=} -a_2 + a_4 = -(U_y - V_x), \quad (26)$$

$$\text{deformation} \overset{\triangle}{=} a_1 - a_5 = (U_x - V_y) \quad (27)$$

In the following examples the estimation of flow within a region is motivated by computing a particular motion of the region, therefore W is designed to be most sensitive to a particular subset of these parameters. For example, if the translation of the region is of most interest then the parameters a_0 and a_3 can be substituted as $||(a_0, a_3)||$ for $||(u,v)||$ in Equation (8).

Figure 15 shows the results of parameterized flow estimation over the hand region of the moving arm over a long sequence (about 540 frames). The parameterized flow is used to automatically track the hand motion throughout the sequence similar to [6] (assuming an initial manual hand segmentation in the first image). The frame numbers are shown with the images. The left graph shows the horizontal and vertical translations (solid

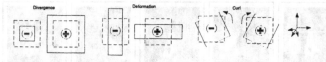

Figure 13. The figure illustrates the motion captured by the various parameters used to represent the motion of the regions. The solid lines indicate the deformed image region and the "–" and "+" indicate the sign of the quantity.

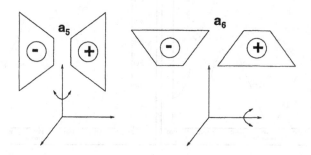

Figure 14. The meaning of the parameters a_5 and a_6

and dashed lines, respectively) and the right graph shows the *curl* of the hand. Notice the smoothness and robustness of these figures.

Parameterized flow models can also be extended to include acceleration. The extension of the affine model requires that the motion parameters across scales be dependent on the scale so that a_i becomes $a_i(s)$. Assuming a constant acceleration for these parameters, the models now become

$$U(x,y) = (a_0 + ax_0's) + (a_1 + ax_1's)x + (a_2 + ax_2's)y \qquad (28)$$
$$V(x,y) = (a_3 + ax_3's) + (a_4 + ax_4's)x + (a_5 + ax_5's)y \qquad (29)$$

where ax_0', ax_3' are the linear horizontal and vertical acceleration components of the motion and the ax_1', ax_2', ax_4' and ax_5' are acceleration components that can be related to angular, divergence and deformation accelerations.

Figure 16 describes an experiment in which the acceleration of a falling book is estimated from an image sequence.[4] Notice that although the book is falling vertically, a small horizontal motion component is present (observe the change of the upper left corner of the book relative to the white stripes). The bottom left graph of Figure 16 shows the horizontal and vertical velocity computed for the sequence (dashed and dotted lines, respectively), and the *predicted* vertical velocity (solid line) based on the velocity computed

---

[4]The book is manually segmented in the first image and tracked automatically afterwords using our multi-scale parameterized flow model.

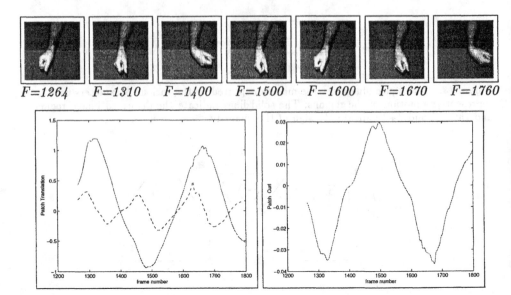

Flow results without acceleration model, vertical and horizontal translation
(left, solid and dashed lines, respectively) and curl (right).

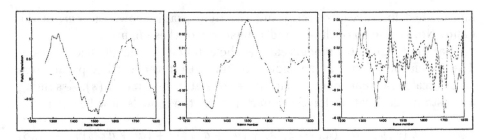

Figure 15.   Seven images (frames indicated below each image) of a long sequence of
the arm in motion (top).Flow results without acceleration model, vertical and horizontal
translation(left, solid and dashed lines, respectively) and curl (center row left to right).
Flow results with an acceleration model, vertical and horizontal translation (left, solid
and dashed lines, respectively), curl (bottom left to right) and vertical and horizontal
acceleration (bottom right).

at the first frame and the *average* acceleration in the first ten frames. The
graphs suggest that the inclusion of acceleration in the image motion model
is valuable in predicting the real motion for a significant amount of time.

   Figures 15 and 17 show the acceleration of the hand and the foot in
motion in addition to the flows for long image sequences. In Figure 17 the
foot performs a pendulum-like motion. The maximal velocity is achieved
midway in its trajectory. The acceleration of the foot does not satisfy the
constant acceleration model, thus the estimates are not very smooth.

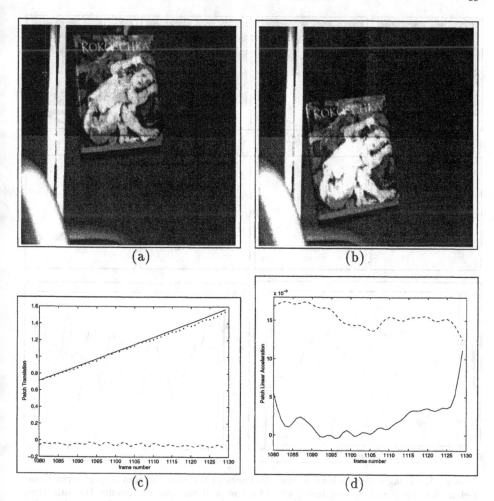

Figure 16. Two images (60 frames apart) of a falling book (a)-(b), the vertical and horizontal velocities (dashed and dotted lines, respectively (c)), and the predicted vertical velocity from the starting velocity and average acceleration, $ax_3{}'$, in the first 10 frames (solid line, (c)) and the vertical and horizontal accelerations (dashed and solid lines respectively, (d))

8. Discussion of the Multi-scale Framework

The proposed multi-scale approach for computing optical flow and acceleration introduced explicit temporal models for image intensity and flow changes. As demonstrated in several image sequences, a multi-scale framework can increase the accuracy of instantaneous motion estimates and recover simultaneously both flow and acceleration.

Algorithms for motion estimation can be quite noisy since they are based on local operators applied over very small neighborhoods between

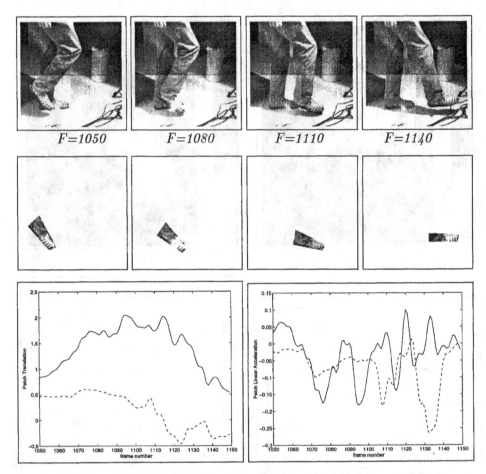

Figure 17. Images from a long image sequences of tracking the foot region using parametric flow (top and middle rows) and the flow and acceleration results (bottom row, left and right respectively) where the horizontal and vertical values are drawn using solid and dashed lines respectively.

two images. Temporal smoothing was proposed by [5] in a regularization framework; in contrast our multi-scale approach employs well-understood scale-space concepts [11, 12] to create smooth estimates. Due to the integrative nature of the multi-scale estimation, motion smoothing is achieved through the estimation process.

In this chapter we developed a new multi-temporal framework for computing flow and acceleration in images. Both dense and parameterized representations were employed and demonstrations on long image sequences were provided. This approach is an extension of the popular brightness-constancy assumption to a temporal scale-space domain. It provides for higher accuracy over a wider range of flows in image sequences. The performance of the approach was demonstrated on long image sequences of human motion and other rapid motions.

Future work can directly employ these models to build representations of image change. The modeling of image acceleration may support qualitative and quantitative reasoning about image change as demonstrated by the falling book example.

Acknowledgments

We would like to thank Michael Black for providing the code for optical flow estimation from two images.

References

1. Adiv G. Determining three-dimensional motion and structure from optical flow generated by several moving objects. *IEEE PAMI*, Vol. 7(4), July 1985, pp. 384-401.
2. S.S. Beauchemin and J.L. Barron. *The Computation of Optical Flow.* ACM Computing Surveys, Vol. 27, No. 3, September 1995, 433-467.
3. J.R. Bergen, P. Anandan, K.J. Hanna and R. Hingorani. *Heirarchical model-based motion estimation.* In G. Sandini, editor, Proc. of Second European Conference on Computer Vision, ECCV-92, Vol. 588 of LNCS-Series, 237-252, Springer-Verlag, May 1992.
4. M.J. Black and P. Anandan. *A Frame-work for Robust Estimation of Optical Flow.* ICCV 1993, Berlin, 231-236.
5. M.J. Black and P. Anandan. *The Robust Estimation of Multiple Motions: Parametric and Piecewise-Smooth Flow Fields.* 1994 Revision of Technical Report P93-00104, Xerox PARC, December 1993.
6. M.J. Black and Y. Yacoob. *Recognizing facial expressions in image sequences using local parameterized models of image motion.* Proceedings of the International Conference on Computer Vision, 1995, Boston, MA, 374-381.
7. A. Blake and A. Zisserman. *Visual Reconstruction* The MIT Press, Cambridge, Massachusetts, 1987.
8. D.J. Fleet and A.D. Jepson. *Computation of Component Image Velocity from Local Phase Information.* IJCV, Vol. 3, No. 4, 77-104.
9. S. Geman and D.E. McClure. *Statistical Methods for Topographic Image Reconstruction.* Bulletin of the International Statistical Institute, LII-4:5-21, 1987.
10. D.J. Heeger. *Optical Flow Using Spatio-temporal Filters.* IJCV, Vol. 1, 279-302.
11. T. Lindeberg. *Scale-Space Theory in Computer Vision.* Kluwer Academic Publishers, 1994.
12. T. Lindeberg. *A Scale Selection Principle for Estimating Image Deformations.* Technical Report, Stockholm University, CVAP 196, 1996.
13. A. Singh. *Incremental Estimation of Image Flow Using a Kalman Filter.* IEEE Proceedings of the Workshop on Visual Motion 1991, , Princeton, 36-43.

LEARNING DEFORMABLE MODELS FOR TRACKING THE HUMAN BODY

ADAM BAUMBERG AND DAVID HOGG
School of Computer Studies
University of Leeds
Leeds LS2 9JT, UK

1. Tracking Human Motion

The analysis and automatic interpretation of images containing moving non-rigid objects, such as walking people, has been the subject of considerable research in the field of computer vision and pattern recognition [1, 2, 3, 4, 5]. In order to build fast and reliable systems some kind of prior model is generally required. A model enables the system to cope with situations where there is considerable background clutter or where information is missing from the image data. This may be due to imaging errors (e.g. blurring due to motion) or due to part of an object becoming hidden from view.

Conventional approaches to the problem of tracking non-rigid objects, (e.g. [1, 3]), require complex hand-crafted models which are not easily adapted to different problems. A more recent approach, the Point Distribution Model (PDM), uses training information to build models for image analysis [6, 7, 8]. The PDM is generated by performing a statistical analysis on a training set of examples of the object of interest. Each example is represented by a set of landmark points placed at equivalent locations on each of the training examples.

In this chapter, we show how this approach can be extended by building flexible 2D contour models, *automatically*, from sequences of training images. Efficient methods are described for using the resulting models for real time tracking using optimal linear filtering techniques, based on a statistical tracking framework [9]. The resulting object-specific tracker produces a very compact representation of the object shape which could be used directly to say something about the posture of the object. Sequences of recovered shape parameters could also be used for motion-based recognition

39

M. Shah and R. Jain (eds.), Motion-Based Recognition, 39–60.
© 1997 *Kluwer Academic Publishers.*

tasks (such as gesture and gait analysis). The model described here is essentially 2D but is trained on a selection of arbitrary views. The variation in shape due to different viewpoints is treated as flexibility in 2D shape, allowing the model to be used for tracking over the range of viewpoints for which it was trained.

2. Building a shape model

The "Point Distribution Model" has proven a useful mechanism for building a compact shape model from training examples of a class of shapes [10, 7, 11]. In our work, the class of shapes of interest are the 2D silhouettes of walking pedestrians viewed from a variety of angles. The conventional PDM requires a human operator to hand generate a set of labeled points (corresponding to particular features) from training images of the object of interest. The training shapes are aligned using an iterative technique, Generalised Procrustes Analysis [12]. The aligned training set is then analysed using "Principal Component Analysis" to obtain the mean shape and a relatively small set of significant modes of variation.

A natural extension of this approach is to automate the whole process, extracting a training set and building the model automatically. The problem is to extract a reasonably consistent shape-vector from real training images containing examples of the object. A simple approach to this problem is described in this chapter. By processing large amounts of data, the effects of noise due to occlusion and mis-segmentation are reduced and a relatively simple segmentation scheme can be employed. In order to extract a large training set of shape-vectors, the processing of image data needs to be reasonably fast.

The control points of a uniform cubic B-spline are used as a shape-vector, since a spline is convenient for data approximation and fast to render. One of the advantages of this approach over the conventional PDM method is that there is no need to estimate positions of features that do not appear in a particular training image. For example, the left arm may be hidden behind the person's body (self-occlusion) and estimating the appropriate feature points becomes difficult and prone to error. By regarding the silhouette as an abstract closed continuous shape (with no landmark features) an automatic procedure can be applied.

2.1. OUTLINE OF THE METHOD

Our method utilises a simple segmentation scheme to extract training shapes of the object of interest. In our experiments image sequences were taken with a fixed camera allowing the use of a background subtraction scheme to segment the moving objects. Other segmentation schemes (e.g.

using colour) could be used for different applications. Once the system has been trained and a contour model generated, the tracker can be applied to more difficult images where a simple segmentation scheme is unreliable.

The system we have implemented takes live video images from a static camera, processes them and extracts fixed length shape-vectors representing the moving objects in the scene. The data is then analysed off-line to generate a model.

There are four main stages to model generation :-

- *Image preprocessing* to obtain a binary background-foreground image.
- *Outline extraction* to obtain the boundary of each foreground shape.
- *Shape vector calculation* to obtain an item of training data.
- *Off-line analysis* to build the shape model.

2.2. IMAGE PREPROCESSING

In order to segment the moving objects from a sequence of images, a background subtraction scheme is used. The background image is continually updated (median filtering over time) to account for changing lighting conditions. An approximation to the median filter can be used (e.g. using a running average [13]).

For a given image frame $I_k(x, y)$, a differenced image $\Delta I_k(x, y)$ is calculated by pixel-wise absolute subtraction from the reference background image $I_{\text{ref}}(x, y)$. i.e.

$$\Delta I_k(x, y) = |I_k(x, y) - I_{\text{ref}}(x, y)|$$

To reduce the effects of noise in the images, the differenced image is blurred using a standard Gaussian blur filter and the resulting blurred difference image $\Delta I'_k$ thresholded to produce a binary image.

The threshold value is chosen to be fairly low to ensure the foreground objects are well defined connected regions in the binary image, although this increases the effects of noise. These regions correspond to moving objects in the scene (in this case, walking pedestrians).

Results of this processing are shown in figure 1.

2.2.1. *Further noise reduction*
When there is poor contrast between the moving object and the background, fragmentation can occur, resulting in several foreground regions where there should only be one connected region. This effect can be reduced by applying morphological filters (see for example [14]) to fill these "gaps". In order to join regions separated by k pixels along an extended boundary, the following operations are performed on the binary image

- k successive dilation operations

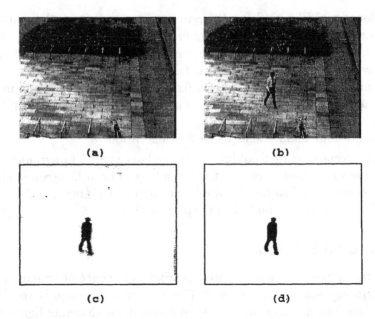

Figure 1. Image Preprocessing: (a) background image, (b) video input image, (c) differenced image, (d) blurred and thresholded image

— k successive erosion operations

2.3. SHAPE VECTOR CALCULATION

2.3.1. *Extracting silhouettes*

Each connected component in the binary image is potentially the silhouette of moving object within the scene. The connected pixels are segmented from the binary image using a standard "flood-fill" algorithm (see for example [15]). Constraints on the region size are used to filter out artifacts due to noise (for instance, a sudden change in lighting conditions may produce very large regions). Each feasible region that satisfies the size constraints is traced (clockwise) to produce a chain of boundary points which is used as the basis for the calculation of a training shape-vector.

2.3.2. *Finding a point of reference on the boundary*

In order to proceed, a fixed reference point on the closed boundary (which will have an associated parameter value $u = 0$) is required. A consistent method is required which is not highly susceptible to noise.

The method used is to find the principal axis (i.e. the axis through the centroid of the boundary points which minimises the sum of the perpendicular distances to that axis). The reference point is chosen to be the upper

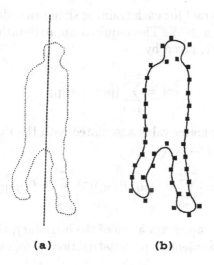

(a) (b)

Figure 2. Extracting a spline:(a) data points with principal axis,(b) resulting spline.

(in terms of image coordinates) of the two points where the axis crosses the boundary.

A more general method may select the intersection point that is nearest to the centroid, or some other suitable choice. In the case where the principal axis may be inappropriate (e.g. vehicles viewed from the side and head on), a very simple method may use the upper-most point (in image coordinates) over the complete contour. The resulting model would be highly specific to a fixed viewing angle (e.g. images of cars taken from a fixed camera).

The boundary points are now reordered so that the first point is the reference point and approximated by a cubic B-spline (see figure 2).

2.3.3. *Approximating with a cubic B-spline*
Each shape will be approximated by a parametric curve $\mathbf{p}(u)$ with $0 \leq u \leq N$ defined using N control points. The control points are combined into a $2N$ dimensional shape-vector, $\mathbf{x}$ given by

$$\mathbf{x} = (R_0, S_0, R_1, S_1, \ldots, R_{N-1}, S_{N-1})^T$$

where (R_i, S_i) are the coordinates of the i'th control point. A point on the curve $\mathbf{p}(u)$ is obtained using a $2 \times 2N$ interpolation matrix $H(u)$ as follows

$$\mathbf{p}(u) = H(u)\mathbf{x}$$

The interpolation matrix $H(u)$ is constructed from a sum of cubic B-spline basis functions (see [16]).

Previous steps extract for each training shape an ordered set of n boundary points, q_i where $n \gg N$. The required approximating spline minimises the error function, erf, given by

$$\text{erf} = \sum_{i=0}^{n-1} |p(u_i) - q_i|^2$$

where u_i is some parameter value associated with the i'th data point. Using standard methods we obtain

$$x = [\sum_{i=0}^{n-1} H^T(u_i)H(u_i)]^{-1} \sum_{j=0}^{n-1} H^T(u_j)q_j \qquad (1)$$

For a reasonably close approximation of the boundary, the parameter values can be set using an arc-length parametrisation as follows:

$$u_k = \begin{cases} 0 & \text{for } k = 0 \\ \lambda \sum_{i=1}^{k} |q_i - q_{(i-1)}| & \text{for } k > 0 \end{cases} \qquad (2)$$

where $q_n \equiv q_0$ and λ is chosen such that $u_n = N$.

Calculating the spline control points requires inverting a $2N \times 2N$ matrix for each shape (equation 1). In order to avoid this computationally expensive step, $n' = wN$ new data points are calculated by linear interpolation (where w is a whole number, typically set to 8). These new data points correspond to the *fixed* uniformly spaced parameter values:

$$u_k \equiv k\left(\frac{N}{n'}\right)$$

The precise method is given in section 2.3.4. Using these new data points and the fixed parameter values, allows us to precalculate $[\sum H^T(u_i)H(u_i)]^{-1}$ once for the whole training set. This efficiently produces a uniform B-spline with the control points placed at approximately uniformly spaced intervals along the contour.

2.3.4. *Selecting Data Points for Spline Approximation*

Conventionally the parameter values associated with data points q_k are based on the Euclidean distances between points (as in equation 2). This leads to a set of values u_k corresponding to the data values q_k. The discrete mapping u_k to q_k can then be extended to a continuous mapping u to $q(u)$ by linear interpolation. Hence given $u_k \leq u \leq u_{k+1}$ it is possible to interpolate $q(u)$ using

$$q(u) = \left(\frac{u - u_k}{u_{k+1} - u_k}\right) q_{k+1} + \left(\frac{u_{k+1} - u}{u_{k+1} - u_k}\right) q_k$$

Regularly spaced parametric values (between 0 and N) are chosen to find n' new data points. These new data points can now be efficiently approximated with a uniform cubic B-spline.

2.4. COMPONENT ANALYSIS OF THE DATA

The training shapes are first rotated and scaled into a "normal" frame so that the principal axis is vertical and the vertical height of each shape is fixed. More generally, a more complex Generalised Procrustes scheme can be used [6, 12].

Principal Component Analysis of the data is performed using a suitable orthogonality condition for the space of B-spline shape-vectors. The mean shape vector, $\overline{\mathbf{x}}$ is calculated in the usual way as well as the covariance matrix, S using

$$\overline{\mathbf{x}} = E(\mathbf{x})$$
$$S = E(\mathbf{x}\mathbf{x}^T) - \overline{\mathbf{x}}\,\overline{\mathbf{x}}^T$$

where $E(\ldots)$ is the standard expectation (or mean) operator over the training set.

A basis of "eigenshapes", $\mathbf{e_i}$ can now be constructed by solving the eigenproblem

$$S\mathcal{H}\mathbf{e_i} = \lambda_i\mathbf{e_i} \qquad (3)$$
$$\mathbf{e_i}^T\mathcal{H}\mathbf{e_i} = 1$$

where S is the training set covariance matrix and by convention the eigenvalues are ordered with $\lambda_0 >= \ldots >= \lambda_{2N-1}$. The matrix, $\mathcal{H}$ is defined by

$$\mathcal{H}_{i,j} = \int_{u=0}^{N} H(u)^T H(u)du$$

$\mathcal{H}$ can be regarded as a finite element mass matrix (assuming unit uniform density) or alternatively as the measurement inverse covariance matrix for an ideal sensor (see [17]). We can easily show that the eigenshapes are linearly independent, $\mathcal{H}$-orthogonal vectors representing the modes of variation of the training data. Writing each training shape vector as a weighted sum of eigenshapes added to the mean

$$\mathbf{x} = \sum_{i=0}^{2N-1} b_i\mathbf{e_i} + \overline{\mathbf{x}}$$

we can also show that the variance of each shape coefficient b_i over the training set is simply λ_i and that the shape coefficients are linearly uncorrelated.

Figure 3.   Some training shape-vectors

Hence, the eigenshapes with relatively small eigenvalues are considered in-significant allowing a considerable reduction in dimensionality.

2.5. MODEL BUILDING RESULTS

Images were taken from 15 minutes of live video of a quiet pedestrian scene containing some moving vehicles. Training shapes were automatically segmented and approximated by a cubic B-spline with 40 control points. Each shape was reflected about the principal axis resulting in over 700 training shapes (corresponding to approximately 50 people). Some of these shapes are shown in figure 3.

2.6. MODES OF VARIATION

Each training shape-vector had 80 parameters (40 control points in 2D). The first 18 modes accounted for 90% of the variance of the training data. The largest 19 eigenvalues are displayed in figure 4. The graph shows that there is a small set of significant eigenvalues and a larger set of relatively small eigenvalues. The small eigenvalues correspond to insignificant modes of variation that can be subsequently ignored.

The first $m = 18$ eigenvectors can thus be used as an orthonormal basis for the model space of allowable shapes. Some of the significant modes of variation of the shape-vectors are shown in figures 5 and 6.

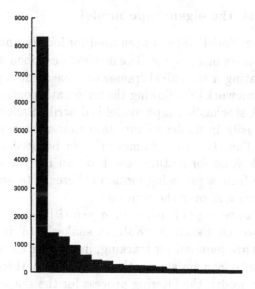

Figure 4. Plot of the first 19 eigenvalues

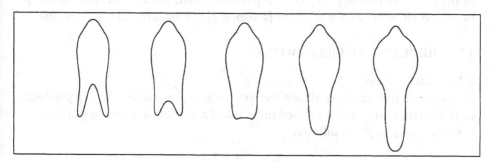

Figure 5. The effect of varying the component of the first mode by ±1.5 standard deviations

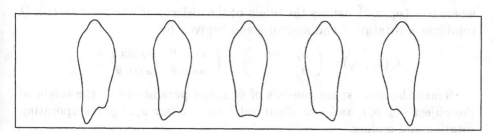

Figure 6. The effect of varying the component of the second mode by ±1.5 standard deviations

3. Tracking with the eigenshape model

The "Active Shape Model" [8] can been used for locally updating shape parameters to fit features in an image. The method described here extends this work by incorporating a statistical framework, based on a general B-spline curve tracking framework [9], allowing the automatic control of spatial (and temporal) scale. A stochastic shape model is described allowing the contour to deform more easily in modes of variation that vary significantly within the training set. The statistical framework can be used to automatically control the search scale for feature search on an individual frame as well as over successive frames (allowing motion coherence to be exploited when "lock" has not been lost over the contour).

A significant advantage of using an *a priori* linear shape model over more general approaches such as "Kalman snakes" [18] is that only a few shape parameters are required for tracking, improving the speed of the system. Furthermore, it can be shown that assuming a theoretical isotropic continuous sensor model, the filtering process for the shape parameters can be decoupled allowing each shape parameter to be filtered independently. In practice the (decoupled) system performs well, even when these assumptions are violated and a discrete (anisotropic) measurement process used.

3.1. THEORETICAL FRAMEWORK

3.1.1. *State Space*

The eigenshape analysis allows the vector $\mathbf{x}$ representing the 2D positions of N control points to be defined in terms of a subset of m shape parameters $\mathbf{b} = (b_0, ..., b_{m-1})^T$ as follows:

$$\mathbf{x} = P\mathbf{b} + \overline{\mathbf{x}}$$

where P is a $2N \times m$ matrix of eigenshapes and $\overline{\mathbf{x}}$ is the mean shape-vector.

A contour in the model frame is projected into the image frame by rotation, scaling and translation using

$$\mathbf{p}(u) = QH(u)\mathbf{x} + \mathbf{o} = QH(u)[P\mathbf{b} + \overline{\mathbf{x}}] + \mathbf{o}$$

where $\mathbf{o} = (o_x, o_y)^T$ defines the origin of the object and the 2 x 2 matrix Q combines a rotation θ and scaling s and is given by

$$Q(a_x, a_y) = \begin{pmatrix} a_x & -a_y \\ a_y & a_x \end{pmatrix} = \begin{pmatrix} s\cos\theta & -s\sin\theta \\ s\sin\theta & s\cos\theta \end{pmatrix}$$

Hence the state space consists of m shape parameters b_i, the origin of the object (o_x, o_y), and two alignment [1] parameters a_x, a_y, incorporating rotation and scaling.

---

[1] where the term alignment refers to rotation and scaling but not translation.

3.2. STOCHASTIC MODEL

3.2.1. *Shape parameters*

The shape part of the state vector is modeled as a simple discrete stochastic process as follows:

$$b_i^{(k)} = b_i^{(k-1)} + w_i^{(k-1)} \quad w_i^k \sim N(0, \mu_i)$$

where b_i^k models the i'th parameter value at frame k and the noise term w_i^k is a zero-meaned, normally distributed random variable with variance μ_i . A dynamic model (assuming constant rate of change) was considered but found to be less stable with no appreciable improvement in performance. The underlying assumption of the shape model is that the shape parameters vary independently (the noise process is isotropic). This is reasonable as over the training set:

$$E(b_i b_j) = 0 \qquad i \neq j$$

As the variance of b_i over the training set is equal to λ_i, it is natural to set the noise terms using

$$\mu_i = \mu \lambda_i$$

where μ is an undetermined shape parameter and is typically set to 0.05. This allows the more significant shape modes to vary more freely. The system parameter μ determines how easily the shape can deform with a value of $\mu = 0$ corresponding to complete rigidity.

3.2.2. *Origin*

The origin of the object is assumed to undergo uniform 2D motion with an additive random noise process (in both velocity and acceleration). This can be expressed by the differential equation:

$$\frac{d}{dt} \begin{pmatrix} o_x \\ \dot{o}_x \end{pmatrix} = \begin{pmatrix} \dot{o}_x \\ 0 \end{pmatrix} + \begin{pmatrix} v_x \\ w_x \end{pmatrix}$$

where $v_x \sim N(0, q_v)$ and $w_x \sim N(0, q_w)$. A corresponding model is used for o_y. Over a walk cycle, changes in shape affect the position of the origin. This can be accommodated by the random velocity term v_x, allowing the underlying "smooth" motion to be recovered. In the absence of sensor measurements this "smoothed" estimate of velocity determines the motion of the origin.

The parameters q_v, q_w may be learned from training sequences of a given scene using standard system identification methods, see for example [19]. However, this may be undesirable for tracking applications where we wish to track *unusual* object motion which is not well represented by a typical training set. In such cases the parameters are set heuristically (e.g. by considering the values required to "break" the filter).

3.2.3. *Alignment parameters*

The alignment parameters a_x, a_y are assumed to be constant with added system noise as described by the equation:

$$\begin{pmatrix} a_x^{(k+1)} \\ a_y^{(k+1)} \end{pmatrix} = \begin{pmatrix} a_x^{(k)} \\ a_y^{(k)} \end{pmatrix} + \mathbf{w}_a^{(k)}$$

where $\mathbf{w}_a^{(k)}$ is a zero-meaned Gaussian random variable with a 2×2 covariance matrix $C_a^{(k)}$. This covariance matrix can be set using a prior stochastic model (for the scale s and orientation θ) of the form

$$\begin{aligned} s^{(k+1)} &= s^{(k)}(1+ds) \\ \theta^{(k+1)} &= \theta^{(k)} + d\theta \end{aligned}$$

where $ds \sim N(0, q_s)$ and $d\theta \sim N(0, q_\theta)$. Assuming $d\theta$, ds are small we can linearise at each frame to obtain

$$C_a^{(k)} \approx \frac{1}{\hat{a}_x^2 + \hat{a}_y^2} Q(\hat{a}_x, \hat{a}_y) \begin{pmatrix} q_s & 0 \\ 0 & q_\theta \end{pmatrix} Q^T(\hat{a}_x, \hat{a}_y)$$

where $\hat{a}_x$, $\hat{a}_y$ are the current estimates for the random variables $a_x^{(k)}$ and $a_y^{(k)}$. The system parameters q_s and q_θ are variances for the relative change in scale and the change in rotation angle between image frames. Again, these can be learned from training data or fixed heuristically.

3.3. THEORETICAL BASIS FOR DECOUPLING SHAPE FILTER

Assume an arbitrarily dense set of isotropic measurements are made for curve points $\mathbf{p}(u)$ with covariance spectral density r with the position and alignment parameters fixed. Denoting the $m \times m$ covariance matrix for the m shape parameters at frame k by $C_b^{(k)}$, the Kalman filter covariance update is given by

$$[C_b^{(k)}]^{-1}(+) = [C_b^{(k)}]^{-1}(-) + \int_{u=0}^{N} P^T H^T(u) Q^T r^{-1} Q H(u) P du \quad (4)$$

The update equation 4 is simplified using the following trivial results
- The alignment matrix is a scaled rotation matrix. Hence $Q^T Q = s^2 I$
- By definition $\int H^T H = \mathcal{H}$

— The matrix of eigenvectors P was derived such that $P^T \mathcal{H} P = I$ (equation 3)

The update equation simplifies to

$$[C_b^{(k)}]^{-1}(+) = [C_b^{(k)}]^{-1}(-) + s^2 r^{-1} I$$

Hence assuming $C_b^{(k)}(-)$ is diagonal, then after applying the measurements, the updated covariance matrix is still diagonal. Assuming $C_b^{(0)}$ is diagonal and noting the diagonal form of the stochastic shape model described in section 3.2.1, the covariance matrix is always diagonal. Thus the system can be decoupled into m independent 1D Kalman filters [2]. Each filter holds an estimate for the shape parameter $\hat{b}_i$ and an associated variance σ_i representing the uncertainty of the estimate.

3.4. DISCRETE ANISOTROPIC MEASUREMENT MODEL

3.4.1. Observed features

Although the object shape is represented by a continuous curve, it is convenient to sample the curve at n_{sub} regularly spaced points with parameter values u_i where u_0 is chosen randomly at each iteration.

At each new frame, measurements are made by searching for suitable image features, along the normal to the estimated contour, n_j, at the sample point $p_j = p(u_j)$. When a reference background image is available a signed contrast measure can be used. Otherwise, a simple sobel edge measure is used. The search is restricted to a specified search window derived from the 2×2 positional covariance matrix C_j obtained from the state estimate covariances. The positional covariance is the sum of a covariance term for each filter. i.e.

$$C_j = C_{\text{align}}(u_j) + C_{\text{origin}}(u_j) + \sum C_{b_i}(u_j)$$

where the terms C_{align}, C_{origin}, C_{b_i} will be determined in subsequent sections.

The search window size ρ_j is now given by

$$\rho_j = 2(n_j^T C_j^{-1} n_j)^{-\frac{1}{2}}$$

The point of maximum contrast or edge strength within the search region is retained as an observation q_j for the j'th sample point. For each point measurement there is an associated measurement variance v_j which

---

[2]i.e. a filter with a 1 dimensional state space

is set proportional to the square size of the search window at that point using

$$v_j = \frac{n_{\mathrm{sub}} n_{\mathrm{iter}}}{N} (c\rho_j)^2$$

where n_{iter} iterations are performed per image frame and c is a system parameter that determines how fast the tracker converges. If there is no significant point of contrast found within the search window (the feature has been lost) then no measurement is made (i.e. $v_j^{-1} = 0$). The normal search direction results in a coupling between the x and y components of the measurements and the 2×2 pointwise *measurement* inverse covariance matrix A_j is thus given by

$$A_j = v_j^{-1} \mathbf{n_j} \mathbf{n_j}^T$$

3.5. TRACKING FILTER

In the interests of speed, the shape, alignment and translation parameters are filtered independently by assuming all other parameters are fixed at their current estimates. For each image frame an iterative scheme is used with a fixed number of iterations. At each iteration measurements are taken and the filters are updated appropriately to get new estimates for the state parameters and new filter covariances.

3.5.1. *The origin filters*
The x and y components of the origin are filtered independently. The measurement model for the x component of the origin, assuming the other parameters are fixed at their current estimates, is given by

$$[\mathbf{p_j}']_x = o_x + [\mathbf{v_j}]_x$$

and similarly for the y component

$$[\mathbf{p_j}']_y = o_y + [\mathbf{v_j}]_y$$

where the noise term $\mathbf{v_j}$ is a zero-meaned random variable with covariance A_j^{-1}. The "origin measurements" $\mathbf{p_j}'$ are calculated from the observed contour points $\mathbf{q_j}$ using

$$\mathbf{p_j}' = \mathbf{q_j} - Q(\hat{a}_x, \hat{a}_y) H(u_j)(P\hat{\mathbf{b}} + \bar{\mathbf{x}})$$

The measurements are applied to each Kalman filter in the usual way and between image frames the filter is used to predict the position of the origin at the next frame. Variances $V(o_x)$ and $V(o_y)$ for the estimates $\hat{o}_x$

and $\hat{o}_y$ are available from the Kalman filter covariance matrices. These variances determine the positional covariance term C_{origin} given by

$$C_{\text{origin}}(u_j) = \begin{pmatrix} V(o_x) & 0 \\ 0 & V(o_y) \end{pmatrix}$$

3.5.2. The alignment filter

If the origin and shape parameters are fixed at their current estimates, the measurement model for the alignment parameters is given by

$$\mathbf{q_j} - \hat{\mathbf{o}} = \begin{pmatrix} s_{jx} & -s_{jy} \\ s_{jy} & s_{jx} \end{pmatrix} \begin{pmatrix} a_x \\ a_y \end{pmatrix} + \mathbf{v_j}$$

where $s_j = H(u_j)(P\hat{\mathbf{b}} + \overline{\mathbf{x}})$.

The estimates $\hat{a}_x$, $\hat{a}_y$ and the 2×2 covariance matrix are updated with the corresponding Kalman filter equations. The alignment parameters are *not* independent. The 2×2 filter covariance C_a determines the positional covariance term for each contour point given by

$$C_{\text{align}}(u_j) = \begin{pmatrix} s_{jx} & -s_{jy} \\ s_{jy} & s_{jx} \end{pmatrix} C_a \begin{pmatrix} s_{jx} & s_{jy} \\ -s_{jy} & s_{jx} \end{pmatrix}$$

3.5.3. The shape filters

The theoretical isotropic sensor model results in a decoupled Kalman filter. This provides a theoretical motivation for filtering each shape parameter independently, even when the anisotropic discrete measurement process is used.

Writing $\Delta \mathbf{p_j} = \mathbf{q_j} - \hat{\mathbf{p}}_j$, the measurement model for the i'th shape filter is given by

$$\Delta \mathbf{p_j} = Q(\hat{a}_x, \hat{a}_y)H(u_j)\mathbf{e_i}(b_i - \hat{b}_i) + \mathbf{v_j}$$

Measurements are combined to obtain an observed change for each shape parameter Δb_i and an associated measurement inverse variance r_i^{-1}. Explicitly

$$r_i^{-1} = \sum_{j=0}^{n_{\text{sub}}-1} \mathbf{e_i}^T H^T(u_j) A_j H(u_j) \mathbf{e_j} \tag{5}$$

$$\Delta b_i = r_i \sum_{j=0}^{n_{\text{sub}}-1} \mathbf{e_i}^T H^T(u_j) A_j \Delta \mathbf{p_j}$$

Each shape filter maintains a variance σ_i for the shape parameter estimate $\hat{b}_i$ which determines the positional covariance term C_{b_i} for a sample point. Explicitly,

$$C_{b_i}(u_j) = Q H(u_j) \mathbf{e}_i \sigma_i \mathbf{e}_i{}^T H^T(u_j) Q^T$$

3.6. INCREASING STABILITY

In the "Active Shape Model" approach, the model space of feasible shapes is constrained by ensuring the vector **b** lies within a hyper-ellipsoid (so that the Mahalanobis distance to the mean shape is constrained). A similar increase in stability can be achieved by applying a virtual input of 0 applied to each shape filter at the start of each image frame with measurement variance for each shape parameter set to λ_i. This approach has several advantages.

- In the absence of image measurements (e.g. due to occlusion) the variance of each shape parameter estimate will rapidly increase. The virtual input ensures each shape variance is bounded. This is valid, because the object shape is assumed to have come from the same (Gaussian) distribution as the training data. Hence the virtual input adds prior knowledge to the system.
- The virtual input will "pull" the solution towards the mean shape before image measurements are made. This discourages *a priori* unlikely solutions but does not prevent them if there is strongly supporting image evidence.

3.7. INITIALISATION

The tracking mechanism requires initial estimates for the state parameters for each tracked object. In our implementation a crude motion detector is used using background subtraction on a subsampled image, updated periodically. The camera is assumed to be fixed (at this initialisation step). The result of this processing is a binary "differenced image" where the foreground pixel regions correspond to moving objects in the scene. Objects that are already being tracked are cleared from this image. The remaining significantly sized connected components are assumed to correspond to new moving objects.

For each of these connected components the bounding box is calculated and the estimated shape is initialised to the mean shape aligned vertically, centred at the origin of the bounding box and scaled to the height of the bounding box.

Figure 7. Results on 1st test sequence

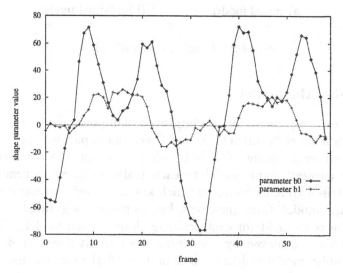

Figure 8. The first 2 shape parameters over successive image frames

3.8. SOME TRACKING RESULTS

An example of running the tracking scheme on a previously unseen test sequence are illustrated in figure 7. The frames are shown left to right top to bottom with every 10th frame displayed. The estimated contour has been superimposed on each image frame. The sequence was processed at 14.75 Hz without dedicated image processing hardware.

The most significant two shape parameters recovered from this sequence are shown in figure 8. From the graph it is clear that the sequence contains two complete walk cycles. These two parameters encapsulate considerable information on the posture of the pedestrian within the walk cycle and this information could be used in a number of recognition tasks.

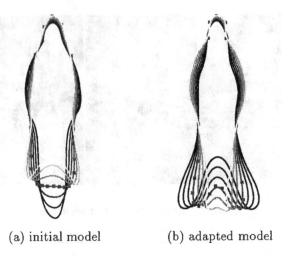

(a) initial model   (b) adapted model

Figure 9.   Principal modes of variation

4. Refining the model

By bootstrapping the system an improved "adapted" model can be generated which is more "compact" in that fewer shape parameters are required
to represent each shape. The basic idea is to use the tracking scheme
to fit the current shape model to each training image to generate new,
reparametrised training contours which are then used to generate a refined
linear shape model. Care must be taken to prevent a gradual loss in accuracy in the contour fit for each training shape as the model is iteratively
refined. This is achieved by combining the current linear model with an
isotropic noise model (adding a constant to all the eigenvalues). This hybrid model can then be used to recover accurate reparametrised training
contours. For more details, see [17].

   A diagram visualising the principal mode of variation for the initial and
adapted shape model for a typical training set is show in figure 9

4.1. TRACKING WITH ADAPTED MODELS

Improved results are obtained with the bootstrapped adapted models. An
image based error measure can be used to compare the tracked contour silhouettes with the "ground truth" segmentation obtained using background
image subtraction. Quantitative results are shown in figure 10. The performance is quantified for image sequences in which varying levels of spatially
structured, temporally uncorrelated noise have been added. Two training
sets were used. The "generic" model is generated using a large training

Performance for temporally uncorrelated noise

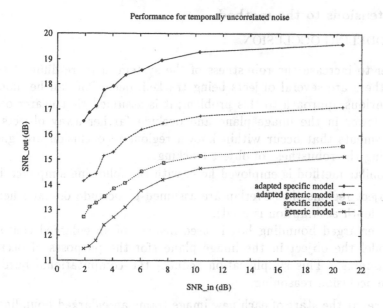

Figure 10. Plot showing accuracy of models

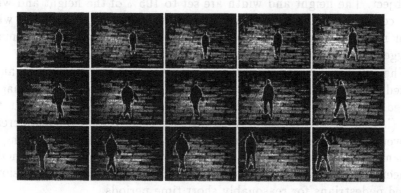

Figure 11. Results on test sequence with zooming camera and adapted model

set containing the silhouette shape of a pedestrian walking in a variety of directions. The "specific" model is generated from a subset of training data in which the pedestrian walks from left to right, parallel to the image plane. The test sequence is similar to this "specific" training sequence.

A further example of tracking is shown in figure 11. This sequence was obtained using a hand held camera in which the camera operator zooms in on the walking person. Image edges were used to drive the tracking mechanism and initialisation was performed by keeping the camera stationary for the first few frames.

5. Extensions to the method

5.1. MODELING OCCLUSION

In order to increase the robustness of the system in more difficult scenes where there are several objects being tracked, occlusion can be modeled. In a previous approach to this problem, it is assumed that nearer objects appear lower in the image plane and occlude farther away objects [20]. Measurements that occur within known regions of occlusion are ignored, improving the robustness of object tracking.

A similar method is employed here, with the following simplifications:-

- Object regions that overlap are assumed to occlude one another (i.e. no depth assumption is used).
- An enlarged bounding box is used instead of an enlarged contour to model the object in the image plane (for the purposes of occlusion reasoning). This simplification reduces the computational burden of the occlusion reasoning.

Hence, at the start of each new image frame an enlarged bounding box is generated for each tracked object centered about the estimated origin of the object. The height and width are set to 105% of the height and width of the contour's bounding box in the previous frame. For every point where two or more rectangles overlap the relevant pixel in an "occlusion image" is flagged.

When measurements are made in the tracking mechanism at an estimated contour point p_i, the associated pixel in the occlusion image is checked. If there is possible occlusion no measurement is made at that point (i.e. the measurement inverse variance is set to zero). This increases the overall measurement variance for each state parameter reducing the Kalman gain and increasing the state parameter uncertainties. The method has been found to improve robustness where there is partial occlusion of tracked pedestrians for reasonably short time periods.

5.2. LEARNING A SPATIO-TEMPORAL MODEL

The purely spatial eigenshape approach works well in many situations. However in extreme clutter when a background image is unavailable a spatio-temporal model is often required to accurately predict changes in shape over time and minimise the effects of distracting image features. Dynamic models for tracking curves can be learned from training data using system identification methods [21, 22]. In this approach, the system state vector is modeled as a general 2nd order stochastic process with a driving noise process. The tracker is trained up on a typical contour motion to allow the optimal system dynamics to be identified. This produces an improved tracker, tuned to a given type of non-rigid motion. We have extended this approach by looking for $\mathcal{H}$-orthogonal physically based vibration modes that best represent the observed dynamics. This new model retains the computational simplicity of the purely spatial approach, whilst increasing the robustness of tracking by improving the accuracy of the prediction step. Full details of the method are given elsewhere [23].

6. Recognition and coding

An important product of the tracking procedure outlined in the body of this chapter is a stream of shape vectors recording the most significant shape parameters that characterise the projection of the object from frame to frame. These shape vectors provide a concise and accurate representation of the projected shape of the object as it moves through the scene. Recognition of individual postures (possibly from a given viewpoint) can be achieved using standard classification procedures operating within the space of shape vectors. In principle, it should also be possible to recognise familiar movements by examining the paths followed by sequences of shape vectors through the 'shape space'.

Finally, this representation of the shape of a moving object could be used as part of a model-based image coding scheme, although it would need to be complemented with coding for the intensity information within an object.

References

1. D. Hogg, "Model-based vision: A program to see a walking person," *Image and Vision Computing*, vol. 1, no. 1, pp. 5–20, 1983.
2. K. Rohr, "Incremental recognition of pedestrians from image sequences," *Computer Vision and Pattern Recognition*, pp. 8–13, 1993.
3. J. M. Rehg and T. Kanade, "Digiteyes: Vision-based hand tracking for human-computer interaction," in *IEEE Workshop on Motion of Non-rigid and Articulated Objects*, pp. 194–199, IEEE Computer Society Press, Nov. 1994. IEEE Catalog No. 94TH0671-8.
4. A. Pentland and B. Horowitz, "Recovery of non-rigid motion and structure," *IEEE Trans. on Pattern Analysis and Machine Intelligence*, vol. 13, pp. 730–742, July 1991.
5. M. K. Leung and H. Y. Yang, "First sight: A human body outline labeling system," *IEEE Trans. on Pattern Analysis and Machine Intelligence*, vol. 17, no. 4, pp. 359–377, 1995.
6. T. J. Cootes, D. H. Cooper, C. J. Taylor, and J. Graham, "A trainable method of parametric shape description," *Image and Vision Computing*, vol. 10, pp. 289–294, June 1993.
7. T. F. Cootes, A. Hill, C. J. Taylor, and J. Haslam, "The use of active shape models for locating structures in medical images," in *Information Processing in Medical Imaging* (B. H.H. and G. A.F., eds.), pp. 33–47, 1993.
8. T. J. Cootes, C. J. Taylor, D. H. Cooper, and J. Graham, "Active shape models - their training and application," *Computer Vision and Image Understanding*, vol. 61,

       pp. 38–59, Jan. 1995.
  9.   A. Blake, R. Curwen, and A. Zisserman, "A framework for spatio-temporal control
       in the tracking of visual contours," *International Journal of Computer Vision*, 1993.
 10.   A. Hill and C. J. Taylor, "Automatic landmark generation for point distribution
       models," in *British Machine Vision Conference*, vol. 2, pp. 429–438, BMVA Press,
       1994.
 11.   A. Lanitis, C. J. Taylor, and T. F. Cootes, "An automatic face identification system
       using flexible appearance models," in *British Machine Vision Conference*, vol. 1,
       pp. 65–74, 1994.
 12.   J. C. Gower, "Generalized procrustes analysis," *Psychometrika*, no. 40, pp. 33–51,
       1975.
 13.   N. J. B. McFarlane and C. P. Schofield, "Segmentation and tracking of piglets in
       images," *Machine Vision and Appications*, vol. 8, pp. 187–193, 1995.
 14.   M. Sonka, V. Hlavac, and R. Boyle, *Image Processing, Analysis and Machine Vision*.
       Chapman and Hall, 1993.
 15.   J. D. Foley and A. Van Dam, *Fundamentals of Interactive Computer Graphics*.
       Addison-Wesley Publishing Co., 1984.
 16.   R. Bartels, J. Beatty, and B. Barsky, *An Introduction to Splines for use in Computer
       Graphics and Geomteric Modeling*. Morgan Kaufmann, 1987.
 17.   A. Baumberg and D. Hogg, "An adaptive eigenshape model," in *British Machine
       Vision Conference* (D. Pycock, ed.), vol. 1, pp. 87–96, BMVA, Sept. 1995.
 18.   D. Terzopoulos and R. Szeliski, "Tracking with kalman snakes," in *Active Vision*
       (A. Blake and A. Yuille, eds.), ch. 1, pp. 3–20, MIT Press, 1992.
 19.   N. K. Sinha and B. Kuszta, *Modeling and Identification of Dynamic Systems*. Van
       Nostrand Reinhold Company, 1983.
 20.   D. Koller, J. Weber, and J. Malik, "Robust multiple car tracking with occlusion
       reasoning," in *European Conference on Computer Vision*, vol. 1, pp. 189–196, May
       1994.
 21.   A. Blake and M. Isard, "3d position, attitude and shape input using video tracking
       of hands and lips," in *Proc. ACM Siggraph*, pp. 185–192, 1994.
 22.   A. Blake, M. Isard, and D. Reynard, "Learning to track the visual motion of con-
       tours," *Artificial Intelligence*, vol. 78, pp. 101–134, 1995.
 23.   A. Baumberg and D. Hogg, "Generating spatiotemporal models from training ex-
       amples," *Image and Vision Computing*, June 1996. in press.

CYCLIC MOTION ANALYSIS USING THE PERIOD TRACE

STEVEN M. SEITZ AND CHARLES R. DYER
Department of Computer Sciences
University of Wisconsin
Madison, WI 53706

1. Introduction

Non-rigid motion analysis is complicated by a lack of general purpose rules or constraints governing how an object or scene evolves over time. In order to design practical algorithms, much of the work to date has focused on object-model-based techniques, such as interpretation of facial expressions, detection of human locomotion, cardiac image analysis, and gesture recognition. In contrast, repeating motions all share common temporal features that can be formally described and interpreted regardless of the particular object or scene that is moving. For instance, it is not necessary to recognize the runner or the motion in Fig. 1 in order to determine his stride frequency. Because many real-world motions repeat, e.g., a heart beating, an athlete running, and a wheel rotating, cyclic motion analysis techniques have broad applicability.

For motions that repeat regularly, detecting the period can yield important information about underlying object or scene properties. For instance, heart-rate relates to activity levels, the period of a hand on a clock to units of time, and a runner's stride frequency to velocity. For repeating motions whose cycle length can change as a function of time, which we call *cyclic* motions, detecting irregularities can illuminate physically important changes in the scene. By analyzing the variance of the period in several cycles of a runner's gait or a swimmer's stroke, anomalies can be located, providing feedback on specific areas which may need improvement [1]. Medical imagery can be analyzed using the same techniques, for instance to detect unevenness in a heart's beating motion.

Analyzing 3D cyclic motions from image sequences is challenging for several reasons, including

M. Shah and R. Jain (eds.), Motion-Based Recognition, 61–85.
© 1997 *Kluwer Academic Publishers.*

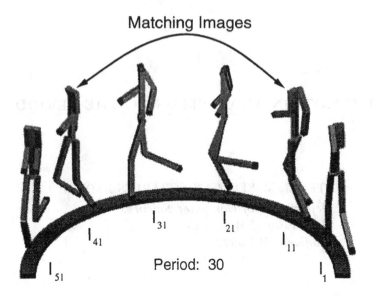

Figure 1. A runner's stride frequency can be determined by detecting *matching* images of the runner at the same point in his cycle. The time between repetitions is the period. Matching images, i.e., 2D views of the same 3D configuration, can be found by tracking the trajectories of five or more image features.

- Irregularities: The motion may not be strictly periodic due to fluctuations in speed from one cycle to the next. These fluctuations must be accounted for in order to robustly detect that a motion is cyclic and to interpret its properties.
- Complexity: The observed motion can be non-rigid and arbitrarily complex. Moreover, motion information must be integrated over long sequences of images.
- Camera motion: Changes in viewpoint affect the apparent motion and therefore complicate the analysis. In order to determine salient features of the underlying motion in the scene, viewpoint-dependent effects must be filtered out.

This chapter addresses the first two difficulties by introducing a compact, strictly temporal motion representation called the *period trace* for the detection and interpretation of cyclic motions. By accounting for changes in speed, uneven cyclic motions can be detected and irregularities located. Computational complexity is minimized by using simple and efficient image comparison operations in the recovery procedure. Using the period trace, an analog of the period is defined for cyclic motions, which we call the *instantaneous period*, giving cycle length as a function of time. Information about motion irregularities can be determined by considering derivatives of the instantaneous period. The issue of camera motion is addressed by incorporation of a novel view-invariant match function for comparing images. The match function incorporates principles of affine invariance to allow for arbitrary changes in viewpoint.

The remainder of the chapter is structured as follows. Section 2 discusses related work on periodic motion analysis. Section 3 describes the *period trace* and its image-based detection using view-invariant techniques. Section 4 introduces several motion features that can be derived from the period trace. Section 5 presents a method for automatically recovering a motion's period trace using an optimization approach. Section 6 describes experiments on real and synthetic image sequences.

2. Related Work

Several researchers have investigated ways of measuring periodicity information from image sequences [2, 3, 4]. Allmen and Dyer [2] described an approach for detecting motion periods under orthographic projection. They used the curvature scale space of point trajectories to detect repeating patterns of curvature maxima and hence infer a period. Polana and Nelson [3] presented a method for detecting periodic motions using Fourier transforms of several point trajectories. In theory, the period of the motion could be detected as well by averaging the fundamental frequencies of the point trajectories, although the authors indicated that determining the period in this way was unreliable. Tsai and Shah [4] described a similar technique, using Fourier transforms of curvature values, where the period was determined from a single point trajectory.

None of these approaches are appropriate for analyzing repeating motions that lack a constant period; in particular, they cannot be used to locate motion irregularities or changes in speed. Furthermore, they rely on detecting periodicity in 2D and fail to account for changes in camera viewpoint and reference frame. For example, the runner's motion in Fig. 1 is periodic in a 3D reference frame that moves with the runner. However, during the course of a cycle, the image motion of points on the runner will not sweep out periodic paths, as a result of the runner's constantly changing attitude relative to the camera.

Motion information has proven useful in a number of related recognition problems. Johansson's pioneering work on moving light displays (MLD's) [5] demonstrated that human motions such as walking and running can be recognized solely based on the trajectories of a small number of bright spots attached to different parts of the body. Since then, motion-based recognition has become an active area of study within computer vision. These efforts include identification of pedestrians [6, 7], MLD-based motion recognition

[8], hand gesture recognition [9, 10], interpretation of facial expressions [11, 12], and temporal textures [13]. Details of many of these efforts are described in other chapters of this book, and in a recent survey article [14].

Cyclic motion analysis is unique among motion-based recognition approaches in that it does not require any type of object- or motion-specific model. Other motion-based recognition techniques [6, 7, 9, 10, 11, 12, 13] require *a priori* models of the underlying object and/or the motion in the scene, although these models can potentially be learned [9, 15]. In contrast, periodicity is a *universal* motion characteristic that can be detected and described without knowledge of the underlying object and without reference to a previous instance of the motion. Therefore, the techniques in this chapter can be used to interpret image sequences of completely unfamiliar objects and motions.

In [16] we introduced a match function that permits view-invariant comparison of images and removes the restriction of a stationary observer. In [17] we described a technique for analyzing irregular cyclic motions using the *period trace*. Here we present a unified framework for the analysis of periodic and cyclic motions from image sequences based on the period trace and view-invariant image matching.

3. Cyclic Motion and the Period Trace

Motion can be described both in spatial and temporal terms, as changes in position or variations in speed. Most existing representations use some combination of spatial and temporal information to describe how an object evolves over time. In many applications, however, we are interested in the temporal evolution of an object or a scene and *not* its instantaneous shape. Consequently, in this section we introduce an alternative, purely temporal approach that is independent of an object's spatial properties. Our temporal motion representation, called the *period trace*, describes how the period changes throughout the course of a motion and contains a wealth of useful motion information. Because no assumptions are made about the spatial structure or representation of the scene, the period trace can be used to analyze both rigidly and non-rigidly moving objects and scenes. To motivate the period trace, we first define what it means for a motion to be *periodic* and, more generally, *cyclic*.

3.1. PERIODIC AND CYCLIC MOTIONS

We define a *motion* $M(t)$ to be a function whose value at time t is the instantaneous configuration of a continuously-moving object or scene[1]. We

---

[1]Although time is implicit, t could represent another quantity.

call a motion P *periodic* if it repeats with period p, i.e.,

$$P(t + p) = P(t) \tag{1}$$

for some constant $p > 0$ and all times t in a given time domain. The smallest such constant p is the *period* and the set $P_{t_0} = \{P(t) \mid t_0 \leq t < t_0 + p\}$ is called the *cycle* beginning at time t_0.

The notion of periodicity given in Eq. (1) has a very restrictive temporal constraint, namely, that the motion is perfectly regular from one cycle to the next. This constraint is relaxed with the introduction of a *period-warping* function, ϕ, as follows:

Definition 1 *A motion C is called* **cyclic** *if*

$$C(\phi(t)) = C(t)$$

for all times t in a given time domain and some increasing continuous function ϕ satisfying $\phi(t) > t$. A function ϕ satisfying these properties is called C-warping.

Intuitively, $\phi(t)$ corresponds to the start of the next cycle after the cycle beginning at time t. The increasing condition on ϕ ensures that a cyclic motion is order preserving, i.e., $t_1 < t_2$ implies $\phi(t_1) < \phi(t_2)$. Notice that all periodic motions are cyclic and that a cyclic motion is periodic when Definition 1 is satisfied for $\phi(t) = t + p$, with p the period. Although not all cyclic motions are periodic, any cyclic motion can be *warped* into a periodic motion by an appropriate temporal transformation (see Section 4.4).

3.2. PROJECTIONS OF 3D CYCLIC MOTIONS

Since we are concerned with image-based analysis of 3D cyclic motions, the effects of camera projection must be considered. For instance, consider the problem of measuring the stride frequency of an athlete filmed while running along a path. Both the relative movement of the runner's limbs and his change in attitude relative to the camera contribute to the motion in the image sequence, but only the former is relevant to the stride frequency. We would like to be able to determine the stride frequency in a way that is not affected by changes in viewpoint. This is accomplished using a view-invariant match function that equates a set of images if and only if they represent views of an object in the same 3D configuration.

3.2.1. Projected Match Criterion

Formally, we want to determine if a set of images *match*, i.e., correspond to views of the same 3D configuration of an object or a scene. Our formulation assumes an *affine* camera model [18, 19] which is a generalization of orthographic, weak perspective, and paraperspective projections. An image sequence $I_1, \ldots, I_m$ is represented as a sequence of $2 \times n$ matrices where each column is the instantaneous position of one of n image features. It is assumed that corresponding columns of I_t and I_s represent projections of the same scene point. Hence, the correspondence between features in different images is assumed to be known. An object or scene, S, is represented as a $3 \times n$ matrix with columns corresponding to the 3D coordinates of features in an object-centered affine reference frame. Without loss of generality, assume that each row of I_t has zero mean; choose the origin of the image coordinates to be the centroid of the feature points. Under an affine projection model a set of images, $\Gamma = \{I_{t_1}, \ldots, I_{t_k}\}$, match if and only if

$$I_{t_i} = \Pi_i S \tag{2}$$

for some fixed $3 \times n$ matrix S, $i = 1, \ldots, k$, and 2×3 matrices $\Pi_1, \ldots, \Pi_k$.

Define the measurement matrix M_Γ of a set of images to be the concatenation of the image measurements:

$$M_\Gamma = \begin{bmatrix} I_{t_1} \\ \vdots \\ I_{t_k} \end{bmatrix}$$

If the images Γ match, then, by Eq. (2), we can express M_Γ as

$$M_\Gamma = \begin{bmatrix} \Pi_1 \\ \vdots \\ \Pi_k \end{bmatrix} S \tag{3}$$

Therefore M_Γ is the product of two matrices, each having rank at most 3. It follows that M_Γ has rank 3 or less. This is the Rank Theorem, due to Tomasi and Kanade [20], and generalized for an affine camera. Conversely, any measurement matrix of rank 3 or less can be decomposed as in Eq. (3) using singular value decomposition [20]. Therefore, we have the following:

Generalized Rank Theorem: *A set of images* Γ *match if and only if* M_Γ *is of rank at most 3.*

Under orthographic projection (the case considered by Tomasi and Kanade), the rank condition alone is not sufficient to determine that a set of images match. However, it is both necessary and sufficient under the more general affine camera model. Hence, the Generalized Rank Theorem can be used to verify if a set of images could have been produced by a 3D cyclic motion. Accordingly, let $I_1, \ldots, I_m$ be a sequence of images, let ϕ be an integer-valued function, and denote $\Gamma_t^\phi = \{I_{\phi^i(t)} \mid 1 \leq \phi^i(t) \leq m\}$. The following result is a consequence of the Generalized Rank Theorem and Eq. (1):

Projected Match Criterion: *An image sequence, $I_1, \ldots, I_m$, is the affine projection of a 3D cyclic motion if and only if there exists an increasing function satisfying $\phi(t) > t$ such that $rank(M_{\Gamma_t^\phi}) \leq 3$ for $t = 1, \ldots, m$.*

Such a function ϕ is called I-*warping*. The Projected Match Criterion suggests a provably-correct way of detecting cyclic motions: check all possible I-warping functions. Under an assumption of periodicity, this brute-force strategy is feasible [16] but exhaustive search is too costly in general. A more efficient method is described in Section 5.

3.2.2. *View-Invariant Image Matching*

Because rank measurements are highly sensitive to numerical errors, a more robust measure of match quality is needed. We can characterize the residual error of a sequence of images by the amount by which we have to perturb features in order to make the images match. Define the match error of a set of images as follows:

$$dist_{\mathcal{A}}(\Gamma) = min \left\{ \|E\|_{rms} \mid rank(M_\Gamma + E) \leq 3 \right\}$$

$\|E\|_{rms}$ is the root-mean-squared norm of the matrix E defined by $\|E\|_{rms} = \sqrt{\frac{1}{2kn} \sum_{i,j} E_{ij}^2}$. Alternatively, $dist_{\mathcal{A}}$ can be expressed in terms of the singular values of M_Γ:

$$dist_{\mathcal{A}}(\Gamma) = \sqrt{\frac{1}{2kn} \sum_{i=4}^{n} \sigma_i^2} \tag{4}$$

where $\sigma_1 \ldots \sigma_n$ are the singular values of Γ. To derive Eq. (4), observe that singular value decomposition gives

$$M_\Gamma = U\Sigma V$$

where U and V are orthogonal matrices and the singular values of M_Γ appear along the diagonal of Σ (a $2k \times n$ diagonal matrix). The above equation can be rewritten:

$$M_\Gamma = U\tilde{\Sigma}V + U\Sigma'V \tag{5}$$

where $\tilde{\Sigma}$ and Σ' are structured as

$$\tilde{\Sigma} = \begin{bmatrix} \sigma_1 & & & \\ & \sigma_2 & & 0 \\ & & \sigma_3 & \\ & 0 & & 0 \end{bmatrix} \qquad \Sigma' = \begin{bmatrix} 0 & & & 0 \\ & & \sigma_4 & \\ & 0 & & \ddots \\ & & & & \sigma_n \end{bmatrix}$$

Eq. (5) can be expressed as

$$M_\Gamma = \tilde{M}_\Gamma + M'_\Gamma$$

where $\tilde{M}_\Gamma = U\tilde{\Sigma}V$ and $M'_\Gamma = U\Sigma'V$.

$\tilde{M}_\Gamma$ is the optimal (in an RMS sense) rank-3 approximation of M_Γ [21]. Hence, M'_Γ is the minimal perturbation of M_Γ that produces a match. Therefore, $dist_A(\Gamma) = \|M'_\Gamma\|_{rms}$. The latter term is just $\|\Sigma'\|_{rms}$ since U and V are orthogonal, and the result follows. □

The measure $dist_A$ gives the average amount (in pixels) necessary to additively perturb the coordinates of each image feature in order to produce a set of matching images. In the case where there are only two images we abbreviate $dist_A(\{I_s, I_t\})$ as $d_A(I_s, I_t)$. $dist_A$ has the following properties:

- $dist_A(\Gamma) = 0$ if and only if the images Γ match exactly.
- $dist_A$ is well-behaved with respect to noise since it is defined in terms of feature measurement perturbations. See, for example, [21].
- $dist_A$ is defined in image coordinates and can be directly related to measurement errors.
- $dist_A$ is always zero when less than five features are considered.
- $dist_A$ may be non-zero when the features are co-planar. Therefore co-planarity of the feature set is not a limitation, in contrast to related techniques [18, 19, 20].
- For n features, $d_A(X, Y) = \frac{\sigma_4}{2\sqrt{n}}$ and the evaluation cost is $O(n)$ arithmetic operations. For m images, the evaluation cost is the smaller of $O(nm^2)$ and $O(mn^2)$.

3.2.3. Occlusions and Lost Features

Matching several images at once is appropriate when a large number of features are continuously visible throughout an image sequence. For real world applications however, this is seldom the case due to factors such as noise and occlusion. Therefore, in practice we compute only pairwise correlations between images that have at least 5 features in common. In terms of the Projected Match Criterion, this amounts to labeling a motion as cyclic if $d(I_t, I_{\phi(t)}) = 0$ for all image pairs I_t and $I_{\phi(t)}$ having at least 5 features in common. Note that this condition is weaker than the Projected Match Criterion; it is only a necessary condition for a motion to be cyclic.

3.3. THE PERIOD TRACE

A C-warping function ϕ contains information concerning the temporal variation of a cyclic motion C. Unfortunately, Definition 1 does not determine a unique C-warping ϕ for a given cyclic motion C. For instance, if C is periodic with period p, $\phi(t) = t + kp$ for any positive integer k satisfies Definition 1. For definiteness we introduce the notion of *instantaneous period*:

Definition 2 *Let C be a cyclic motion. Let*

$$\phi_1 = \text{pointwise-infimum } \{\phi \ C\text{-warping}\}$$

Define $\tau_1(t) = \phi_1(t) - t$. τ_1 is called the **instantaneous period** *of C.*

The fact that ϕ_1 is C-warping follows from the continuity of C: Let $\mathcal{A} = \{\phi \ C\text{-warping}\}$ and let $\{\psi_i\}_{i=1}^{\infty}$ be a subset of $\mathcal{A}$ converging to ϕ_1 at t. Then $C(\psi_i(t)) = C(t)$ for each i, so $\lim_{i \to \infty} C(\psi_i(t)) = C(t)$. But $\lim_{i \to \infty} C(\psi_i(t)) = C(\phi_1(t))$ by continuity of C. Hence $C(\phi_1(t)) = C(t)$ for all values of t, i.e., ϕ_1 is C-warping.

Intuitively, $\tau_1(t)$ is the length of the cycle beginning at time t. For instance, if C is periodic, τ_1 is the period. From τ_1, several useful quantities are computable, including τ_n, the instantaneous combined length of the next n cycles. Accordingly, let

$$\phi_n(t) = \overbrace{(\phi \circ \phi \circ \ldots \circ \phi)}^{n}(t)$$

for integers $n > 0$. Then the *nth instantaneous period*, τ_n, is defined as

$$\tau_n(t) = \phi_n(t) - t \tag{6}$$

In addition to being continuous, τ_n has the following property:

$$\frac{\tau_n(t+h) - \tau_n(t)}{h} > -1 \tag{7}$$

which follows from Eq. (6) and the increasing condition on ϕ_1. If τ_n is differentiable, Eq. (7) is equivalent to the condition $\tau_n' > -1$.

The inverse functions exist and are defined as

$$\phi_{-n} = (\phi_n)^{-1}$$
$$\tau_{-n}(t) = \phi_{-n}(t) - t$$

for $n > 0$. $\tau_{-n}(t)$ has the intuitive interpretation as the combined length of the previous n cycles ending at time t. Its value is always negative.

We refer to the set of functions $\{\tau_n \mid n \neq 0\}$ as the *period trace* of a cyclic motion C. For instance, the period trace of a periodic motion is a set of constant functions: $\tau_n = np$, where p is the period, as shown in Fig. 2. The period trace is a comprehensive map of the cycles in a repeating motion and identifies all corresponding configurations in different cycles. These properties make the period trace a useful tool for comparing different cycles and tracking changes in a repeating motion over time.

4. Trends and Irregularities

The period trace of a cyclic motion describes how the motion varies from one cycle to the next. Consequently, many important attributes of a motion can be computed directly from the period trace, without reference to the spatial structure of the underlying scene. In this section we present several features that quantify temporal trends and irregularities in cyclic motions.

4.1. LOCAL FEATURES

Because the period trace reflects the cumulative motion of n cycles, it is not obvious that local temporal information can be derived from it. In this section we show that local temporal irregularities can indeed be computed from derivatives of the instantaneous periods. Any interval $[t, t+h]$ in which τ_n is not constant, for some n, is said to be an *irregular interval*. The quantity $\tau_n(t+h) - \tau_n(t)$ gives the cumulative change in the nth period, and $\frac{\tau_n(t+h) - \tau_n(t)}{h}$ is the mean rate of change of τ_n in the interval $[t, t+h]$. Letting $h \to 0$, the mean rate of change converges to the instantaneous rate of change, $\tau_n'(t)$. Henceforth, we denote τ_n' as the one-sided derivative defined by $\tau_n'(t) = \lim_{h \to 0} \frac{\tau_n(t+h) - \tau_n(t)}{h}$.

Values of t for which $\tau_n'(t) \neq 0$ are *irregular points*, i.e., points where the motion is faster or slower at $\phi_n(t)$ than at t. Moreover, τ_n' provides both the sign and magnitude of the irregularity. For instance, $\tau_1'(t) = 1$ indicates that the period at time t is increasing at a rate of 1 unit, i.e., motion is faster at t than at $\phi_1(t)$ by 1 unit. Because the motion is faster at the beginning of the cycle than at the end, it must slow down during the course of the cycle, thereby causing a net increase in the period.

Generally, not all points t where $\tau_n'(t) \neq 0$, for some $n \neq 0$, correspond to times where the motion is irregular relative to the norm. A nonzero value of $\tau_n'(t)$ only indicates that motion is irregular at time t *relative* to time $\phi_n(t)$. A point t where the motion is globally irregular satisfies $\tau_n'(t) \neq 0$, for all $n \neq 0$ (see Fig. 2(f)).

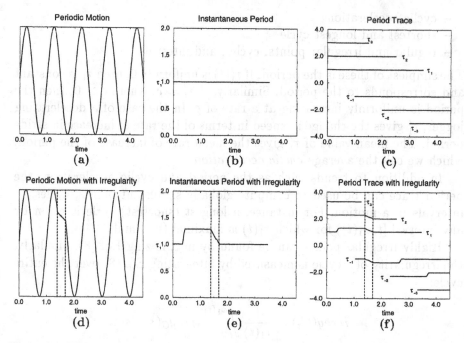

Figure 2. Effects of an irregularity on the instantaneous period. A periodic sinusoid signal (a) has a constant period (b) and a period trace consisting of multiples of the period (c). A small irregularity is introduced (d) by temporarily slowing by a factor of 10. The irregularity shows up in two places in the instantaneous period (e), where a ramp occurs once when the irregularity enters the current cycle and once where it leaves. The actual location of the irregularity is detected using the period trace (f) where the ramps "line-up" only at the true irregularity (i.e., in a small interval around time 1.5).

The total first-order irregularity of a cyclic motion C at a point in time t can be defined as

$$irreg_C(t) = mean \{|\tau_n'(t)| \mid n \neq 0\} \qquad (8)$$

Higher-order irregularities may also be relevant in certain situations. For instance, consider a jogger whose stride frequency is steadily decreasing. Due to the changes in speed, τ_n' will be nonzero and second-order irregularities may be more interesting, i.e., where $\tau_n''(t) \neq 0$. Unless otherwise qualified, the term *irregularity* henceforth refers to a first-order irregularity.

4.2. GLOBAL FEATURES

The period trace can also be used to compute various global features of a cyclic motion. These include

- period

- cyclic acceleration
- shortest and longest cycle
- regular and irregular points, cycles, and intervals

The simplest of these is the period. If $\tau_1'(t)$ is uniformly 0, then τ_1 is constant and corresponds to the period. Similarly, if $\tau_1' = c$, where $c > 0$, then the period is uniformly increasing at a rate of c. In the case of a decelerating jogger, τ_1' gives the change in speed in terms of the rate of increase in stride period. The mean value of τ_1' gives the mean rate of increase of the period, which we call the average *cyclic acceleration*.

In addition to trends such as the period and cyclic acceleration, the period trace can be used to compute globally significant points, cycles, or intervals in a motion. For instance, a longest (shortest) cycle is given by any interval $[t, \phi_1(t))$ for which $\tau_1(t)$ is maximal (minimal).

Highly irregular points can be found by maximizing Eq. (8). Similarly, the irregularity of a cycle is measured by integrating Eq. (8) over the entire cycle:

$$irreg(C_t) = \frac{1}{\tau_1(t)} \int_t^{\phi_1(t)} irreg_C(s) \, ds$$

Another useful feature is the median[2] cumulative irregularity of an interval:

$$irreg_C(t_1, t_2) = median \; \{|\tau_n(t_2) - \tau_n(t_1)| \; \mid \; n \neq 0\}$$

4.3. MOTION SIGNATURES

Some types of cyclic motion produce telltale temporal signatures. Examples include periodic motions and motions with uniform cyclic acceleration, both of which have instantaneous periods that are linear. The form of a motion's period trace may also indicate something about the distribution of irregularities. Cyclic motions that contain isolated irregularities have very distinct signatures (see Figs. 2 and 7); the instantaneous period of a motion of this sort is piecewise-constant except at isolated intervals whose locations and extents correspond to the irregularities.

The locations of irregularities in the period trace provide other important qualitative clues about the behavior of a moving object. Any cyclic motion containing at least one irregularity has a period trace with ramps in various places, e.g., see Fig. 2(f). There are two categories of ramps; those that line-up, recurring at the same place in each cycle, and isolated ramps that don't recur. Generally, irregular motions will have both types of ramps,

---

[2] We have found the *median* to be more robust than the *mean* in this context.

but certain motions have only ramps that line-up. These are motions where irregularities occur at roughly the same point in each cycle. For instance, a heart may contract at a different rate in each cycle, resulting in a series of ramps that line-up at each contraction.

4.4. MEASURING RELATIVE SPEED AND REMOVING IRREGULARITIES

The period trace provides detailed information about cycles and changes in speed but does not explicitly give instantaneous speed as a function of time. Consider the motion of a jogger whose stride frequency varies through time. Suppose we wish to describe these variations not as changes in stride frequency, but in terms of differences in speed relative to the norm. For instance, we might say that the jogger is currently running twice as fast as normal (e.g., 15 mph versus 7.5 mph). This can be achieved by recovering a *time-warping* function that maps a perfectly regular, e.g. periodic, motion into the uneven motion of the runner.

Any cyclic motion can be warped into a periodic motion by *flattening* its period trace. This flattening transformation and its inverse are the time-warping functions we're after. The first step is to find a function that maps a cyclic motion into a periodic motion. In other words, given a cyclic motion, $C(t)$, we wish to find a reparameterization $s = \sigma(t)$ such that $\tilde{\tau}_1(s)$ is constant, where

$$\tilde{\tau}_1(s) = (\sigma \circ \phi_1 \circ \sigma^{-1})(s) - s$$

The result is that $C(s)$ is periodic, satisfying Eq. (1).

For any cyclic motion $C(t)$, there exist innumerable reparameterizations, $s = \sigma(t)$, such that $C(s)$ is periodic. A natural choice is one that synchronizes the motion with respect to a particular cycle. The result is that the reference cycle remains intact and all other cycles are aligned to match the reference cycle. This approach is attractive because a cycle may be chosen that satisfies a particular criterion and the whole motion will be rectified to conform to the criterion. For instance, irregularities may be removed by choosing a *regular* reference cycle, using the techniques of Section 4.2. Similarly, choosing a cycle that has a particular anomaly will result in a motion with a corresponding anomaly in every cycle.

Let $C(t)$ be a cyclic motion and choose a cycle, C_{t_0}, starting at time t_0. Define time warping function σ and its inverse as

$$\sigma(t) \quad = \quad \phi_{-n}(t) + n\tau_1(t_0) \quad \text{for } t \in [\phi_n(t_0), \phi_{n+1}(t_0))$$
$$\sigma^{-1}(s) \quad = \quad \phi_n(s - n\tau_1(t_0)) \quad \text{for } \sigma^{-1}(s) \in [\phi_n(t_0), \phi_{n+1}(t_0))$$

Claim: $C(s)$ *is periodic, where* $s = \sigma(t)$.

Proof: Fix a value of s. Suppose, without loss of generality, that $\sigma^{-1}(s) \in [\phi_n(t_0), \phi_{n+1}(t_0))$. Then

$$
\begin{aligned}
\tilde{\tau}_1(s) &= (\sigma \circ \phi_1 \circ \phi_n)(s - n\tau_1(t_0)) - s \\
&= (\phi_{-n-1} \circ \phi_1 \circ \phi_n)(s - n\tau_1(t_0)) + (n+1)\tau_1(t_0) - s \\
&= s - n\tau_1(t_0) + (n+1)\tau_1(t_0) - s \\
&= \tau_1(t_0)
\end{aligned}
$$

Therefore $\tilde{\tau}_1(s)$ is constant, so $C(s)$ is periodic with period $\tau_1(t_0)$. □

An image sequence is rectified using σ^{-1} as follows: given an initial image sequence $I_1, I_2, \ldots, I_m$, the rectified sequence is $I_{\sigma^{-1}(1)}, I_{\sigma^{-1}(2)}, \ldots, I_{\sigma^{-1}(k)}$, where $k = \lfloor \sigma(m) \rfloor$. The effect is that any irregularities are removed and there is no variation in speed from one cycle to the next.

Whereas σ^{-1} can be used for making a sequence periodic, σ is useful for describing how a cyclic motion deviates from periodic. The derivative of σ gives the factor by which motion is locally faster with respect to the reference cycle. Contrast this with τ_1', which gives the change in cycle length as a function of time; $\frac{d\sigma}{dt}$ gives relative speed directly, but requires a reference cycle. For instance, $\frac{d\sigma}{dt}(t_1) = 2$ indicates that the motion at time t_1 is twice as fast as normal. In Section 6 these techniques are used to measure relative speed and to demonstrate the rectification process for a real image sequence.

5. Recovering the Period Trace

In order to detect cyclic motion features such as those introduced in Section 4, we must be able to recover functions $\tau_n(t)$ for each integer $n \neq 0$ from a cyclic motion C. Because the complete period trace can be computed from τ_1, it is sufficient to determine τ_1. For periodic motions, τ_1 is constant and can be determined using unconstrained scalar optimization methods. More generally, the recovery of τ_1 is posed as a constrained functional minimization problem.

5.1. THE PERIODIC CASE

If a cyclic motion is known to have a certain form, this information may be used to simplify the recovery of its period trace. Particularly, if an *a priori* motion model is available, e.g., linear or piecewise constant, the model can ameliorate the task of determining τ_1 by reducing the search space and making the recovery procedure less sensitive to noise. Existing techniques for recovering periodicity information [2, 3, 4, 16]. implicitly use this strategy, exploiting the assumption of a constant period. The case where such a model is not available is treated in the next section.

The instantaneous periods of a periodic motion are constant and correspond to the multiples of the period. Therefore, the period trace of a periodic motion can be found by optimizing some function over the set of candidate periods. Towards this end, we introduce a tool for assessing the significance of a candidate period or period trace.

5.1.1. *A Significance Measure*

In the presence of noise and quantization effects it is unlikely that corresponding images of a 3D cyclic motion will match precisely, as dictated by the Projected Match Criterion. Still, any candidate period $\tilde{p}$ induces a correspondence between images spaced $\tilde{p}$ frames apart. In order to discriminate between cyclic and non-cyclic motions, it is necessary to evaluate the quality of this correspondence, based on how well the images match. For this purpose we use a statistical measure based on the Kolmogorov-Smirnov (K-S) test [22].

The quality of a computed period trace can be assessed by considering the problem within the context of sampling theory. The operation of comparing two images $\mathbf{I}_s$ and $\mathbf{I}_t$ can be thought of as taking a random point sample from *match space*: $\mathcal{M} = \{d_{\mathcal{A}}(\mathbf{I}_s, \mathbf{I}_t) \mid 1 \leq s < t \leq m\}$. A period trace $\pi = \{\tau_n \mid n \neq 0\}$ determines a specific sample $\mathcal{S}_\pi \subset \mathcal{M}$ given by

$$\mathcal{S}_\pi = \{d_{\mathcal{A}}(\mathbf{I}_t, \mathbf{I}_{t+\tau_n(t)}) \mid n > 0\}$$

For example, a period of p corresponds to the sample:

$$\mathcal{S}_p = \{d_{\mathcal{A}}(\mathbf{I}_t, \mathbf{I}_{t+np}) \mid n > 0\}$$

$\mathcal{S}_\pi$ contains the match scores of precisely those images that correspond, under the hypothesis that scene motion is cyclic with period trace π. If this hypothesis is correct, the distribution of $\mathcal{S}_\pi$ should differ significantly from that of $\mathcal{M}$, since matching images minimize $d_{\mathcal{A}}$. The K-S test gives the probability, or *P-value*, of a random sample of $\mathcal{M}$ matching the cumulative histogram of $\mathcal{S}_\pi$, i.e., the probability that the motion is not cyclic. For details on how to calculate the K-S statistic, consult [22].

The P-value of a candidate period trace π is evaluated by applying the K-S test to $\mathcal{S}_\pi$ and $\mathcal{M}$. Since the objective is to find a period trace that minimizes $d_{\mathcal{A}}$ it is only necessary to assess P-values of period traces for which

$$mean(\mathcal{S}_\pi) < mean(\mathcal{M}) \tag{9}$$

Candidate period traces for which this inequality is violated are normalized by assigning P-values of 1. The P-value of a motion is defined to be the minimum P-value of all possible period traces.

In practice, a motion's exact P-value is never computed; the P-value of the period trace obtained by the optimization procedure provides an upper bound that is generally sufficient to differentiate cyclic from non-cyclic motions. We have found this test to be quite robust; real cyclic motions yield P-values that are exceedingly small, on the order of 10^{-20} or smaller, whereas non-cyclic motions have P-values at or near 1. In fact, every one of the non-cyclic motions we evaluated had a P-value of 1. The reason is that image sequences are decidedly non-random; adjacent images (i.e., I_t and I_{t+1}) tend to be similar, resulting in a natural bias towards small periods. $\mathcal{M}$ contains match scores for *all* adjacent frames and it is therefore unlikely that Eq. (9) will be satisfied for any period trace π chosen at random. As a result, most period traces will have P-values of 1. Note that this bias helps to differentiate cyclic from non-cyclic motions since the latter tend to have P-values of 1. For cyclic images sequences there is an implicit assumption that the instantaneous period is uniformly greater than 1. In other words, the frame rate (images per second) should be greater than the frequency (cycles per second), or else the period trace cannot be detected.

5.1.2. *Detecting Periods Using the K-S Test*

Using the K-S test it is straightforward both to calculate the probability that a motion is periodic and to compute the most likely period. Given an image sequence $I_1, \ldots, I_m$, evaluate the K-S test for every candidate period $\tilde{p} \le (m-1)/2$. The value of $\tilde{p}$ with smallest P-value constitutes the most likely period and the corresponding P-value gives an upper bound on the probability that the motion is not cyclic.

A potential complication arises in discriminating between the true period and its multiples. If p is the true period, note that $S_{kp} \subset S_p$ for any positive integer multiple k of p. Hence, it is possible that $mean(S_{kp}) \le mean(S_p)$, for some value of k. However, the K-S test assigns greater significance to larger samples, adding a built-in bias in favor of smaller periods. Because S_{kp} represents a smaller sample than S_p, the latter is automatically assigned more significance, resulting in a smaller P-value. Consequently, the P-value of the true period will tend to be more significant than those of its multiples due to the natural bias of the K-S test. This phenomenon can be seen in Fig. 5, where the true period has a much smaller P-value than its multiple.

5.2. THE CYCLIC CASE

When nothing is known *a priori* about a cyclic motion, the recovery of its period trace can be posed as a constrained functional optimization problem. Because the complete period trace can be computed from τ_1, it is sufficient to determine $\tau_1(t)$ for each value of t. Given an image sequence $\mathbf{I}_1, \ldots, \mathbf{I}_m$, we seek a τ_1 that minimizes the energy function E, comprised of two terms:

$$E_1(\tau_1, t) = mean\ \{d_A(\mathbf{I}_t, \mathbf{I}_{t+\tau_1(t)}) \mid 1 \le t,\ t + \tau_1(t) \le m\}$$
$$E_2(\tau_1, t) = |\tau_1'(t)|$$
$$E(\tau_1) = \int_{t_1}^{t_2} E_1(\tau_1, t) + \alpha E_2(\tau_1, t)\ dt \qquad (10)$$

The first term seeks to maximize correlations between corresponding images, whereas the second term is added as a continuity constraint on τ_1. α is a scalar weighting factor that balances E_1 and E_2. τ_1 is also subject to the following hard constraints, as imposed in Section 3.3:

$$\tau_1(t) > 0 \qquad (11)$$
$$\tau_1(t + 1) - \tau_1(t) \ge -1 \qquad (12)$$

Note that the strict inequality in Eq. (7) is relaxed above since it cannot be enforced on the discrete domain $t \in [1, m]$.

We used a multiscale *snake* algorithm adapted from [23, 24] for recovering the period trace by iteratively minimizing the energy function E. The snake is initialized to a rough estimate of τ_1 and is then incrementally refined so as to reach a state of locally minimal energy. At each iteration a value of t is selected and $\tau_1(t)$ is adjusted by 0, 1, or -1 unit so as to globally decrease $E(\tau_1)$ while checking that Eqs. (11) and (12) are satisfied. A pass consists of iterating over all successive values of $t = 1, 2, \ldots, m$. The algorithm performs repeated passes until E converges. Convergence is guaranteed since energy decreases monotonically in each iteration. To avoid problems with local minima, the optimization approach is repeated at three scales of increasing resolution using the output of each stage as the initialization for the next.

Initialization is performed by setting τ_1 to the constant function that minimizes the K-S test, as described in Section 5.1. In combination with the multiscale approach, we found this simple strategy to provide adequate initialization for the optimization procedure to converge correctly for the sequences that we tested. However, this strategy is designed for near-periodic sequences; highly irregular cyclic sequences may require a more sophisticated initialization procedure, perhaps incorporating user-interaction, as in [23].

5.3. THE PERIOD TRACE RECOVERY ALGORITHM

The following sequence of operations is used to compute the period trace and its P-value from a sequence of images $\mathbf{I}_1, \ldots, \mathbf{I}_m$:

1. Compute P-values for each candidate period $\tilde{p} = 2, \ldots, \frac{m-1}{2}$. The candidate p with smallest P-value is the most likely period. The associated P-value gives the probability that the motion is not periodic.
2. Initialize the snake to $\tau_1 = p$ at the coarsest scale and apply energy minimization passes until the snake converges. Upon convergence, the new snake configuration is used to initialize energy minimization at the next finer scale. The procedure repeats until convergence is reached at the finest scale.
3. Obtain the P-value of the computed period trace using the K-S test, giving the probability that the motion is not cyclic.

As noted in Section 5.1.1, this algorithm cannot detect periods of one frame or less. Such motions will generally be found to be *non-cyclic* with a P-value of 1.

6. Experiments

One synthetic and two real image sequences are used in this section to show the performance of the optimization methods described in Section 5 and the detection of temporal motion features. Selected images from the sequences are shown in Figs. 3 and 4. Additional experiments with cardiac X-ray images are presented in [17, 25].

6.1. SYNTHETIC JOGGER

In order to study the effects of noise and perspective distortion on the algorithm, we created a simulation of a human jogging around a short track using a volumetric model and real motion data (see Fig. 4). To model the human torso, we used nine parallelepipeds connected by revolute joints. The model was roughly 250 units tall and the track had a diameter of 200 units. The camera was fixed at an angle of 30 degrees from horizontal. The periodic motion consisted of a sequence of joint angles extrapolated from real motion data of a human running, provided by N. Goddard [8]. The projected positions of vertices of the parallelepipeds were used as input to the algorithm. No attempt was made to distinguish visible from occluded vertices.

We found the algorithm of Section 5.1 to be highly robust to both feature noise and perspective effects (see Fig. 5). At the highest noise level of 10 pixels (the projected jogger was roughly 25 units wide and 250 tall), the true period was correctly detected with a P-value of 10^{-41}. At the highest level of perspective distortion (shown in Fig. 4), the detected period had a P-value of 10^{-47}. In all 123 trials the period with lowest P-value matched the

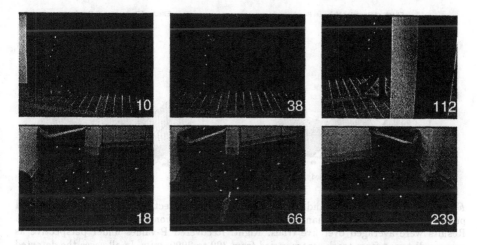

Figure 3. Selected images from image sequences of cyclic motions. Top: Human subject walking. Bottom: Rotating turntable. The second and third images match in each set shown above. These and other correspondences were computed automatically from the period trace.

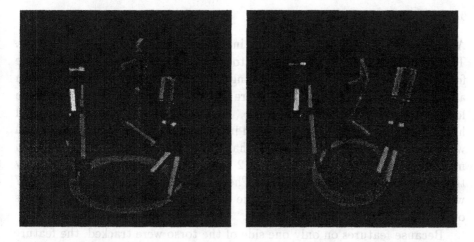

Figure 4. Projections of a simulated jogger moving around a track. Left: The jogger at three different points along the path under orthographic projection. Right: Same scene under perspective projection with the camera 300 units from the track center. Note the projective distortions in the perspective image that are not modeled by orthographic projection.

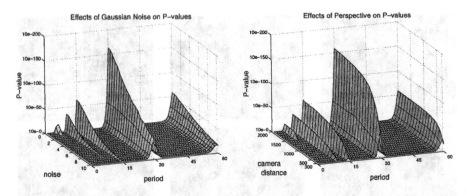

Figure 5. Effects of noise and perspective on period detection. Left: Plot of recovered period P-values under Gaussian noise with standard deviation varying from 0 to 10 pixels. Results were averaged over five trials. Right: Recovered P-values under perspective as camera distance from track center varied from 300 to 2000 units. In all cases the detected period of 30 frames exactly matched the true period.

true period of 30 frames. The most evident effect of noise and perspective was a gradual degradation of P-values, as seen in the Fig. 5.

6.2. HUMAN WALKER

A human subject was filmed walking in an arc subtending about 70 degrees. The camera was manually rotated to keep the subject in view, and the focal length was slowly changed during filming. To aid in feature detection and tracking, reflective markers were placed in areas which were visible for the duration of the sequence (i.e., right arm, right leg, mid torso, and head). We used the method described in Section 5.1 to determine the period with smallest P-value. Fig. 3 shows that the algorithm successfully detected matching images; images 38 and 112 are views of the subject at roughly the same point in her walk cycle. The algorithm detected a period of 37 with a P-value of 10^{-18}. The P-value of the detected period clearly beats out all other candidates (see Fig. 6).

Because features on only one side of the torso were tracked, the feature set was nearly planar. Co-planarity of the feature set is known to be problematic for many shape-from-motion algorithms [18, 19, 20]. Note, however, that it does not cause a problem for our algorithm because an explicit 3D scene representation is never computed. Note also that the entire motion is nearly affine; the features in each image can be roughly approximated by a horizontal shear and/or reflection of the features in the first image. In fact, we have found this property to be true of other human locomotory motions such as running, skipping and jumping. Since the match function d_A equates all images that are related by a 3D affine transformation, purely

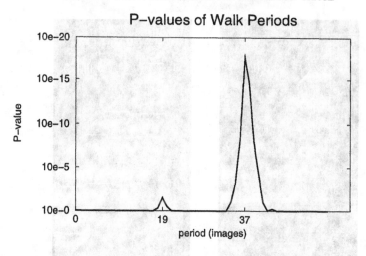

Figure 6. Period P-values of a walking human. The period with the highest level of significance is 37, with a P-value of 10^{-18}.

affine motions appear stationary, and period detection fails. The periods of nearly-affine motions, however, can be reliably detected with our approach, but the resulting P-values tend to be larger.

6.3. ROTATING TURNTABLE

A rotating turntable was filmed using a moving, hand-held video camera. Reflective markers were placed both on the turntable and elsewhere in the static scene. Note that although the motion of the turntable is rigid, the scene as a whole moves non-rigidly. Custom software was used to track the markers as the camera moved around the scene and the turntable simultaneously rotated. Twice the turntable was briefly touched to temporarily slow the rotation and produce an irregularity.

Fig. 7 shows the recovered period trace. The figure illustrates that the optimization procedure successfully located low-energy "valleys" in the match space, corresponding to the dark contours in the temporal correlation plot. The two regions with the highest irregularity values correspond to the two brief intervals in which the turntable was touched. The width of each interval indicates how long the turntable was touched and the irregularity value determines the extent to which the rotation was slowed. Notice that there are ramps in several locations in the period trace. A ramp's presence at point (t,s) indicates that the turntable's speed at time t is different than at time s, since the period is locally changing. Therefore, a motion irregularity occurs at time t *or* at time s. In order to determine where

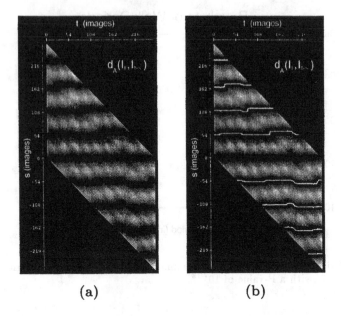

(a)   (b)

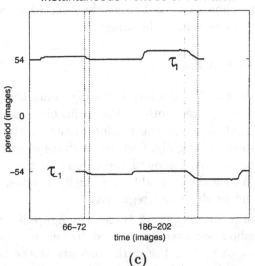

(c)

Figure 7. Period trace of the turntable motion sequence. (a) Temporal correlation plot of $d_A(I_t, I_{t+s})$ with values shown as intensity (light – low correlation, dark – high correlation). (b) The recovered period trace (white) is superimposed on the correlation plot. (c) A graph of τ_1 and τ_{-1} with the two most irregular intervals marked. Notice that discontinuities appear in various places but "line up" only at actual motion irregularities. The turntable frequency is $33\frac{1}{3}$ revolutions per minute, which corresponds to 54 frames per revolution at NTSC video rate.

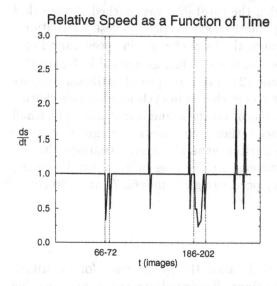

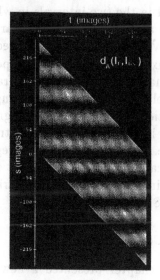

Figure 8. Image sequence rectification using the period trace. The turntable sequence was rectified, producing a periodic sequence with irregularities automatically removed. Left: Speed relative to the first cycle. Touching the turntable slowed the rotational speed, producing momentary reductions in $\frac{ds}{dt}$. Other fluctuations are present but are not significant since the local variations in speed cancel. Right: Temporal correlation plot of the rectified sequence. Observe that the period trace (dark lines) consists of constant functions, indicative of a periodic motion. The period with smallest P-value is 54, exactly matching the ground truth of $33\frac{1}{3}$ rpm at NTSC video rate.

specifically the irregularity occurs, simply look for ramps that "line-up", i.e., that occur at the same interval in $\tau_{-n}, \ldots, \tau_n$. The only ramps that line-up are the two intervals marked in Fig. 7, both of which correspond to intervals where the turntable was touched. Image 66 in Fig. 3 is contained in a detected irregular interval and shows the turntable being touched.

The rectification procedure of Section 4.4 was applied to the image sequence, producing a periodic image sequence (see Fig. 8) comprised of a subset of the images in the original turntable sequence. Notice that the period trace has been *flattened* and no longer contains signs of irregularities. The first complete cycle was automatically detected and used as the reference cycle. In the rectified sequence, the turntable rotates at a constant rate of 54 frames per revolution, exactly matching the ground truth rate of $33\frac{1}{3}$ rpm at NTSC video rate. The turntable sequence was found to be cyclic with a P-value of 10^{-229}. The rectified sequence was determined periodic with a P-value smaller than 10^{-324}, the smallest number representable with 8-byte double floating point precision.

The left graph in Fig. 8 shows instantaneous speed relative to the reference cycle. The slowing of the turntable is reflected by a reduction of speed in the two intervals in which the turntable was touched. Notice that there are four other momentary fluctuations in speed. These fluctuations, however, are not significant because the local changes in speed cancel out. For example, the rotational speed is twice as fast as normal in frame 126, and half as fast as normal in frame 127. These reciprocal variations exactly cancel each other out, producing no net change in cycle length. This characteristic pattern is common in graphs of this type and is indicative of a small error in the recovered period trace rather than a motion irregularity in the scene. In contrast, motion irregularities generally leave variations that do not cancel, resulting in a net increase or decrease in cycle length during the time in which the irregularity occurs (e.g., frames 66-72 and 186-202 in Fig. 8).

7. Conclusion

This chapter presented a new tool called the *period trace* for describing temporal properties of cyclic motions. By delimiting cycles and locating inter-cycle image correspondences, the period trace provides a framework for tracking changes in an object's motion over time. From this representation, several types of cyclic-motion features were derived, relating to motion trends and irregularities.

A second contribution was an affine-invariant image match function for comparing images under variable viewpoint. Using this function, we derived necessary and sufficient conditions for determining whether an uncalibrated image sequence is the projection of a 3D cyclic motion in the scene. Whereas previous work on affine-invariance was limited to static scenes or rigid-body motion, our results indicate that affine-invariant techniques have application to objects and scenes whose motion is arbitrarily non-rigid.

An energy-based optimization technique was described for recovering the period trace from image sequences obtained with a moving uncalibrated camera. Results on both real and synthetic image sequences indicate that the technique detects the period and other features of cyclic motions with high reliability and is robust with respect to measurement errors and perspective effects.

Acknowledgments

The support of the National Science Foundation under Grant Nos. IRI-9220782 and CDA-9222948 is gratefully acknowledged.

References

1. T. S. Perry, "Biomechanically engineered athletes," *IEEE Spectrum*, pp. 43–44,

April 1990.

2. M. Allmen and C. R. Dyer, "Cyclic motion detection using spatiotemporal surfaces and curves," in *Proc. 10th Int. Conf. on Pattern Recognition*, pp. 365–370, 1990.

3. R. Polana and R. Nelson, "Detecting activities," in *Proc. Computer Vision and Pattern Recognition Conf.*, pp. 2–7, 1993.

4. P. Tsai, M. Shah, K. Keiter, and T. Kasparis, "Cyclic motion detection for motion based recognition," *Pattern Recognition*, vol. 27, no. 12, pp. 1591–1603, 1994.

5. G. Johansson, "Visual perception of biological motion and a model for its analysis," *Perception and Psychophysics*, vol. 14, no. 2, pp. 201–211, 1973.

6. D. Hogg, "Model-based vision: A program to see a walking person," *Image and Vision Computing*, vol. 1, no. 1, pp. 5–20, 1983.

7. K. Rohr, "Incremental recognition of pedestrians from image sequences," in *Proc. Computer Vision and Pattern Recognition Conf.*, pp. 8–13, 1993.

8. N. H. Goddard, *The Perception of Articulated Motion: Recognizing Moving Light Displays*. PhD thesis, University of Rochester, Rochester, NY, 1992.

9. T. Darrell and A. Pentland, "Space-time gestures," in *Proc. Computer Vision and Pattern Recognition Conf.*, pp. 335–340, 1993.

10. J. W. Davis and M. Shah, "Visual gesture recognition," *IEE Proc. Vision, Image and Signal Processing*, vol. 141, no. 2, pp. 101–106, 1994.

11. Y. Yacoob and L. Davis, "Computing spatio-temporal representations of human faces," in *Proc. Computer Vision and Pattern Recognition Conf.*, pp. 70–75, 1994.

12. I. A. Essa and A. Pentland, "A vision system for observing and extracting facial action parameters," in *Proc. Computer Vision and Pattern Recognition Conf.*, pp. 76–83, 1994.

13. R. Polana and R. Nelson, "Recognition of motion from temporal texture," in *Proc. Computer Vision and Pattern Recognition Conf.*, pp. 129–134, 1992.

14. C. Cedras and M. Shah, "Motion-based recognition: A survey," *Image and Vision Computing*, vol. 13, no. 2, pp. 129–155, 1995.

15. A. M. Baumberg and D. C. Hogg, "An efficient method for contour tracking using active shape models," in *Proc. Workshop on Motion of Non-Rigid and Articulatea Objects*, pp. 194–199, 1994.

16. S. M. Seitz and C. R. Dyer, "Affine invariant detection of periodic motion," in *Proc. Computer Vision and Pattern Recognition Conf.*, pp. 970–975, 1994.

17. S. M. Seitz and C. R. Dyer, "Detecting irregularities in cyclic motion," in *Proc. Workshop on Motion of Non-Rigid and Articulated Objects*, pp. 178–185, 1994.

18. J. J. Koenderink and A. J. van Doorn, "Affine structure from motion," *J. Opt. Soc. Am. A*, vol. 8, pp. 377–385, 1991.

19. L. S. Shapiro, A. Zisserman, and M. Brady, "3D motion recovery via affine epipolar geometry," *Int. J. of Computer Vision*, vol. 16, pp. 147–182, 1995.

20. C. Tomasi and T. Kanade, "Shape and motion from image streams under orthography: A factorization method," *Int. J. of Computer Vision*, vol. 9, no. 2, pp. 137–154, 1992.

21. G. W. Stewart, *Introduction to Matrix Computations*. New York, NY: Academic Press, 1973.

22. W. H. Press, B. P. Flannery, S. A. Teukolsky, and W. T. Vetterling, *Numerica Recipes in C*. Cambridge, MA: Cambridge University Press, 1988.

23. M. Kass, A. Witkin, and D. Terzopoulos, "Snakes: Active contour models," *Int. J of Computer Vision*, vol. 1, no. 4, pp. 321–331, 1988.

24. D. J. Williams and M. Shah, "A fast algorithm for active contours and curvature information," *CVGIP: Image Understanding*, vol. 1, no. 55, pp. 14–26, 1992.

25. S. M. Seitz and C. R. Dyer, "View-invariant analysis of cyclic motion," *Int. J. o Computer Vision*, 1997. To appear.

TEMPORAL TEXTURE AND ACTIVITY RECOGNITION

RAMPRASAD POLANA
MicroStrategy Inc.
8000 Towers Crescent Drive
Vienna, VA 22182

AND

RANDAL NELSON
Department of Computer Science
University of Rochester
Rochester, New York 14627

1. Introduction

Who has not watched ripples spread across a pool and known water thereby? Or seen leaves shimmer their silver backs in a summer breeze and known a tree? Who has not known the butterfly by her fluttering? Or seen a distant figure walking and known there goes a man? The motion recognition ability of the human visual system is remarkable. People are able to distinguish both highly structured motions, such as those produced by walking, running, swimming or flying birds, and more statistical patterns such as those due to blowing snow, flowing water or fluttering leaves. We have demonstrated similar recognition capabilities in an automated machine vision system using efficient low-level techniques that can be implemented in real-time.

Visual motion has, of course, long been considered as a vital source of information in natural vision systems. In an unpredictable environment, survival often depends on the ability of the visual system to detect moving objects and quickly identify potential sources of danger. Many comparatively unsophisticated systems, such as those possessed by insects and lower vertebrates, are essentially blind to anything that is not moving. While other visual cues essentially provide information about a static world, motion provides the information necessary to interact with a dynamic en-

M. Shah and R. Jain (eds.), Motion-Based Recognition, 87–124.

vironment. Moving objects in a scene are typically the first attended to. Even in the more sophisticated systems of higher vertebrates, including man, motion in the visual field retains its important role. A wide variety of (semi)quantitative information relating to object segmentation, depth, three dimensional shape, and object and observer motion, seems to be derived from the visual motion field.

1.1. VISUAL MOTION ANALYSIS

Understanding visual motion is important for distinguishing the sources of different motions, identifying objects moving relative to the surrounding environment, and also for navigational tasks such as obstacle detection and collision avoidance. Even in the absence of moving objects in the scene, movements of the head and eye cause apparent motion in the images projected on the retina. It is thus important to be able to distinguish quickly objects that are moving relative to the environment from objects whose retinal projections are moving due to head and eye movements.

The potential wealth of information derivable from retinal motion has inspired a large body of work on computation of the exact geometric quantities such as the 3-D shape of objects, their location, and the motion of the observer. This reconstruction problem is sometimes referred to as the *structure-from-motion problem*. Much of the research to date has focussed on finding robust solutions to this structure-from-motion problem. Important navigational problems such as detection of independently moving objects, passive navigation, target pursuit, and related problems, can be thought of as simple applications of a structure-from-motion module of a machine vision system. Though a number of important results have been reported in this area, the high-level shape-from-motion algorithms have proven to be sensitive to the accuracy of the underlying motion information, which is often low. For this reason, their use in real-world situations is rather limited. The structure from motion problem entails (semi)-complete recovery of the relative depth map and the three-dimensional motion parameters. Many nontrivial visual tasks however, do not require complete knowledge of structure of the surrounding environment or three-dimensional motion of the moving objects.

The emphasis on visual motion as a means of quantitative reconstruction of world geometry has tended to obscure the fact that motion can also be used for recognition. In fact, in biological systems the use of motion information for recognition is often more evident than its use in reconstruction. A simple example occurs in the case of the common toad *Bufo bufo*, where any elongated object of a certain size exhibiting motion along the longitudinal axis, is identified as a potential food item, and elicits an orienting response [7]. Another example is the recognition of the female grayling butterflies by males. Tinbergen [25] reported that male butterflies fly towards crude paper models moving overhead, and that their response was not affected by the color or shape of the model. The key stimulus provoking males to fly towards the model turned out to be the pattern of movement: the flickering and up-and-down movements of a butterfly. Scout bees notify other bees the direction of a feeding place they have discovered by means of a 'waggling dance', and the other bees recognize the dance and fly directly toward the food without deviating [27]. It has also been noted that bees approach oscillating flowers almost twice as fast as still flowers [28]. Many organisms have elaborate courting ceremonies in which recognition of the type of movement or dance of the opposite sex, is crucial. For instance, wing flapping in ceratin female insects will cause the males to be attracted more strongly [4]. Birds pay no attention to leaves and branches moving in the breeze, or during a storm, but they immediately observe movement of a living person or animal in the midst of such an environment [23]. In general, a moving object is often distinguished better than a motionless one.

Humans have a remarkable ability to recognize different kinds of motion, both of discrete objects, such as animals or people, and in distributed patterns as in windblown leaves, or waves on a pond. The classic demonstration of pure motion recognition by humans is provided by Moving Light Display (MLD) experiments [14], where human subjects are able to distinguish activities such as walking, running or stair climbing, from lights attached to the joints of an actor. More subtle movement characteristics can be distinguished as well. For example, human observers can identify the actor's gender, and even identify the actor if known to them, by his or her gait. Similar discrimination abilities using motion alone have been observed in non-human animals as well [7].

Such abilities suggest that, in the case of machine vision, it is possible to use motion as a means of recognition directly, rather than indirectly through a geometric reconstruction.

1.2. RELATED WORK

The everyday experience of visual motion incorporates a considerable element of recognition; this may even be its dominant attribute. Yet surprisingly, this aspect of motion has been neglected by the research community on computational motion analysis. A recognition system using motion information can recognize a wide variety of objects and motions. An example is distinguishing two persons by their gait, or more broadly, discriminating between running and walking.

There is a large body of psychological literature addressing the perception of motion, most of it concerned with primitive percepts. A modest amount of this work addresses motion recognition issues at an abstract level [5], [8].

Other researchers have worked towards obtaining higher-level descriptions, usually employing a linguistic approach for representing motion concepts [3]. Human motion, specifically walking, has been studied extensively using model-based approaches [16]. Computational approaches have also been proposed to understand biological ambulatory patterns utilizing specific anatomical constraints [13].

Motion recognition is well studied in the context of Johansson's MLDs. The demonstrations using MLDs show that trajectories of a few specific points corresponding to joints of an actor performing an activity, can be used as a key for recognizing the activity. Consequently, Rashid [20] addresses the correspondence of the points in an MLD sequence between successive frames and the problem of obtaining trajectories of those points. Goddard [10] recognizes MLDs involving single actors moving parallel to the image plane using a connectionist approach utilizing the lower-level features of trajectories.

Trajectories of specific points belonging to an object have also been used outside the context of MLDs. Gould and Shah [11] build a trajectory primal sketch that represents significant changes in motion, with the purpose of identifying objects using the trajectories of a few representative points [12]. In other studies by Allman and Dyer [1] and by Tsai et al. [26], the curvature features of the trajectories are used to detect cyclic motion.

A few studies in speech recognition using visual cues (lip-reading) are relevant to motion recognition. Petajan et al. [18] use images of lip-movements over time. Finn and Montgomery [9] and Mase and Pentland [17] track points around the mouth and use features of specific points to characterize the spoken words.

Only a few researchers have attempted motion recognition directly using purely low-level features of retinal motion information. One example is the study by Anderson et al. [2] on spatiotemporal energy measures to characterize different sources of motion. Following our research on temporal textures, Szummer [24] attempted temporal texture modelling using ARMA models. Steven Seitz [22] developed an affine-invariant technique for detecting periodic motion at an early stage.

1.3. APPLICATIONS OF MOTION RECOGNITION

Motion recognition algorithms, both for temporal texture and activity, have potential applications in several important areas. The motion of many natural objects can be modeled as a periodic activity. A machine algorithm for recognizing such motions would be useful in applications, such as automated surveillance. Motion detection via image differencing can be used for intruder detection; however such systems are subject to false alarms, especially in outdoor environments, since the system is triggered by anything that moves, whether it is a person, a dog, or a tree blown by the wind. Motion recognition techniques can be used disambiguate such situations, and to build more reliable and accurate detection systems.

Another application is in industrial monitoring. Many manufacturing operations involve a long sequence of simple operations, each performed repeatedly at high speed, by a specialized mechanism at a particular location. It should be possible to set up one or more fixed cameras that cover the area of interest, and to characterize the allowed motions in each region of the image(s). Such automated systems can be used in hazardous plant conditions and can also be a cheap alternative compared to manual systems. They could also detect critical breakdowns immediately rather than after a number of defective units have been produced.

Lip-reading systems based on motion recognition can be used for improving the recognition rate of utterances, by characterizing motion of the lips and motion surrounding the mouth area for different utterances. For instance, employing the feature vectors computed by the algorithms described here alone, Virginia de Sa [6] obtained a recognition rate of approximately 70 percent in this application. Recognition of hand gestures can aid in designing better man-machine interfaces. Rhyne and Wolf [21] study the use of gestures for editing operations and menu-oriented operations of pointing, selecting etc. On-line handwritten character recognition can be made easier by recognizing the motion trajectories of the hand for different characters. In medical diagnostics, motion recognition can help in detecting abnormal motions of heart. In meteorological applications, temporal texture recognition techniques can be used for classifying motion patterns of clouds and of ocean waves. Another potential application is the analysis of traffic flow in busy highways and intersections.

1.4. CLASSES OF MOTION

As a first step towards motion recognition by a machine, we define three common classes of visual motion. Our classification is based on the spatial and temporal regularity exhibited by the two-dimensional motion field observed in a sample image sequence (the two-dimensional motion field is the projection of the three-dimensional motion in the scene onto the image plane). We define the first class, *temporal textures*, to be motion patterns that exhibit statistical regularity but have indeterminate spatial and temporal extent. Examples of temporal textures include wind blown trees or grass, turbulent flow in cloud patterns, ripples on water, the motion of a flock of birds etc. The second class, *activities*, consists of motion patterns that are temporally periodic and possess compact spatial structure. Examples of activities include walking, running, rotating or reciprocating machinery, etc. A third class, *motion events*, consists of isolated simple motions that do not exhibit any temporal or spatial repetition. Examples of motion events are opening a door, starting a car, throwing a ball etc.

Because of the differences in their motion characteristics, different recognition algorithms are needed for different classes. For example, recognition techniques for activities and motion events have to explicitly undo the spatial translation to achieve spatial translation invariance. But the techniques for temporal textures do not require compensation of spatial translation, because of the spatial uniformity inherent in temporal textures.

In the following section, we present the experiments and results related to the classification of temporal textures. We describe an algorithm for detecting periodic activities in image sequences exploiting its temporal periodicity in section 3. In section 4, we demonstrate the recognition of activities while the object of interest remains stationary in the image frame. In the same section, we also describe tracking and normalization with respect to scale changes for non-stationary objects. In section 5, we conclude the chapter with a discussion of some extensions and possible future research directions.

2. Recognition of Motion from Temporal Texture

Despite the differences in domain, some techniques of spatial texture analysis are applicable to temporal textures. Classical texture analysis is concerned with the identification of spatial invariances in the gray-level patterns in an image region. These invariances may be either structurally or statistically defined. The basic idea is to characterize different sorts of "stuff" of indeterminate spatial extent in terms of such invariances. We extend this basic idea into the temporal dimension with the idea of recognizing similar "stuff" in dynamic scenes. This is motivated in part by the existence of a large class of natural phenomena that seem to have characteristic motions, but indeterminate spatial extent. Examples include wind-blown trees or grass, turbulent flow in cloud patterns, ripples on water, falling snow, the motion of a flock of birds or a crowd of people, etc. As with spatial textures, the main criteria for selecting features are that they change little within a given texture (i.e. an area of the same stuff), and that they vary significantly between different textures.

Since most changes along the time dimension are due to motion in the image, it makes sense to preprocess the time-varying image to obtain motion information, as it is in object motion that the physical invariances lie. In this case, a natural choice is the optic flow field. The basic source of information is thus a time varying vector field representing an approximation to the two dimensional motion field induced by movement in the world. Such a field contains considerably more information than the scaler valued field associated with gray level texture analysis. In addition, the direction and magnitude of motion have a more direct relationship to typically salient events in the world than the gray-level of a single pixel. A problem with using optic flow is that it is difficult to compute accurately. One solution is to devise measures that are insensitive to inaccuracy. Another is to utilize partial, albeit more reliable information. An example is *normal flow*, the gradient parallel component of the optic flow, which is simpler to compute locally from an image sequences than the full motion field.

The simplest local motion measures are the magnitude and direction of the normal flow. We examine several statistical features based on the distribution of these first-order quantities. The direction and magnitude can be combined locally, both spatially and temporally to obtain second order local motion measures. We also examine features based on the distribution of some second order measures. All these are described below.

2.1. TEMPORAL TEXTURE FEATURES

A useful first order statistic can be derived from the distribution of flow directions. Intuitively, what is being measured is the non-uniformity in direction of motion. Our non-uniformity statistic was computed by discretizing the direction into 8 possible values, computing a histogram over the relevant neighborhood of the image, and summing the absolute deviation from a uniform distribution. It should be noted that the normal flow direction at a pixel is parallel (or anti-parallel) to the gradient direction. Thus measures based on the normal flow direction alone depend on the underlying intensity texture. To reduce this dependence, the normal flow directions in the histogram are normalized by the 4-way histogram of gradient directions. This feature is invariant under translation, rotation, and temporal and spatial scaling.

A useful statistic based on the distribution of the normal flow magnitude is the average flow magnitude divided by its standard deviation. The scaling by the standard deviation has the effect of making the measure robust under scaling changes. One way to think of this statistic is as a measure of "peakiness" in the velocity distribution. It is also invariant under translation, rotation, and temporal and spatial scaling.

Second order statistics of the normal flow direction distribution can be derived from the difference statistics, which give the number of pixels pairs at a given offset which differ in their values by a given amount. These difference statistics can be represented by a co-occurrence matrix of the normal flow direction surrounding a pixel. Co-occurrence matrices are computed for four directions, (horizontal, vertical, positive diagonal and negative diagonal), at a distance proportional to the average flow magnitude. This yields invariance with respect to scaling. In each direction the ratio of the number of pixel pairs differing in direction by at most one to the number of pixel pairs differing by more than one is computed. This ratio is the sum of the first two difference statistics to the sum of the last three difference statistics. Logarithms of the resulting ratios are used as a feature in each of the four directions, and represent a measure of the spatial homogeneity of the flow. These features are invariant under translation, rotation and scaling.

Finally, we considered statistics of some second order flow features, namely, estimates of the divergence and curl of the motion field obtained from the normal flow. Positive and negative divergence, and positive and negative curl were taken as separate features to give four different second order features. The features used are the mean values of these quantities over the region of interest. They are invariant with respect to rotation and translation, but not scaling. If scale invariant features are desired, ratios of the differential measures can be used.

2.2. CLASSIFICATION EXPERIMENTS

A set of image sequences representing both oriented temporal textures such as flowing water and non-oriented textures such as leaves fluttering in the wind was digitized. In addition, sequences representing uniform expansion and rotation of a textured scene were obtained. These were used in classification experiments utilizing the features described above. Seven different texture samples, listed below, were used for the experiments.

A: fluttering crepe paper bands
B: cloth waving in the wind
C: motion of leaves in the wind
D: flow of water in a river.
E: turbulent motion of water
F: uniformly expanding image produced by forward observer motion
G: uniformly rotating image produced by observer roll

For each sample texture, two image sequences consisting of 16 256x256 pixel frames taken at 30 Hertz were split into quadrants to obtain eight independent sample image sequences of 128x128 pixels. The normal flow field was computed between each consecutive pair of image frames using a

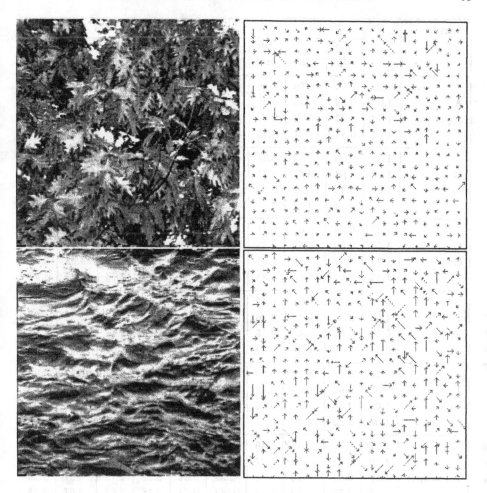

Figure 1. Sample texture and normal flows of turbulent water and leaves

multi-resolution flow computation, with the direction of normal flow quantized to one of eight directions. Sample images and normal flows in the form of needle diagrams for leaves in the wind and turbulent water are illustrated in 1. The end result of the processing was a sample of 8 normal flow sequences of fifteen frames each for each texture.

Classification experiments were run using a nearest centroid classifier. More elaborate classifiers could be used, but the nearest centroid method gives a fairly direct indication of the utility of the features. The features used were those described in the previous section, namely

a: nonuniformity of normal flow direction
b: mean flow magnitude divided by its standard deviation
c: directional difference statistics in four directions

texture	b	c				a	d			
		pos	*neg*	*pos*	*neg*		*hor*	*vert*	*pos*	*neg*
	mag	*div*	*div*	*curl*	*curl*	*dir*			*diag*	*diag*
	1.09	0.46	-1.00	0.26	-0.23	0.94	2.17	3.16	2.66	2.24
A	1.07	0.39	-0.95	0.26	-0.20	0.92	2.66	3.32	3.79	2.78
approach	1.02	0.33	-0.94	0.24	-0.18	0.92	2.36	3.17	3.27	2.49
	1.01	0.42	-1.01	0.33	-0.25	0.91	2.59	3.45	3.37	2.68
	1.08	0.59	-0.69	0.27	-0.19	0.48	5.01	9.08	5.19	5.17
B	1.00	0.64	-0.57	0.24	-0.22	0.66	4.67	8.35	4.96	5.05
bands	1.08	0.46	-0.62	0.21	-0.20	0.83	4.35	7.87	4.51	4.40
	1.22	0.54	-0.54	0.18	-0.20	0.69	5.31	8.95	5.54	5.45
	1.41	0.64	-0.62	0.31	-0.32	0.92	4.26	6.17	4.80	4.28
C	1.52	0.53	-0.55	0.32	-0.33	0.91	4.93	5.68	6.14	4.82
cloth	1.28	0.61	-0.59	0.31	-0.30	0.94	3.39	4.97	3.88	3.33
	1.39	0.61	-0.64	0.33	-0.34	0.90	3.59	4.73	4.27	3.56
	0.96	0.70	-0.21	0.19	-0.29	0.94	1.48	2.10	2.27	1.46
D	1.06	0.30	-0.43	0.26	-0.17	0.95	1.55	2.28	2.35	1.57
leaves	0.88	0.52	-0.43	0.25	-0.27	0.96	1.26	1.86	2.05	1.23
	0.95	0.38	-0.39	0.29	-0.26	0.97	1.30	1.87	2.05	1.24
	1.18	0.43	-0.39	0.09	-0.58	0.92	2.95	4.07	3.52	3.02
E	1.02	0.62	-0.42	0.08	-0.66	0.94	3.25	4.18	4.07	3.39
rotate	1.03	0.38	-0.35	0.05	-0.66	0.93	2.92	3.97	3.62	3.07
	1.08	0.52	-0.33	0.11	-0.72	0.94	2.78	3.78	3.59	2.90
	1.12	0.72	-0.63	0.40	-0.39	0.94	2.45	2.97	3.96	2.52
F	1.20	0.81	-0.58	0.37	-0.40	0.92	2.61	3.05	4.30	2.69
turbulence	1.10	0.59	-0.76	0.42	-0.39	0.94	2.18	2.73	3.67	2.25
	1.06	0.79	-0.52	0.43	-0.42	0.94	2.16	2.61	3.67	2.15
	1.29	0.44	-0.55	0.19	-0.16	0.86	4.63	5.14	7.15	4.79
G	1.49	0.48	-0.38	0.17	-0.18	0.81	5.02	5.61	7.03	5.11
water	1.25	0.51	-0.58	0.20	-0.18	0.88	4.29	4.77	6.21	4.50
	1.51	0.44	-0.52	0.22	-0.22	0.88	3.86	4.17	6.07	3.87

TABLE 1. Sample texture features showing interclass and intraclass variations

d: positive and negative curl and divergence estimates

The first four samples of each texture are used as training set to compute the centroid of the cluster corresponding to that texture in the feature space. The different feature values are converted into common units by mapping the average of the resulting centroids to a unit vector. Table 1 contains the values of these features for each flow sample. It can be seen

Feature Combination	Correct Classification	Percent Success
all	28	100
c,d	28	100
b,d	24	85
a,c	21	75
d	21	75
c	20	71

TABLE 2. Nearest centroid classification

comp.	b		c		a		d			eigenval
1	.29	.51	.95	.01	-.29	-.20	.15	.20	.17	.16	54.95
2	-.24	.42	.37	.78	.07	.10	-.82	-.37	-.89	-.69	5.82
3	-.08	.78	-.50	.11	-.15	0.0	.15	.05	0.0	.13	2.54

TABLE 3. First three principal components

that, overall, the within sample variation is smaller than the between sample variation as desired. No single feature is sufficient to distinguish all the textures, but for each texture, there is at least one feature that clearly separates it from the others. For example, as would be expected, texture A containing an approaching object is distinguished by high divergence. For texture B, containing moving vertical bands, the second order difference feature in the vertical direction clearly separates it from the rest.

The remaining four samples are tested using a nearest centroid classification scheme. The results of classification are summarized in Table 2. Note that none of the features alone is sufficient to separate all the textures, but the combination gives one hundred percent success in the classification of the test cases. In fact, the second order features alone are sufficient for successful classification in all cases.

We also performed a principal component analysis of these features to gauge the relative importance of different features in producing the variation in the sample values. The first three principal components of the entire data set are shown in Table 3. Note that the first principle component has a high eigenvalue, and relatively high proportions of the second order features, particularly positive and negative divergence. This is consistent with the finding that the second order features alone are sufficient for classification in this case. The principal components within each sample contain small absolute coefficients for the same second order features, showing that these features are most useful in classification.

3. Detecting Activities

Activities involve a regularly repeating sequence of motion events. If we consider an image sequence as a spatiotemporal solid with two spatial dimensions x, y and one time dimension t, then repeated activity tends to give rise to periodic or semi-periodic gray level signals along smooth curves in the image solid. We refer to these curves as *reference curves*. If these curves could be identified and samples extracted along them over several cycles, then frequency domain techniques could be used in order to judge the degree of periodicity.

First we shall formalize the concept of a periodic object. An object is defined as a set of points P. Associated with each $p \in P$ is a function $X_p(t)$ giving its location (in a fixed 3D coordinate system) as a function of time. A stationary periodic object (ie. a stationary object exhibiting periodic activity) has the property that $X_p(t) = X_p(t + \tau)$ for all $p \in P$, where τ is the time period for one cycle of the activity and is independent of p. We now define a translating periodic object. Such an object has the property that $X_p(t) = Y_p(t) + Z(t)$, where Y_p satisfies $Y_p(t) = Y_p(t + \tau)$ and $Z(t)$ is a path in 3D space independent of p. It can be assumed that $Z(0) = \mathbf{0}$ so that $X_p(0) = Y_p(0)$. Intuitively, a periodic object characterized by $Y_p(t)$ is translated along the path $Z(t)$ (we are assuming the object does not undergo any rotation and the viewing angle does not change). If we compensate for the translation of the object, we would be looking at a stationary periodic object as shown by the equation: $X_p(t) - Z(t) = Y_p(t) = Y_p(t + \tau) = X_p(t + \tau) - Z(t + \tau)$. Note that $Z(t)$ is not necessarily periodic. Note also that a stationary periodic object is a special case of translating periodic object with no translation, or in other words $Z(t) = \mathbf{0}$ for all t.

Corresponding to each point p of a translating periodic object, we define a 3D-reference curve $R_p(t)$ to be the path $X_p(0) + Z(t)$. We also define a 2D-reference curve $r_p(t)$ corresponding to a point p of the object, to be the projection of $R_p(t)$ onto the image plane over time (hence $r_p(t)$ is a curve in (x, y, t) space). The gray-level signal along the 2D-reference curve $r_p(t)$ is determined by the set of points of the object that appear along the 3D-reference curve $R_p(t)$. It can be shown that the same set of points of the object recur periodically along each reference curve $R_p(t)$. For example, the point p is on the reference curve $R_p(t)$ at time zero, and it coincides with the reference curve at regular intervals of τ (since

$X_p(\tau) = Y_p(\tau) + Z(\tau) = Y_p(0) + Z(\tau) = X_p(0) + Z(\tau)$). Similarly, every other point of the object on the reference curve $R_p(t)$ recurs along $R_p(t)$ at intervals of τ.

Consider the case of human walking. This is an example of a nonstationary activity; that is, if we attach a reference point to the person walking, that point does not remain at one location in the image. If the person is walking with constant velocity, however, and is not too close to the camera, then the reference point moves across the image on a path composed of a constant velocity component modulated by whatever periodic motion the reference point undergoes. Thus, if we know the average velocity of the person over several cycles, we can compute the spatiotemporal line of motion along which the periodicity can be observed. If the person moves with average velocity (u, v) the spatiotemporal line of motion will be determined by the equations $(x, y) = (u, v) * t + (x_0, y_0)$, where (x, y) is the position of the object in space at time t and (x_0, t_0) is the position at time zero. This applies to any object undergoing constant velocity locomotion.

3.1. PERIODICITY DETECTION

From Fourier theory we know that any periodic signal can be decomposed into a fundamental and harmonics. That is, we can consider the energy of a periodic signal to be concentrated at frequencies which are integral multiples of some fundamental frequency. This implies that if we compute the discrete Fourier transform of a sampled periodic signal, we will observe peaks at the fundamental frequency and its harmonics. Hence, in theory, the periodicity of a signal can be detected by obtaining its Fourier transform and checking whether all the energy in the spectrum is contained in a fundamental frequency and its integral multiples.

The real-world signals, however are seldom perfectly periodic. In the case of signals arising from activity in image sequences, disturbances can arise from errors in the uniform translation assumption, varying background and lighting behind a locomoting actor, and other sources. In addition, for computational purposes, we need to truncate the signal at some finite length which may not be an exact integral multiple of its period. Nevertheless, the frequency defined by the highest amplitude often represents the fundamental frequency of the signal. Hence we can get an idea of the periodicity in a signal by summing the energy at the highest amplitude frequency and its multiples, and comparing that quantity to the energy at the remaining frequencies. In practice, since peaks in a Fourier transform tend to be slightly broadened for a variety of reasons caused by factors like the finite length of the sample, we define the periodicity measure p_f of a signal f as a normalized difference of the sum of the power spectrum values at the highest amplitude frequency and its multiples, and the sum of the power spectrum values at the frequencies halfway between. That is,

$$p_f = (\sum_i F_{iw} - \sum_i F_{(iw+w/2)})/(\sum_i F_{iw} + \sum_i F_{iw+w/2})$$

where F is the energy spectrum of the signal f and w is the frequency corresponding to the highest amplitude in the energy spectrum.

The measure is normalized with respect to the total energy at the frequencies of interest so that it is one for a completely periodic signal and zero for a flat spectrum. In general, if a signal consists of frequencies other than one single fundamental and its multiples, its periodicity measure will be low.

Because the signal along any given reference curve in the image solid may be ambiguous, we need a way of combining periodicity measures of a number of signals from reference curves associated with the same actor. Each frequency w is assigned a value equal to the sum of the periodicity measures P_w from all the signals whose highest amplitude occurred at that frequency. The result is the same as suppressing all but the maximum frequency in each transform, weighting each by the periodicity measure of the signal, and summing them. The maximum value of this combined signal is taken as the fundamental frequency, and the associated periodicity measure is the average of the periodicity measures of the contributing signals. Thus, the periodicity measure P for an entire image sequence is defined as

$$P = \max_w(P_w/n_w)$$

where n_w and P_w are the number of pixels at which the highest amplitude frequency is w and the sum of periodicity measures at those pixels respectively.

In the following, we assumed that any activity that existed in the data would be either stationary, or locomotory in a manner that produced an overall translating motion. We also assumed that there was at most one actor in the scene, though a certain amount of background motion could be tolerated. A third assumption is that the viewing angle and the scene illumination does not change significantly so that the intensity along the reference curves remains periodic. The first assumption turns out not to be too restrictive – a large number of natural periodic activities fit into one of the two categories. The second can be relaxed with some additional preprocessing. Refer to the discussions section for how this can be achieved and how the other assumptions can be relaxed as well.

The first step of the algorithm is to identify locations in the scene where movement of any sort is occurring. This is done by computing the normal flow magnitude at each pixel between each successive pair of frames using a spatiotemporal differential method. Those pixels at which significant motion is present are marked, and the centroid of the marked pixels computed in each frame. The mean velocity (if any) of the actor is then computed by fitting a linear trajectory to the sequence of centroids. This is where the one-actor assumption comes into play. If several actors were present, simple clustering techniques could be used to isolate the regions in the scene corresponding to different activities. The reference curves were taken as the lines in the spatiotemporal solid parallel to that generated by the linear-fitted trajectory of the centroid. Signals were extracted along these curves, and those that displayed significant spread over a period of at least half as long as the signal were selected for processing. This had the effect of eliminating the need to process regions in which no motion occurred, as well as regions affected only by an occasional blip. The periodicity measures for all signals extracted is computed and are used in computing periodicity measure P for the entire image sequence as described above.

The periodic activity detection algorithm can be summarized as follows:

- *Input:* The input to the algorithm is a digitized image sequence consisting of 128 frames of resolution 128x128 pixels.
- *Output:* A periodicity measure indicating the amount of periodicity in the image sequence is computed. This is used to decide whether the image sequence contains a periodic activity and if so, to locate the region of the activity.
- *Step 1.* Compute normal flow magnitude at each pixel between each successive pair of frames using the differential method.
- *Step 2.* Mark pixels corresponding to significant motion in the scene by thresholding the normal flow magnitude. Compute centroid of the marked pixels in each frame. Compute the mean velocity (if any) of the actor by fitting a linear trajectory to the sequence of centroids. Take reference curves to be the lines in the spatiotemporal solid parallel to the linear trajectory of centroids of motion.
- *Step 3.* Extract gray-level signals along the reference curves. Compute the dominant frequency w and the periodicity measure P_w for each individual signal extracted.
- *Step 4.* Compute overall periodicity measure P for the image sequence using formula given in the last section.

3.2. EXPERIMENTS

We ran experiments on four different activities, and a number of non-periodic motions. The sequences were first recorded on video and then digitized later with suitable temporal sampling so that at least four cycles of the activity were captured in 128 frames. Following is a description of each activity and the conditions under which they were digitized.

- Walk: A person walking across a room viewed in profile. Six sequences of 128 frames of size 128x128 pixels were obtained. Half the sequences contained one person and the other half a second.
- Exercise: A person performing jumping jacks. Four sequences of 128 frames of 128x128 pixels, two each of two different people.
- Swing: A person swinging viewed from the side. Six sequences of 128 frames of 128x128 pixels, three each of two different people.
- Frog: A toy frog simulating swimming activity viewed from above. Four sequences of 128 frames of 64x256 pixels.
- Nonperiodic: Various sequences taken from television shows and live outdoor shots: splashing water, closeup of crowd at a political rally, a plane flying overhead, a robot hand picking up and manipulating objects (2 sequences), the input to an eye tracker (eyeball movements), leaves fluttering in the wind, turbulent flow in a stream. In all, 8 sequences of 128 frames of 128x128 pixels.

The swing and exercise activities were shot outdoors and contained background motion as well. Among the periodic activities, a single sequence of uniform rotation is included as well. Sample images of these activities are shown in figure 2.

The periodicity measures computed using the above algorithm are plotted for all 20 periodic and all 8 non-periodic sequences in figure 3. As is evident from the graphs and the projected scatter plot, the technique separates complex periodic from non-periodic motion nicely. The requirement that an empirically determined threshold be used is not a great drawback in this case, nor is it particularly surprising, since even the the intuitive notion of periodic activity falls on a continuum. Is the motion of a branch waving somewhat irregularly in the wind periodic or non-periodic? Here, we classified it as non-periodic, but it had one of the higher periodicity measures, as might be expected.

The method we described satisfies the several desirable invariances. It is invariant to image illumination, contrast, translation, rotation and scale. It is also invariant to the magnitude of locomotory motion and the speed of the activity. It is also fairly robust with respect to small changes in viewing angle. The periodicity measure does not depend on the number of pixels involved in the activity. If desired, a restriction on the minimum number of pixels can be imposed so that only activities of a minimum size can be recognized. The swing and exercise sequences were taken outdoors where there is a small amount of background motion. This comprises not only moving trees and plants, but also moving people and occasional crossing of a car. The thresholding stage on motion magnitude in step 2 of the algorithm (in our implementation one-half pixel per frame is used) eliminates small background motion, but it can not eliminate larger background motion such as produced by a car passing. That periodicity can be detected even in this case demonstrates that the technique is reasonably tolerant of background clutter and an occasional disturbance. The technique also provides a method for localizing activity in the scene by back-projecting the reference curves having high periodicity measures into the image solid.

The complexity of detection is proportional to the number of pixels involved in the activity. About half the work is computing the fast Fourier transforms at each of the pixels. Most of the rest of the time is occupied by the motion detection process. The detection procedure currently runs on an SGI machine using four processors and it take approximately 15 seconds to process a 128 frame sequence of 128x128 images.

3.3. EXTENSIONS

So far, we have assumed that the actors giving rise to the activity move with constant velocity along linear paths. The case of nonlinearly moving objects can be handled by tracking the object of interest given a coarse estimate of its initial location and velocity. This would generate reference curves that were not straight lines. We have already demonstrated the usefulness of the centroid of motion for computing the velocity of linearly moving objects. It could also be used for tracking the actors moving on more complex trajectories. Use of the motion centroid can be unreliable in estimating the centroid of the object if the shape of the object changes as it moves. In this case use of a prediction and correction mechanism using past values over a sufficiently long period can help.

The detection scheme also assumes that there is only one activity in the scene except for some background clutter. If there are multiple activities in the scene, this detection technique can still be applied provided the activities can be spatially isolated so that they do not interfere with each other. In this case they can segmented using the motion information and later tracked separately. Even an occasional crossing of different activities can be tolerated as long as the regions can be separated again later. In our experiments, the periodic activity samples consist of at least four cycles of the activity. Minimum four cycles were used to detect the actual frequency given that there is a considerable amount element of non-repetitive structure from the background in the case of translating actors.

4. Recognizing Activities

In this section, we describe a robust method for recognizing activities, including ones, such as walking, that involve simultaneous translation of the actor. The recognition scheme is based on low-level features of motion, and does not require the recognition or tracking of specific parts of the actor. We make use of the motion field computed between successive frames as a basis to segment and track the actor, detect scale changes and compensate for translation and scaling of the actor in the image sequence. The resulting gray-level image sequence consists of the activity at a constant distance from camera while the actor remains stationary in the image frame. We combine this with methods for detecting stationary activities and classifying them to obtain a real-time implementable system of activity recognition.

4.1. SPATIAL AND TEMPORAL NORMALIZATION

We used a spatiotemporal motion magnitude template as a basis for the recognition of activities. In order for this to work, the motion to be identified must be normalized with respect to spatial scale, spatial translation and temporal scale and translation. Template matching is a well studied and frequently effective method of recognition. It fails when sufficiently rigid normalization cannot be carried out. It turns out that periodicity inherent in motion such as walking or running is a sufficiently strong cue to allow normalization to be performed.

Given a gray valued image sequence, we first detect pixels corresponding to independently moving objects. These pixels are grouped using spatial clustering methods and an object of interest (the actor performing the activity) is selected and tracked. The subsequent activity detection and recognition can be applied to each independently moving object. For each selected object, a spatiotemporal template of motion features is obtained from the motion of the corresponding pixels, and it is used to match the test sample with the reference motion templates of known activities.

Spatial translation invariance is achieved by tracking the selected object through successive frames, estimating the spatial translation parameters and compensating for the translation of the object. The spatial translation parameters are estimated by assuming the object is moving along a locally linear trajectory and using a least squares technique. Spatial scale invariance is achieved by measuring the spatial size of the object through successive frames, estimating the spatial scale parameters and compensating for the changes in scale. This method is described in detail in the next subsection.

Temporal scale invariance is achieved by detecting the frequency of the activity and obtaining one cycle of activity by averaging motion information of multiple cycles. A more complete discussion of the periodicity detection and frequency estimation can be found in [19]. Temporal translation has turned out to be hard to estimate from the motion information, but it was handled in the matching stage by matching the test template with reference template at all possible temporal translations.

The following sections describe the steps of the algorithm in greater detail.

4.2. DETECTING INDEPENDENTLY MOVING OBJECTS

In order to apply the motion recognition algorithm, the moving actor must be segmented out of the scene. If there are multiple actors in the scene, the recognition algorithm should be applied to each of them separately. If the motions do not interfere with each other, they can be segmented and tracked separately. By using a predictive tracker, an occasional crossing of different activities can be tolerated as long as the regions can be separated again later. For this, it is important to initially detect each actor in the scene and spatially isolate them.

Fortunately, independent motion provides an exceptionally strong segmentation cue. Nelson [15] has demonstrated a real-time method of detecting independently moving objects even in the case that the observer is itself moving. Using such a method, we can detect the pixels in an image sequence that exhibit motion independent of that of the background and segment the image frames into distinct regions corresponding to different moving objects. Some other common methods of segmenting multiple moving objects are: using color cues, using distance obtained from a range sensor, or selecting objects moving with a certain image velocity. The exact method is not of particular importance as long as there is a way of selecting an actor of interest.

4.3. TRACKING IN THE PRESENCE OF OTHER MOVING OBJECTS

Once an actor has been detected, the object must be tracked. There is a lot of excellent work on tracking that gives good results in the presence of temporary occlusion, noise, pixel dropout, and other sorts of interference. For the purposes of our experiments, however, the motion signal was sufficiently strong that fairly simple techniques sufficed. If there is a single moving object in the scene, the object can be effectively tracked by following the centroid of the moving pixels corresponding to that object, and fitting a locally linear trajectory. For the scenes involving multiple actors, we used a simple predicitive filter, where the region expected to be occupied was predicted from past frames, and only pixels within that frame allowed to contribute to the new position estimate. This allows us to deal with the sort of temporary occlusion produced, e.g., by walking persons whose visual paths cross, and to exclude outliers due to other motion in the scene. This allowed us to deal with outdoor scenes involving a few actors and occasional road traffic and wind movement of vegetation. For scenes with more severe motion clutter, there is a large suite of better tracking algorithms that could be applied.

A demonstration of our simple tracking algorithm in the presence of multiple moving objects and occlusions by other objects is shown in figure 4. The illustration shows an image sequence consisting of two persons walking towards and crossing each other. The object of interest in this case is the person walking from right to left. The first eight frames of the 64 frame image sequence given here consist of only the first person walking. The second person temporarily occludes the first person. In the first eight frames, we thresholded the motion to highlight significant motion. Using those locations we estimated the shape of the object of interest in the form of a rectangle surrounding the object, which is illustrated in the figure. The estimated positions are shown with a plus (+) sign. It can also be seen that the tracking algorithm smoothly tracks the first person even when there is occlusion.

4.4. ELIMINATING BACKGROUND MOTION

Estimating the trajectory of an object and compensating for its translation so that the object remains in the center of the image causes the previously stationary background to appear to be moving. In the compensated image frames, the background will be moving with the same velocity magnitude as the object velocity estimated but in the opposite direction. After computing the flow fields between successive frames of the compensated image sequence, we eliminate any motion that is consistent with the background velocity by making the estimate at that point unknown.

Alternatively, instead of translating the image frames and recomputing the flow frames, we could update the flow field by subtracting the estimated trajectory motion (u, v) of the object. Such a subtraction, however leads to large inaccuracies in the measured flow, primarily because the differential techniques we use for speed have low accuracy, and subtraction can cause the values to lose all significance. Hence, we chose to recompute the motion measurements after translating the image frames.

4.5. CHANGING SCALE

In this section, we show how the changes in spatial scale of the activity can be detected and compensated for. We make a key assumption here: that the height of the object of interest does not change through the image sequence. This is certainly true for the activities of human walking, running etc., and it is a reasonable assumption for a host of other activities. (Even when the height is changing, the periodic repetition of the activity requires that the same height recur through successive cycles of the activity, and hence fitting the model described below over many periods in this case will give good estimates of scale changes).

The model of projected image height of the object is illustrated in figure 5. Let H be the actual height of the object in three-dimensional world. (It is assumed that H is unchanging over time). According to the imaging model illustrated in the figure, the image coordinates (x_t, y_t) in image frame t are related to the three-dimensional world coordinates (X_t, Y_t, Z_t) as $(x_t, y_t) = (X_t/Z_t, Y_t/Z_t)$, where Z_t is the distance of the object from the camera at image frame t. From this we obtain the formula $h_t = H/Z_t$ where h_t is the projected image height of the object at image frame t. Now, if we assume the object is approaching or moving away from camera at a locally constant velocity, say W, then $Z_t = Z_0 + W * t$ is the distance of the object from camera. Using the relation $h_0 = H/Z_0$, we find the image height of the object over time to be

$$h_t = h_0/(1 + w * t)$$

where $w = W/Z_0$ is a constant scale factor. (Notice that w is negative if the object is approaching the camera, positive if the object is moving away, and it is exactly equal to zero if the object's distance from the camera does not change).

By estimating height of the object in each flow frame and using the model described above, we obtain an estimate for the locally constant scale factor w and then compensate for the scale changes by scaling the image frame t so as to match the scale of the activity in the reference database (which is fixed before hand). Unfortunately, the relation $h_t = h_0/(1+w*t)$ is not linear in w and so we can not directly use the least squares technique to estimate w. Instead, we use an approximation to the model $h_t = h_0(1-w*t)$ which is a good approximation if the term $w*t$ is small. We keep $w*t$ small by using the model in small temporal neighborhoods, so that t is very small (also $w = W/Z_0$ is small if the distance of the object from camera is large compared to the speed with which it is approaching or moving afar, which is true in most circumstances). Note that the relation $1/h_t = (1 + w * t)/h_0$ is linear in w, but using least squares to minimize the error between observed and model heights is not same as minimizing the error between observed and model inverted heights, and hence the estimate of w is not same, even though it may produce a reasonable estimate.

The steps involved in detecting and compensating for changes in scale are thus: measure the image height of the object in each flow frame, use the heights in the last K frames to estimate the scale factor w and scale the image frame t to match the reference database scale, and recompute flow frame t.

To demonstrate the above technique, we have digitized an image sequence from the video recorded by NIST (National Institute of Standards & Technology) using a video camera mounted on a van looking straight ahead while the van is being driven on the road at about 30mph around the NIST grounds in Gaithersberg, Maryland. The image sequence is shown in figure 6 which consists of a person walking across the street as the van is approaching. The image heights of the person for this sequence are hand-measured and plotted (dotted-line) in figure 7. As can be seen, a linear fit over the entire image sequence is inappropriate for this data. We estimated the scale factor over the entire sequence using the least squares technique for inverted heights and plotted (solid-line) the resulting fit to the data in the same figure. It gives a reasonably good approximation in the beginning where the distance from camera is large and at the right end a slight deviation from actual heights is seen where the distance of the person from camera is smaller compared to the speed of the vehicle. By using local linear models a better approximation is obtained and when the image frames are scaled and tracked as before, we obtain the stationary walking activity shown in figure 6. The motion magnitude feature vector is computed for this image sequence and classification algorithm applied and it was correctly classified as walking (there being six other choices and unknown).

4.6. FEATURE VECTOR COMPUTATION AND MATCHING

Given a gray-valued image sequence, the actor is detected, tracked and spatial scale changes are estimated. The image sequence is transformed so as to compensate for the spatial translation and scale changes of the actor. The resulting modified image sequence puts the actor at the center of the image frame and at the same scale throughout the image sequence. The image frame is reduced to the size of the object and the motion is computed between successive image frames.

Each flow frame t is divided into a spatial grid of XxY dimension and the motion magnitudes in each spatial cell are summed. Let $M(x, y, t)$ be the motion magnitude in flow frame t corresponding to spatial cell (x, y). According to the definition of a periodic activity, for each fixed (x, y), the signal $M(x, y, t)$ over time should be periodic. For each (x, y), we compute the periodicity index as described in [19] and combine the individual periodicity indices to get a periodicity measure for the whole image frame. By thresholding the resulting periodicity measure it is possible to determine if the motion produced by the object is periodic. If it is found that there is sufficient periodicity in the motion, we proceed to compute the feature vector of motion magnitudes.

The frequency of the activity is found along with the periodicity measure and it is used to divide the entire image sequence into cycles. The flow frames are folded over temporally, so as to obtain a single cycle of the activity and the motion in different cycles is averaged to obtain motion in a single combined cycle of activity. The length of the cycle is divided into T temporal divisions and motion is totalled in each temporal division corresponding to each spatial cell (x, y) separately. The resulting spatiotemporal motion template is used as the feature vector to match against reference motion templates of known activities.

The classification method we have used is the nearest centroid algorithm, which is simple to implement and effectively shows the discriminating power of the feature vector. In this algorithm, feature vectors of a number of reference sample image sequences of each known activity are computed and their centroid is taken as the reference feature vector for the corresponding class. Feature vectors computed from a test sample image sequence are matched against the reference feature vectors of each class and the test sample is classified as the class corresponding to the centroid closest in euclidean distance.

To be able to recognize an activity we need to fix thresholds on the distances between a test feature vector and the centroids of the classes. This allows us to classify a motion as unknown if it is not sufficiently close to any of the known activities. We can fix the thresholds of a reference class as the average distance of reference samples from the centroid. This way, we would be recognizing a test vector as belonging to class k if the test vector falls within a circular region of radius threshold and center as the centroid. To achieve greater accuracy we first find the principal components of the reference vectors in each class and weigh the test vector elements inversely proportional to the corresponding coefficients in the first principal component. Feature vector elements which are more consistent within the class are given higher priority in matching by the above process and the elements whose the variability is greatest are weighed down. The net result of this procedure is to make the recognition regions around each class centroid ellipsoidal instead of circular. This is an approximation to a Mahalanobis distance representing the within-class distribution. Because our reference class contains only a few samples, we use only the first principle component.

The feature vectors computed, while corrected for temporal scale (frequency) are not corrected for temporal translation (phase). For reference samples, we manually selected the phase so that the same time point of the activity is taken as the beginning of the cycle for all reference vectors considered. For test samples, we handle the unknown temporal phase, by simply trying a match at each possible phase and picking the best. Since the pattern matching phase of the algorithm currently represents only a small fraction of the total computational effort, the above method works efficiently compared to alternative methods of finding the temporal phase of the activity. The temporal resolution of the pattern is typically small (i.e., less than 10 samples per cycle),

4.7. EXPERIMENTAL RESULTS

We ran experiments on seven different types of activities. Following is a description of each activity and the conditions under which they were digitized.

- Walk: A person walking on a treadmill.
- Exercise: A person exercising on a machine.
- Jump: A person performing jumping jacks.
- Swing: A person swinging viewed from the side.
- Run: A person running on a treadmill.
- Ski: A person skiing on a skiing machine.
- Frog: A toy frog simulating swimming activity viewed from above.

All samples were digitized at a spatial resolution of 128x128 pixels, except those for walk and run which were digitized at a resolution of 64x128 pixels. Pixels were 8 bit gray levels. The swing and exercise activities were shot outdoors and contained background motion.

The algorithm was used to successfully classify seven different types of activities which included: walking, running, jumping jacks, exercises on a machine, swinging, skiing and swimming activity of a toy frog. Image frames from a sample each of the seven activities are illustrated in figure 8. The image sequences were first recorded on video and then digitized later with suitable temporal sampling so that at least four cycles of the activity were captured in 128 frames of resolution 128x128 (walking and running are performed on a treadmill to satisfy the stationary activity assumption and they were of resolution 128x64 pixels). Four cycles were needed to reliably detect the fundamental frequency given that there is a considerable amount of non-repetitive structure from the background in the case of translating actors.

We first digitized eight samples of each activity by the same person under the same conditions with respect to scene illumination, background, and camera position. We created the reference database taking half of the samples belonging to each activity. In other words, the reference database consists of four samples of each of the seven activities. The remaining four samples of each activity are used to create the test database. In addition, we digitized four samples of walking by a different person and eight samples of the frog under different lighting conditions and different background and foreground gradients. These samples also differed from the reference database in frequency, speed of motion, and spatial scale. These samples were also added to the test database. The samples in the test database were classified by a nearest centroid classification technique using the samples in the reference database as training set.

We conducted experiments with the three feature vectors, described briefly above, and more completely below. In each case the feature vector consists of a local statistic computed over each of a set of cells constituting a partition of the spatiotemporal solid. We divided each spatial dimension into four divisions and the temporal dimension into six divisions, so that we get a feature vector of length 96. We experimented with three different statistics: total motion magnitude in each cell, direction of maximum motion (where the directions are quantized into eight sectors) and the motion magnitude in maximum motion direction. Sample features vectors are illustrated in 9 using the total motion magnitude statistic for a walk and a run sequence. We initially computed the feature vectors by finding a zero phase marker within a cycle using the method described previously. However, more reliable results were achieved by matching each test feature vector with the reference feature vector six times, corresponding to different temporal offsets, and choosing the best match obtained. The results below utilize the later matching. This classification resulted in correct classification of every sample in the test database, including the samples using a different actor and different backgrounds, which were not represented in the reference database. In other words, the algorithm successfully classified all samples achieving 100% correct classification.

The results of classification using different variations are shown in terms of percentage of test cases correctly classified in table 4. The percentage of correct classification does not give a good indication for the quality of classification. Hence, we also illustrate the results by the confusion matrix which shows how closely test samples belonging to various classes match the reference samples of those classes. The confusion matrix using the total motion magnitude feature vector is shown in Figure 10.

4.8. CLUTTER TOLERANCE

A one hundred percent correct classification does not reveal much about the robustness of the algorithm. To test the degree of robustness of the activity recognition algorithm, we degraded the samples of walking by adding motion clutter due to leaves blowing in the wind. The motion of leaves was chosen instead of random noise because the degradation because it is a realistic example of structured motion clutter that is commonly present in the background.

To understand how much background clutter can be tolerated by this

Feature vector	Total Test Samples	Correct Classified	Percent Success	Failures
maximal motion direction	40	32	80	walk by different actor and frog under different gradients
maximal motion magnitude	40	39	97.5	walk by different actor
total motion magnitude	40	40	100	None

TABLE 4. Classification results

Added Clutter Percentage	Total Test Samples	Successfully Detected	Correctly Classified
25	4	3	3
50	4	3	3
75	4	2	0
100	4	2	0
150	4	1	0
200	4	0	0

TABLE 5. Classification results with motion clutter (samples are of walk)

technique, we experimented with the walk samples by adding motion clutter produced by blowing leaves This structured motion clutter is added in a controlled fashion so that its mean magnitude represents a varying percentage of the mean magnitude of the signal, and the resulting samples are classified using a feature vector of total motion magnitudes. The motion clutter is added at increasing magnifications until the recognition algorithm could no longer classify the activity correctly. The results are tabulated in 5. The results show that the recognition scheme can tolerate motion clutter whose magnitude is equal to one half that of the activity, and it displays degraded, but still useful performance for even higher clutter magnitudes.

4.9. REAL-TIME IMPLEMENTATION

We have implemented the above algorithm on SGI architecture with multiple processors. The complexity is proportional to the number of pixels involved in the activity. The majority of the work involved is the normal flow computation at the original resolution of the image sequence. With four processors, the flow computation at 128x64 resolution takes 40 to 50 milliseconds between two successive image frames. The remaining processing involves frames of much reduced resolution and it takes from 20-30 milliseconds. The implementation includes displaying the original image frame, gradient, flow, and the XxY grid motion magnitudes and the classification result at every frame of the image sequence. A sample screen through a run of the program is illustrated in figure 11. The total computation for each frame takes 60-80 milliseconds. Of course, more processors can be used for faster running times.

The recognition algorithm described above is robust to varying image illumination and contrast because the method uses only motion information which is invariant to these. It is also invariant to spatial and temporal translation and scale due to the normalization of the feature vectors, and the multiple temporal matching. It is also fairly robust with respect to small changes in viewing angle (i.e on the order of 20 degrees). The swing and exercise sequences were taken outdoors where there is a small amount of background motion. This comprises not only moving trees and plants, but also moving people and an occasional crossing of a car. That the activities can be detected even in these cases demonstrates that the technique is tolerant of the usual background clutter and an occasional disturbance.

5. Conclusions

Our goal was to obtain a robust low-level alternative for the recognition of complex motion patterns. We have computed the normal flow between successive image frames using a simple differential technique and used it as a basis to obtain meaningful features. The issue of the accuracy of the motion computation is not of significance, as it is sufficient to show that useful features can be extracted from sufficiently reliable estimates of motion. The use of accurate motion estimation algorithms would only improve the reliability of the techniques described.

We have studied two classes of motion patterns namely, temporal textures and activities, and proposed efficient robust algorithms for their recognition using low-level statistical features based on motion. For both temporal textures and activities we have identified a set of features derived from the optic flow computed from successive frames of an image sequence, and demonstrated recognition using simple classification schemes. We have also shown that periodicity is an inherent low-level motion cue that can be exploited for robust detection of periodic phenomenon without prior structural knowledge. The recognition algorithms were implemented in near real-time using real-world image sequence samples.

We also described a general technique for periodic activity recognition.

The affect of changing viewing angle in activity recognition has not been addressed in our implementation. It was shown that a change in viewing angle of up to 30 degrees can be tolerated in the recognition of walking. However, motion templates derived from views having wider angular separation (compare walking towards the camera with walking perpendicular to the viewing axis), would exhibit substantial differences. A motion template consisting of three-dimensional motions might be used to handle this, but this requires computation of three-dimensional motion vectors which is, as yet, an unreliable and expensive process. The proposed method of handling different viewing angles is to incorporate multiple reference classes corresponding to different views of the same activity.

References

1. M. Allmen and C.R. Dyer. Cyclic motion detection using spatiotemporal surface and curves. In *Proc. Int. Conf. on Pattern Recognition*, pages 365–370, 1990.
2. C. H. Anderson, P. J. Burt, and G. S. van der Wal. Change detection and tracking using pyramid transform techniques. In *Proc. SPIE Conference on Intelligent Robots and Computer Vision*, pages 300–305, 1985.
3. N.I. Badler. *Temporal Scene Analysis: Conceptual Descriptions of Object Movements*. PhD thesis, Univ of Toronto, 1975.
4. J. Crane. Imaginal behaviour of trinidad butterfly. *Zoologica*, 40:167–196, 1955.
5. J.E. Cutting. Six tenets for event perception. *Cognition*, pages 71–78, 1981.
6. Virginia R. de Sa. *Unsupervised Classification Learning from Cross-Modality Structure in the Environment*. PhD thesis, Computer Science Department, Univ of Rochester, 1994.
7. J.P. Ewart. Neuroethology of releasing mechanisms: Prey-catching in toads. *Behavioral and Brian Sciences*, 10:337–405, 1987.
8. J.E. Feldman. Time, space and form in vision. Technical Report 244, University of Rochester, Computer Science Department, 1988.
9. K.E. Finn and A.A. Montgomery. Automatic optically-based recognition of speech. *Pattern Recognition Letters*, 8:159–164, 1988.
10. N.H. Goddard. Representing and recognizing event sequences. In *Proc. AAAI Workshop on Neural Architectures for Computer Vision*, 1989.
11. K. Gould and M. Shah. The trajectory primal sketch: A multi-scale scheme for representing motion characterestics. In *IEEE Conf. Computer Vision and Pattern Recognition*, pages 79–85, 1989.
12. K. Gould, K. Rangarajan, and M.A. Shah. Detection and representation of events in motion trajectories. In Gonzalez and Mahdavieh, editors, *Advances in Image Processing and Analysis*. SPIE Optical Engineering Press, 1992.
13. D.D. Hoffman and B.E. Flinchbuagh. The interpretation of biological motion. *Biological Cybernatics*, pages 195–204, 1982.
14. G. Johansson. Visual perception of biological motion and a model for its analysis. *Perception and Psychophysics*, 14:201–211, 1973.
15. R.C. Nelson. Qualitative detection of motion by a moving observer. In *Proc. of IEEE CVPR*, pages 173–178, 1991.
16. J. O'Rourke and N.I. Badler. Model-based image analysis of human motion using

constraint propagation. *PAMI*, 3(4):522–537, 1980.

17. A. Pentland and K. Mase. Lip reading: Automatic visual recognition of spoken words. Technical Report 117, M.I.T. Media Lab Vision Science, 1989.

18. E.D. Petajan, B. Bischoff, and N.M. Brooke. An improved automatic lipreading system to enhance speech recognition. In *SIGCHI'88: Human Factors in Computing Systems*, pages 19–25, 1988.

19. R. Polana and R.C. Nelson. Detecting activities. *Journal of Visual Communication and Image Representation*, 5(2):172–180, 1994.

20. R.F. Rashid. *LIGHTS: A System for Interpretation of Moving Light Displays*. PhD thesis, Computer Science Dept, University of Rochester, 1980.

21. J.R. Rhyne and C.G. Wolf. Gestural interfaces for information processing applications. Technical Report 12179, IBM Research Report, 1986.

22. S.M. Seitz and C.R. Dyer. Affine invariant detection of periodic motion. In *Proceedings of CVPR*, 1994.

23. R.H. Smythe. *Vision in the Animal World*. St. Martin's Press, NY, 1975.

24. Marcin Olaf Szummer. *Temporal Texture Modeling*. PhD thesis, Department of Electrical Engineering and Computer Science, Massachusetts Institute of Technology, 1995.

25. N. Tinbergen. *The Study of Instinct*. Oxford: Clarendon Press, 1951.

26. Tsai, Ping-Sing, Keiter, K., Kasparis, T., and Shah, M. Cyclic motion detection. *Pattern Recognition*, 27(12), 1994.

27. K. 'Frisch von'. *Bees: Their Vision, Taste, Smell and Language*. Moscow,IL, 1955.

28. E. Wolf and G. Zerrahn-Wolf. Flicker and the reactions of bees to flowers. *Journal of Gen. Physiol.*, 20:511–518, 1936.

Figure 2. Sample images from four periodic sequences, walk, exercise, swing and rotation; and two non-periodic sequences, plane and leaves

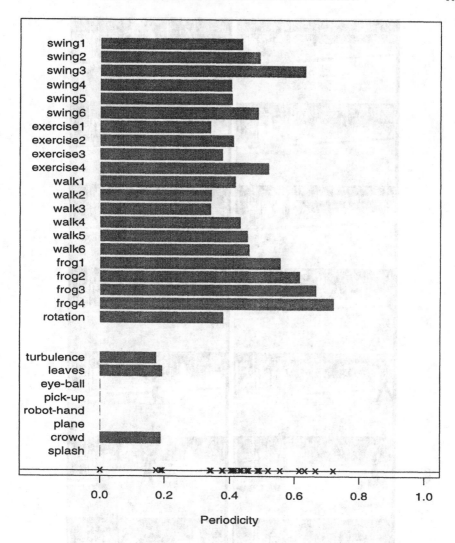

Figure 3. Periodicity measure for Periodic and Nonperiodic sequences

Figure 4. Image frames of walking in the presence of multiple moving objects and result of tracking

Figure 6. Image tra are of a person walking across the street taken by a camera mounted on an approaching van, and the result after stacking and compensating for scale changes.

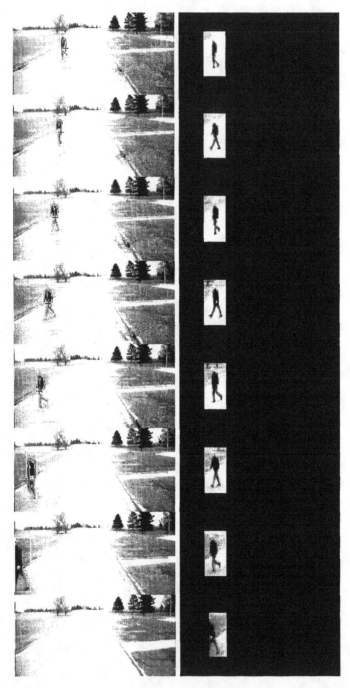

Figure 6. Image frames of a person walking across the street taken by a camera mounted on an approaching van, and the result after tracking and compensating for scale changes

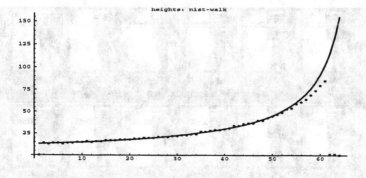

Figure 7. Actual image heights of the person (dotted) and the fitted model (solid)

Figure 8. Sample images from periodic activities: walk, run, swing, jump, ski, exercise and toy frog

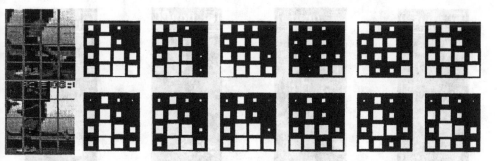

Figure 9. Sample total motion magnitude feature vector for a sample of walk (top) and a sample of run (bottom), one cycle of activity is divided into six time divisions shown horizontally, each frame shows spatial distribution of motion in a4x4 spatial grid (size of each square is proportional to the amount of motion in the neighborhood).

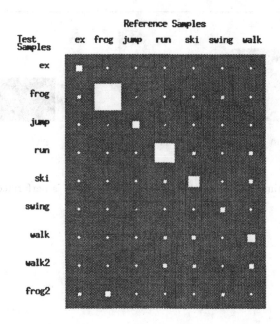

Figure 10. Confusion matrix for the feature vector using total motion magnitude

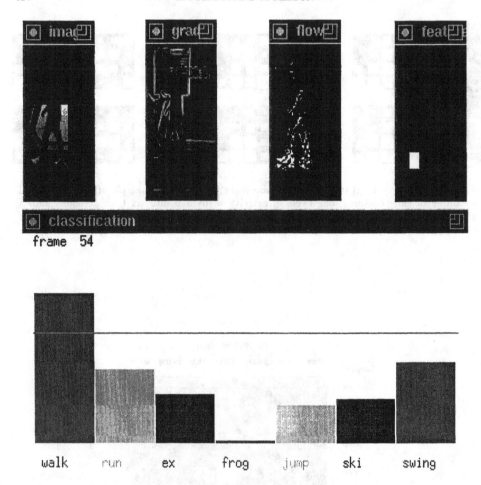

Figure 11. Real-time implementation screen: classification is performed at every frame

ACTION RECOGNITION USING TEMPORAL TEMPLATES

AARON F. BOBICK AND JAMES W. DAVIS
MIT Media Laboratory
20 Ames St., Cambridge, MA 02139
E-mail: bobick,jdavis@media.mit.edu

1. Introduction

The recent shift in computer vision from static images to video sequences has focused research on the understanding of *action* or behavior. In particular, the lure of wireless interfaces (e.g. [13]) and interactive environments [11, 3] has heightened interest in understanding human actions. Recently a number of approaches have appeared attempting the full three-dimensional reconstruction of the human form from image sequences, with the presumption that such information would be useful and perhaps even necessary to understand the action taking place (e.g. [22]). This chapter presents an alternative to the three-dimensional reconstruction proposal. We develop a view-based approach to the representation and recognition of action that is designed to support the direct recognition of the motion itself.

In previous work [4, 6] we described how people can easily recognize action in even extremely blurred image sequences such as shown in Figure 1 and in `lowres_action.mov`[1]. Such capabilities argue for recognizing action from the motion itself, as opposed to first reconstructing a 3-dimensional model of a person, and then recognizing the action of the model as advocated in [1, 7, 16, 22, 23, 9, 28]. In [4] we proposed a representation and recognition theory that decomposed motion-based recognition into first describing *where* there is motion (the spatial pattern) and then describing *how* the motion is moving. The approach is a natural extension of Black and Yacoob's work on facial expression recognition [2].

In this chapter we continue to develop this approach. We review the construction of a binary *motion-energy* image (MEI) which represents where motion has occurred in an image sequence (Section 3.1). We next generate a

---

[1]`http://vismod.www.media.mit.edu/vismod/demos/actions/lowres_action.mov`

M. Shah and R. Jain (eds.), Motion-Based Recognition, 125–146.

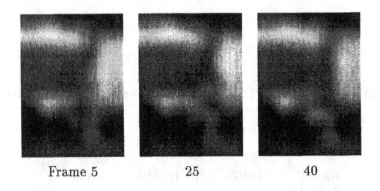

Frame 5   25   40

Figure 1. Selected frames from video of someone performing an action. Even with almost no structure present in each frame people can trivially recognize the action as someone sitting.

motion-history image (MHI) which is a scalar-valued image where intensity is a function of recency of motion (Section 3.2). Taken together, the MEI and MHI can be considered as a two component version of a *temporal template*, a vector-valued image where each component of each pixel is some function of the motion at that pixel location. These view-specific templates are matched against the stored models of views of known actions. To evaluate the power of the representation we evaluate the discrimination power on a set of 18 aerobics exercises (Section 4). Finally we present a recognition method which automatically performs temporal segmentation, is invariant to linear changes in speed, and runs in real-time on a standard platform (Section 5).

2. Prior work

The number of approaches to recognizing motion and action has recently grown at a tremendous rate. For an excellent survey on the machine understanding of motion (particularly human motion) see the work of Cédras and Shah [8]. Their review article details methods for extracting motion information (e.g. optic flow) and for performing matching. That survey article also discusses recent work in motion recognition (e.g. lip-reading and gesture recognition). In this chapter we divide the immediately relevant prior work into two general areas: configuration-based tracking and recognition and motion-based recognition.

2.1. CONFIGURATION-BASED TRACKING AND RECOGNITION OF ACTION

2.1.1. *Tracking*

The first and most obvious body of relevant work includes the approaches using structural or appearance-based representations to tracking and understanding human action. Some believe that a 3-D description is necessary and sufficient for understanding action (e.g. [16, 7, 23, 14, 22, 15]), while others choose to analyze the 2-D appearance as a means of interpretation (e.g. [9, 10, 1, 28]). We now take a closer look at these approaches.

The most common method for attaining the 3-D information in the action is to recover the pose of the object at each time instant using a 3-D model of the object. A common method for model fitting in these works is to use a residual measure between the projected model and object contours (e.g. edges of body in the image). This generally requires a strong segmentation of foreground/background and also of the individual body parts to aid the model alignment process. It is difficult to imagine such techniques could be extended to the blurred sequence of Figure 1.

For example, Rehg and Kanade [22] used a 27 degree-of-freedom (DOF) model of a human hand in their system called "Digiteyes". Local image-based trackers are employed to align the projected model lines to the finger edges against a solid background. The work of Goncalves et al. [15] promoted 3-D tracking of the human arm against a uniform background using a two cone arm model and a single camera. Though it may be possible to extend their approach to the whole body as claimed, it seems unlikely that it is appropriate for non-constrained human motion with self-occlusion. Hogg [16] and Rohr [23] used a full-body cylindrical model for tracking walking humans in natural scenes. Rohr incorporates a 1 DOF pose parameter to aid in the model fitting. All the poses in a walking action are indexed by a single number. Here there is only a small subset of poses which can exist. Gavrila and Davis [14] also used a full-body model (22 DOF, tapered superquadrics) for tracking human motion against a complex background. For simplifying the edge detection in cases of self-occlusion, the user is required to wear a tight-fitting body suit with contrasting limb colors.

One advantage of having the recovered model is the ability to estimate and predict the feature locations, for instance edges, in the following frames. Given the past history of the model configurations, prediction is commonly attained using Kalman filtering [23, 22, 15] and velocity constraints [21, 14].

Because of the self-occlusions that frequently occur in articulated objects, some employ multiple cameras and restrict the motion to small regions [22, 14] to help with projective model occlusion constraints. A single camera is used in [16, 15, 23], but the actions tracked in these works had little deviation in the depth of motion. Acquiring the 3-D information from image sequences is currently a complicated process, many times necessitating human intervention or contrived imaging environments.

2.1.2. *Recognition*

As for action recognition, Campbell and Bobick [7] used a commercially available system to obtain 3-D data of human body limb positions. Their system removes redundancies that exist for particular actions and performs recognition using only the information that varies between actions. This method examines the relevant parts of the body, as opposed to the entire body data. Siskind [25] similarly used known object configurations. The input to his system consisted of line-drawings of a person, table, and ball. The positions, orientations, shapes, and sizes of the objects are known at all times. The approach uses support, contact, and attachment primitives and event logic to determine the actions of dropping, throwing, picking up, and putting down. These two approaches address the problem of recognizing actions when the precise configuration of the person and environment is known while the methods from the previous section concentrate on the recovery of the object pose.

In contrast to the 3-D reconstruction and recognition approaches, others attempt to use only the 2-D appearance of the action (e.g. [1, 10, 9, 28]). View-based representations of 2-D statics are used in a multitude of frameworks, where an action is described by a sequence of 2-D instances/poses of the object. Many methods require a normalized image of the object (usually with no background) for representation. For example, Cui et al. [9], Darrell and Pentland [10], and also Wilson and Bobick [26] present results using actions (mostly hand gestures), where the actual grayscale images (with no background) are used in the representation for the action. Though hand appearances remain fairly similar over a wide range of people, with the obvious exception of skin color, actions that include the appearance of the total body are not as visually consistent across different people due to obvious natural variations and different clothing. As opposed to using the actual raw grayscale image, Yamato et al. [28] examines body silhouettes, and Akita [1] employs body contours/edges. Yamato utilizes low-level silhouettes of human actions in a Hidden Markov Model (HMM) framework, where binary silhouettes of background-subtracted images are vector quantized and used as input to the HMMs. In Akita's work [1], the use of edges and some simple 2-D body configuration knowledge (e.g. the arm is a protrusion out from the torso) are used to determine the body parts in a hierarchical manner (first find legs, then head, arms, trunk) based on stability. Individual parts are found by chaining local contour information. These two approaches help alleviate *some* of the variability between people but introduce other problems such as the disappearance of movement that happens to be within the silhouetted region and also the varying amount of contour/edge information that arises when the background or clothing is high versus low frequency (as in most natural scenes). Also, the problem of examining the entire body, as opposed to only the desired regions, still exists, as it does in much of the 3-D work.

Whether using 2-D or 3-D structural information, many of the approaches discussed so far consider an action to be comprised of a sequence of poses of an object. Underlying all of these techniques is the requirement that there be individual features or properties that can be extracted and tracked from each frame of the image sequence. Hence, motion understanding is really accomplished by recognizing a sequence of static configurations. This understanding generally requires previous recognition and segmentation of the person [21]. We now consider recognition of action within a motion-based framework.

2.2. MOTION-BASED RECOGNITION

Direct motion recognition [21, 24, 20, 2, 27, 25, 12, 4, 6] approaches attempt to characterize the motion itself without reference to the underlying static poses of the body. Two main approaches include the analysis of the body region as a single "blob-like" entity and the tracking of predefined *regions* (e.g. legs, head, mouth) using motion instead of structural features.

Of the "blob-analysis" approaches, the work of Polana and Nelson [21], Shavit and Jepson [24], and also Little and Boyd [20] are most applicable. Polana and Nelson use repetitive motion as a strong cue to recognize cyclic walking motions. They track and recognize people walking in outdoor scenes by gathering a feature vector, over the entire body, of low-level motion characteristics (optical-flow magnitudes) and periodicity measurements. After gathering training samples, recognition is performed using a nearest centroid algorithm. By assuming a fixed height and velocity of each person, they show how their approach is extendible to tracking multiple people in simple cases. Shavit and Jepson also take an approach using the gross overall motion of the person. The body, an animated silhouette figure, is coarsely modeled as an ellipsoid. Optical flow measurements are used to help create a phase portrait for the system, which is then analyzed for the force, rotation, and strain dynamics. Similarly, Little and Boyd recognize people walking by analyzing the motion associated with two ellipsoids fit to the body. One ellipsoid is fit using the motion region silhouette of the person, and the other ellipsoid is fit using motion magnitudes as weighting factors. The relative phase of various measures (e.g. centroid movement, weighted centroid movement, torque) over time for each of the ellipses characterizes the gait of several people.

There is a group of work which focuses on motions associated with facial expressions (e.g. characteristic motion of the mouth, eyes, and eyebrows) using region-based motion properties [27, 2, 12]. The goal of this research is to recognize human facial expressions as a dynamic system, where the motion of interest regions (locations known *a priori*) is relevant. Their approaches characterize the expressions using the underlying motion properties rather than represent the action as a sequence of poses or configurations. For Black and Yacoob [2], and also Yacoob and Davis [27], optical flow measurements are used to help track predefined polygonal patches placed on interest regions (e.g. mouth). The parameterization and location relative to the face of each patch was given *a priori*. The temporal trajectories of the motion parameters were qualitatively described according to positive or negative intervals. Then these qualitative labels were used in a rule-based, temporal model for recognition to determine expressions such as anger or happiness. Recently, Ju, Black, and Yacoob [19] have extended this work with faces to include tracking the legs of a person walking. As opposed to the simple, independent patches used for faces, an articulated three-patch model was needed for tracking the legs. Many problems, such as large motions, occlusions, and shadows, make motion estimation in that situation more challenging than for the facial case. We extended this expression recognition approach by applying a similar framework to the domain of full-body motion [5].

Optical flow, rather than patches, was used by Essa [12] to estimate muscle activation on a detailed, physically-based model of the face. One recognition approach classifies expressions by a similarity measure to the typical patterns of muscle activation. Another recognition method matches motion energy templates derived from the muscle activations. These templates compress the activity sequence into a single entity. In this chapter, we develop similar templates, but our templates incorporate the temporal motion characteristics.

3. Temporal templates

Our goal is to construct a view-specific representation of action, where action is defined as motion over time. For now we assume that either the background is static, or that the motion of the object can be separated from either camera-induced or distractor motion. At the conclusion of this chapter we discuss methods for eliminating incidental motion from the processing.

In this section we define a multi-component image representation of action based upon the observed motion. The basic idea is to construct a vector-image which can be matched against stored representations of known

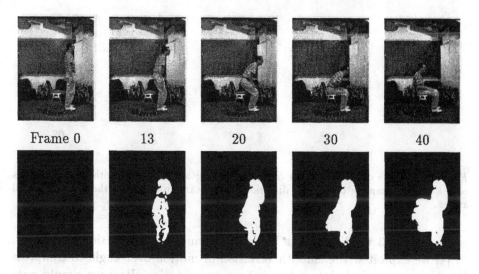

| Frame 0 | 13 | 20 | 30 | 40 |

Figure 2. Example of someone sitting. Top row contains key frames; bottom row is cumulative motion images starting from Frame 0.

actions; this image is used as a temporal template.

3.1. MOTION-ENERGY IMAGES

Consider the example of someone sitting, as shown in Figure 2. The top row contains key frames in a sitting sequence. The bottom row displays cumulative binary motion images — to be described momentarily — computed from the start frame to the corresponding frame above. As expected the sequence sweeps out a particular region of the image; our claim is that the shape of that region (*where* there is motion) can be used to suggest both the action occurring and the viewing condition (angle).

We refer to these binary cumulative motion images as *motion-energy images* (MEI). Let $I(x,y,t)$ be an image sequence, and let $D(x,y,t)$ be a binary image sequence indicating regions of motion; for many applications image-differencing is adequate to generate D. Then the binary MEI $E_\tau(x,y,t)$ is defined

$$E_\tau(x,y,t) = \bigcup_{i=0}^{\tau-1} D(x,y,t-i)$$

We note that the duration τ is critical in defining the temporal extent of an action. Fortunately, in the recognition section we derive a backward-looking (in time) algorithm which can dynamically search over a range of τ.

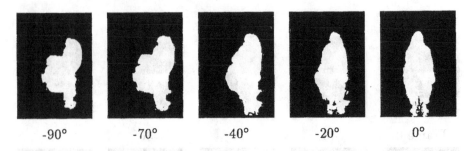

$$-90°\qquad\quad -70°\qquad\quad -40°\qquad\quad -20°\qquad\quad 0°$$

Figure 3. MEIs of sitting action over 90° viewing angle. The smooth change implies only a coarse sampling of viewing direction is necessary to recognize the action from all angles.

In Figure 3 we display the MEIs of viewing a sitting action across 90°. In [4] we exploited the smooth variation of motion over angle to compress the entire view circle into a low-order representation. Here we simply note that because of the slow variation across angle, we only need to sample the view sphere coarsely to recognize all directions. In the evaluation section of this chapter we use samplings of every 30° to recognize a large variety of motions (Section 4).

3.2. MOTION-HISTORY IMAGES

To represent *how* (as opposed to where) motion the image is moving we form a *motion-history* image (MHI). In an MHI H_τ, pixel intensity is a function of the temporal history of motion at that point. For the results presented here we use a simple replacement and decay operator:

$$H_\tau(x,y,t) = \begin{cases} \tau & \text{if } D(x,y,t) = 1 \\ \max\left(0, H_\tau(x,y,t-1) - 1\right) & \text{otherwise} \end{cases}$$

The result is a scalar-valued image where more recently moving pixels are brighter. Examples of MHIs are presented in Figure 4 and the dynamic construction of an MHI is illustrated in `mhi_generation.mov`[2]. Note that the MEI can be generated by thresholding the MHI above zero.

One possible objection to the approach described here is that there is no consideration of optic flow, the direction of image motion. In response, it is important to note the relation between the construction of the MHI and direction of motion. Consider the waving example in Figure 4 where the arms fan upwards. Because the arms are isolated components — they do not occlude other moving components — the motion-history image implicitly represents the direction of movement: the motion in the arm down

[2]`http://vismod.www.media.mit.edu/vismod/demos/actions/mhi_generation.mov`

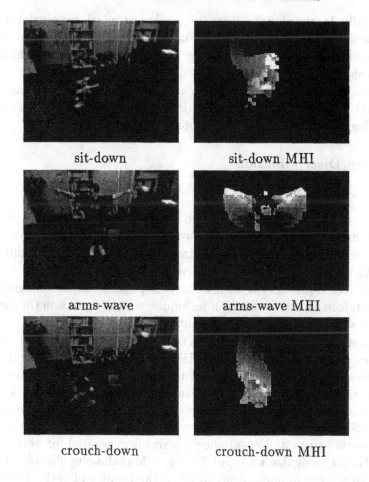

sit-down   sit-down MHI

arms-wave   arms-wave MHI

crouch-down   crouch-down MHI

Figure 4. Action moves along with their MHIs used in a real-time system.

position is "older" than the motion when the arms are up. For these types of articulated objects, and for simple movements where there is not significant motion self-occlusion, the direction of motion is well represented using the MHI. As motions become more complicated the optic flow is more difficult to discern, but is typically not lost completely.

3.3. EXTENDING TEMPORAL TEMPLATES

The MEI and MHI are two components of a vector image designed to encode a variety of motion properties in a spatially indexed manner. Other possible components of the temporal templates include power in directional motion integrated over time (e.g. "in this pixel there has been a large amount of motion in the down direction during the integrating time window") or the spatially localized periodicity of motion (a pixel by pixel version of Polana and Nelson [21]). The vector-image template is similar in spirit to the vector-image based on orientation and edges used by Jones and Malik [18] for robust stereo matching.

For the results in this chapter we use only the two components derived above (MEI and MHI) for representation and recognition. We are currently considering other components to improve our performance.

4. Action Discrimination

4.1. MATCHING TEMPORAL TEMPLATES

To construct a recognition system, we need to define a matching algorithm for the temporal template. Because we are using an appearance-based approach, we must first define the desired invariants for the matching technique. As we are using a view sensitive approach, it is desirable to have a matching technique that is as invariant as possible to the imaging situation. Therefore we have selected a technique which is rotation (in the image plane), scale, and translation invariant.

We first collect training examples of each action from a variety of viewing angles. Given a set of MEIs and MHIs for each view/action combination, we compute statistical descriptions of the these images using moment-based features. Our current choice are 7 Hu moments [17] which are known to yield reasonable shape discrimination in a translation- and scale-invariant manner (See appendix). For each view of each action a statistical model of the moments (mean and covariance matrix) is generated for both the MEI and MHI. To recognize an input action, a Mahalanobis distance is calculated between the moment description of the input and each of the known actions. In this section we show results using this distance metric in terms of its separation of different actions.

Note that we have no fundamental reason for selecting this method of scale- and translation-invariant template matching. The approach outlined has the advantage of not being computationally taxing making real-time implementation feasible; one disadvantage is that the Hu moments are difficult to reason about intuitively. Also, we note that the matching methods for the MEI and MHI need not be the same; in fact, given the distinction we make between where there is motion from how the motion is moving one might expect different matching criteria.

4.2. TESTING ON AEROBICS DATA: ONE CAMERA

To evaluate the power of the temporal template representation, we recorded video sequences of 18 aerobics exercises performed several times by an expe-

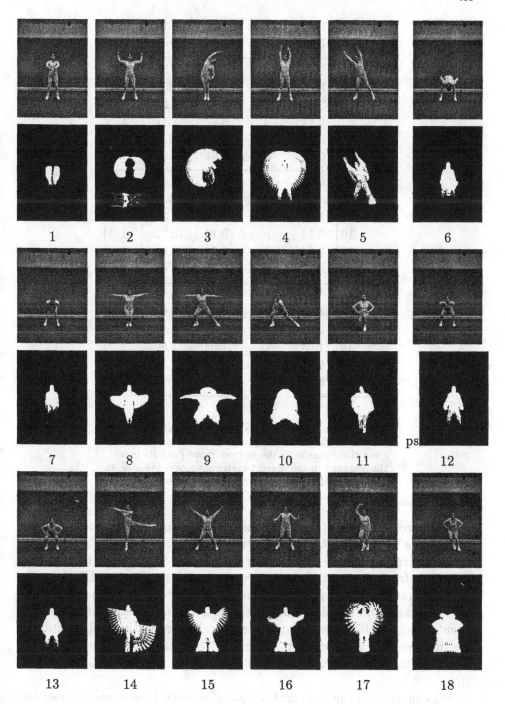

Figure 5. A single key frame and MEI from the frontal view of each of 18 aerobics exercises used to test the representation.

	Closest Dist	Closest Move	Correct Dist	Median Dist	Rank
Test 1	1.43	4	1.44	2.55	2
2	3.14	2	3.14	12.00	1
3	3.08	3	3.08	8.39	1
4	0.47	4	0.47	2.11	1
5	6.84	5	6.84	19.24	1
6	0.32	10	0.61	0.64	7
Test 7	0.97	7	0.97	2.03	1
8	20.47	8	20.47	35.89	1
9	1.05	8	1.77	2.37	4
10	0.14	10	0.14	0.72	1
11	0.24	11	0.24	1.01	1
12	0.79	12	0.79	4.42	1
Test 13	0.13	6	0.25	0.51	3
14	4.01	14	4.01	7.98	1
15	0.34	15	0.34	1.84	1
16	1.03	15	1.04	1.59	2
17	0.65	17	0.65	2.18	1
18	0.48	10	0.51	0.94	4

TABLE 1. Test results using one camera at 30° off frontal. Each row corresponds to one test move and gives the distance to the nearest move (and its index), the distance to the correct matching move, the median distance, and the ranking of the correct move.

rienced aerobics instructor. The sequence `aerobic_action.mov`[3] provides an example. Seven views of the action — +90° to −90° in 30° increments in the horizontal plane — were recorded. Figure 5 shows the frontal view of one key frame for each of the moves along with the frontal MEI. We take the fact that the MEI makes clear to a human observer the nature of the motion as anecdotal evidence of the strength of this component of the representation. For this experiment the temporal segmentation and selection of the time window over which to integrate were performed manually. Later we will detail a self-segmenting, time-scaling recognition system (Section 5).

We constructed the temporal template for each view of each move, and

[3]http://vismod.www.media.mit.edu/vismod/demos/actions/aerobic_action.mov

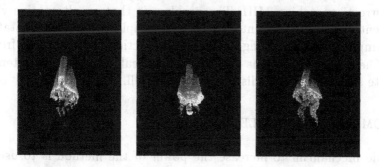

Figure 6. An example of MHIs with similar statistics. (a) Test input of move 13 at 30°.
(b) Closest match which is move 6 at 0°. (c) Correct match.

then computed the Hu moments on each component. To do a useful Mahalanobis procedure would require watching several different people performing the same actions; this multi-subject approach is taken in the next
section where we develop a recognition procedure using a full covariance
(Section 5). Instead, here we design the experiment to be a measurement
of confusion. A new test subject performed each move and the input data
was recorded by two cameras viewing the action at approximately 30° to
left and 60° to the right of the subject. The temporal template for each of
the two views of the test input actions was constructed, and the associated
moments computed.

Our first test uses only the left (30°) camera as input and matches
against all 7 views of all 18 moves (126 total). We select as a metric a *pooled*
independent Mahalanobis distance using a diagonal covariance matrix to
accommodate variations in magnitude of the moments. Table 1 displays
the results. Indicated are the distance to the move closest to the input (as
well as its index), the distance to the correct matching move, the median
distance (to give a sense of scale), and the ranking of the correct move in
terms of least distance.

The first result to note is that 12 of 18 moves are correctly identified
using the single view. This performance is quite good considering the com-
pactness of the representation (a total of 14 moments from two correlated
motion images) and the large size of the target set. Second, the typical
situation in which the best match is not the correct move, the difference
in distances from the input to the closest move versus the correct move is
small compared to the median distance. Examples of this include test moves
1, 9, 13, 16, 18. In fact for moves 1, 16, 18 the difference is negligible.

To analyze the confusion difficulties further consider the example shown
in Figure 6. Displayed here, left to right, are the input MHI (move 13 at
view angle 30°), the closest match MHI (move 6 at view angle 0°), and

the "correct" matching MHI. The problem is that an alternative view of a different action projects into a temporal template with similar statistics. For example, consider sitting and crouching actions when viewed from the front. The observed motions are almost identical, and the coarse temporal template statistics do not distinguish them well.

4.3. COMBINING MULTIPLE VIEWS

A simple mechanism to increase the power of the method is to use more than one camera. Several approaches are possible. For this experiment, we use two cameras and find the minimum sum of Mahalanobis distances between the two input templates and two stored views of an action that have the correct angular difference between them, in this case 90°. The assumption embodied in this approach is that we know the approximate angular relationship between the cameras.

Table 2 provides the same statistics as the first table, but now using two cameras. Notice that the classification now contains only 3 errors. The improvement of the result reflects the fact that for most pairs of this suite of actions, there is some view in which they look distinct. Because we have 90° between the two input views the system can usually correctly identify most actions.

We mention that if the approximate calibration between cameras is not known (and is not to be estimated) one can still logically combine the information by requiring consistency in labeling. That is, we remove the inter-angle constraint, but do require that both views select the same action. The algorithm would be to select the move whose Mahalanobis sum is least, regardless of the angle between the target views. If available, angular order information — e.g. camera 1 is to the left of camera 2 — can be included. When this approach is applied to the aerobics data shown here we still get similar discrimination. This is not surprising because the input views are so distinct.

To analyze the remaining errors, consider Figure 7 which shows the input for move 16. Left to right are the 30° MHIs for the input, the best match (move 15), and the correct match. The test subject performed the move much less precisely than the original aerobics instructor. Because we were not using a Mahalanobis variance across subjects, the current experiment could not accommodate such variation. In addition, the test subject moved her body slowly while wearing low frequency clothing resulting in an MHI that has large gaps in the body region. We attribute this type of failure to our simple (i.e. naive) motion analysis; a more robust motion detection mechanism would reduce the number of such situations.

		Closest Dist	Closest Move	Correct Dist	Median Dist	Rank
Test	1	2.13	1	2.13	6.51	1
	2	12.92	2	12.92	19.58	1
	3	7.17	3	7.17	18.92	1
	4	1.07	4	1.07	7.91	1
	5	16.42	5	16.42	32.73	1
	6	0.88	6	0.88	3.25	1
Test	7	3.02	7	3.02	7.81	1
	8	36.76	8	36.76	49.89	1
	9	5.10	8	6.74	8.93	3
	10	0.68	10	0.68	3.19	1
	11	1.20	11	1.20	3.68	1
	12	2.77	12	2.77	15.12	1
Test	13	0.57	13	0.57	2.17	1
	14	6.07	14	6.07	16.86	1
	15	2.28	15	2.28	8.69	1
	16	1.86	15	2.35	6.72	2
	17	2.67	8	3.24	7.10	3
	18	1.18	18	1.18	4.39	1

TABLE 2. Results using two cameras where the angular interval is known and any matching views must have the same angular distance.

Figure 7. Example of error where failure is caused by both the inadequacy of using image differencing to estimate image motion and the lack of the variance data in the recognition procedure.

5. Segmentation and recognition

The final element of performing recognition is the temporal segmentation and matching. During the training phase we measure the minimum and maximum duration that an action may take, τ_{min} and τ_{max}. If the test actions are performed at varying speeds, we need to choose the right τ for the computation of the MEI and the MHI. Our current system uses a backward looking variable time window. Because of the simple nature of the replacement operator we can construct a highly efficient algorithm for approximating a search over a wide range of τ.

The algorithm is as follows: At each time step a new MHI $H_\tau(x, y, t)$ is computed setting $\tau = \tau_{max}$, where τ_{max} is the longest time window we want the system to consider. We choose $\Delta\tau$ to be $(\tau_{max} - \tau_{min})/(n - 1)$ where n is the number of temporal integration windows to be considered.[4] A simple thresholding of MHI values less than $(\tau - \Delta\tau)$ followed by a scaling operation generates $H_{(\tau - \Delta\tau)}$ from H_τ. Iterating we compute all n MHIs at each time step. Binarization of the MHIs yields the corresponding MEIs.

After computing the various MEIs and MHIs, we compute the Hu moments for each image. We then check the Mahalanobis distance of the MEI parameters against the known view/action pairs; the mean and the covariance matrix for each view/action pair is derived from multiple subjects performing the same move. Any action found to be within a threshold distance of the input is tested for agreement of the MHI. If more than one action is matched, we select the action with the smallest distance.

The aerobics data were generated from only two individuals who performed the actions precisely, without adequate variation to generate a statistical distribution. To test the real-time recognition system, we created a new, smaller action set using multiple people to provide training examples. Our experimental system recognizes 180° views of the actions *sitting*, *arm waving*, and *crouching* (See Figure 4). The training required four people and sampling the view circle every 45°. The system performs well, rarely misclassifying the actions. The errors which do arise are mainly caused by problems with image differencing and also due to our approximation of the temporal search window $n < (\tau_{max} - \tau_{min} + 1)$. The sequence two_cam_demo.mov[5] provides a demonstration of this two camera technique. An avatar shown in the bottom window changes based upon the recognition state of the system.

The system runs at approximately 9 Hz using 2 CCD cameras connected to a Silicon Graphics 200MHz Indy; the images are digitized at a

---

[4]Ideally $n = \tau_{max} - \tau_{min} + 1$ resulting in a complete search of the time window between τ_{max} and τ_{min}. Only computational limitations argue for a smaller n.

[5]http://vismod.www.media.mit.edu/vismod/demos/actions/two_cam_demo.mov

size of 160x120. For these three moves $\tau_{max}=19$ (approximately 2 seconds), $\tau_{min} = 11$ (approximately 1 second), and we chose $n = 6$. The comparison operation is virtually no cost in terms of computational load, so adding more actions does not affect the speed of the algorithm, only the accuracy of the recognition.

6. Extensions, problems, and applications

We have presented a novel representation and recognition technique for identifying actions. The approach is based upon temporal templates and their dynamic matching in time. Initial experiments in both measuring the sensitivity of the representation and in constructing real-time recognition systems have shown the effectiveness of the method.

There are, of course, some difficulties in the current approach. Several of these are easily rectified. As mentioned, a more sophisticated motion detection algorithm would increase robustness. Also, as developed, the method assumes all motion present in the image should be incorporated into the temporal templates. Clearly, this approach would fail when two people are in the field of view. To implement our real-time system we use a tracking bounding box which attempts to isolate the relevant motions.

A worse condition is when one person partially occludes another, making separation difficult, if not impossible. Here multiple cameras is an obvious solution. Since occlusion is view angle specific, multiple cameras reduce the chance the occlusion is present in all views. For monitoring situations, we have experimented with the use of an overhead camera to select which ground based cameras have a clear view of a subject and where the subject would appear in each image.

6.1. INCIDENTAL MOTION

A more serious difficulty arises when the motion of part of the body is not specified during an action. Consider, for example, throwing a ball. Whether the legs move is not determined by the action itself, inducing huge variability in the statistical description of the temporal templates. To extend this paradigm to such actions requires some mechanism to automatically mask away regions of this type of motion. Our current thinking is to process only the motion signal associated with the dominant motions.

Two other examples of motion that must be removed are camera motion and locomotion (if we assume the person is performing some action while locomoting and what we want to see is the underlying action). In both instances the problem can be overcome by using a body centered motion field. The basic idea would be to subtract out any image motion induced by camera movement or locomotion. Of these two phenomena, camera motion

Figure 8. The KIDSROOM interactive play-space. Using a modified version of temporal templates the room responds to the actions of the children. All sensing is performed using vision from 3 cameras.

elimination is significantly easier because of the over constrained nature of estimating egomotion. Our only insight at this point is that because the temporal template technique does not require accurate flow fields it may be necessary only to approximately compensate for these effects and then to threshold the image motion more severely than we have done to date.

6.2. THE KIDSROOM: AN APPLICATION

We conclude by mentioning a recent application we developed in which we employed a version of the temporal template technique described. On October 30th, 1996 we debuted The KidsRoom, an interactive play-space for children [3]. The basic idea is that the room is aware of the children (maximum of 4) and takes them through a story where the responses of the room are affected by what the children do. Computers control the lighting, sound effects, performance of the score, and illustrations projected on the two walls of the room that are actually video screens. The current scenario is an adventurous trip to Monsterland; a snapshot is shown in Figure 8.

In the last scene the monsters appear and teach the children to dance — basically to perform certain actions. Using a modified version of the MEIs[6] the room can compliment the children on well performed moves (e.g. spinning) and then turn control of the situation over to them: the

---

[6]The MEIs were computed from background subtracted images instead of binary motion images. This change was necessary because of the high variability of incidental body motion. By using the background subtracted images the body was always included.

monsters follow the children if the children perform the moves they were taught. The interactive narration coerces the children to room locations where occlusion is not a problem. Of all the vision processes required, the modified temporal template is one of the more robust. We take the ease of use of the method to be an indication of its potential.

References

1. Akita, K. Image sequence analysis of real world human motion. *Pattern Recognition*, 17, 1984.
2. Black, M. and Y. Yacoob. Tracking and recognizing rigid and non-rigid facial motion using local parametric models of image motion. In *Proc. Int. Conf. Comp. Vis.*, 1995.
3. A. Bobick, J. Davis, S. Intille, F. Baird, L. Campbell, Y. Ivanov, C. Pinhanez, A. Schutte, and A. Wilson. Kidsroom: Action recognition in an interactive story environment. PerCom TR 398, MIT Media Lab, 1996.
4. Bobick, A. and J. Davis. An appearance-based representation of action. In *Proc. Int. Conf. Pat. Rec.*, August 1996.
5. Bobick, A. and J. Davis. An appearance-based representation of action. In *Proc. Int. Conf. Pat. Rec.*, August 1996.
6. Bobick, A. and J. Davis. Real time recognition of activity using temporal templates. In *IEEE Workshop on Applications of Computer Vision*, Sarasota, December 1996.
7. Campbell, L. and A. Bobick. Recognition of human body motion using phase space constraints. In *Proc. Int. Conf. Comp. Vis.*, 1995.
8. Cédras, C., and Shah, M. Motion-based recognition: A survey. *Image and Vision Computing*, 13(2):129–155, March 1995.
9. Cui, Y., D. Swets, and J. Weng. Learning-based hand sign recognition using shoslif-m. In *Proc. Int. Conf. Comp. Vis.*, 1995.
10. Darrell, T. and A. Pentland. Space-time gestures. In *Proc. Comp. Vis. and Pattern Rec.*, 1993.
11. Darrell, T., P. Maes, B. Blumberg, and A. Pentland. A novel environment for situated vision and behavior. In *IEEE Wkshp. for Visual Behaviors (CVPR-94)*, 1994.
12. Essa, I. and A. Pentland. Facial expression recognition using a dynamic model and motion energy. In *Proc. Int. Conf. Comp. Vis.*, June 1995.
13. Freeman, W., and M. Roth. Orientation histogram for hand gesture recognition. In *Int'l Workshop on Automatic Face- and Gesture-Recognition*, 1995.
14. Gavrila, D. and L. Davis. Tracking of humans in actions: a 3-d model-based approach. In *ARPA Image Understanding Workshop*, Feb 1996.
15. Goncalves, L., E. DiBernardo, E. Ursella, P. Perona. Monocular tracking of the human arm in 3d. In *Proc. Int. Conf. Comp. Vis.*, June 1995.
16. Hogg, D. Model-based vision: a paradigm to see a walking person. *Image and Vision Computing*, 1(1), 1983.
17. Hu, M. Visual pattern recognition by moment invariants. *IRE Trans. Information Theory*, IT-8(2), 1962.
18. D. Jones and J. Malik. Computational framework for determining stereo correspondence from a set of linear spatial filters. *Image and Vision Computing*, 10(10):699–708, 1992.
19. Ju, S., Black, M., and Y. Yacoob. Cardboard people: a parameterized model of articulated image motion. In *Submitted to the Second International Conference on Automatic Face and Gesture Recognition*, 1996.
20. Little, J., and J. Boyd. Describing motion for recognition. In *International Symposium on Computer Vision*, pages 235–240, November 1995.

21. Polana, R. and R. Nelson. Low level recognition of human motion. In *IEEE Workshop on Non-rigid and Articulated Motion*, 1994.
22. Rehg, J. and T. Kanade. Model-based tracking of self-occluding articulated objects. In *Proc. Int. Conf. Comp. Vis.*, 1995.
23. Rohr, K. Towards model-based recognition of human movements in image sequences. *CVGIP, Image Understanding*, 59(1), 1994.
24. Shavit, E. and A. Jepson. Motion understanding using phase portraits. In *IJCAI Workshop: Looking at People*, 1995.
25. Siskind, J. M. Grounding language in perception. In *SPIE*, September 1993.
26. Wilson, A. and A. Bobick. Learning visual behavior for gesture analysis. In *Proc. IEEE Int'l. Symp. on Comp. Vis.*, Coral Gables, Florida, November 1995.
27. Yacoob, Y. and L. Davis. Computing spatio-temporal representations of human faces. In *Proc. Comp. Vis. and Pattern Rec.*, 1994.
28. Yamato, J., J. Ohya, and K. Ishii. Recognizing human action in time sequential images using hidden markov models. In *Proc. Comp. Vis. and Pattern Rec.*, 1992.

A. General moments

The two-dimensional $(p + q)$th order moments of a density distribution function $\rho(x, y)$ (e.g. image intensity) are defined in terms of Riemann integrals as

$$m_{pq} = \int_{-\infty}^{\infty} \int_{-\infty}^{\infty} x^p y^q \rho(x, y) dx dy, \tag{1}$$

for $p, q = 0, 1, 2, \cdots$.

B. Moments invariant to translation

The central moments μ_{pq} are defined as

$$\mu_{pq} = \int_{-\infty}^{\infty} \int_{-\infty}^{\infty} (x - \bar{x})^p (y - \bar{y})^q \rho(x, y) d(x - \bar{x}) d(y - \bar{y}), \tag{2}$$

where

$$\bar{x} = m_{10}/m_{00},$$
$$\bar{y} = m_{01}/m_{00}.$$

It is well known that under the translation of coordinates, the central moments do not change, and are therefore invariants under translation. It is quite easy to express the central moments μ_{pq} in terms of the ordinary moments m_{pq}. For the first four orders, we have

$$\mu_{00} = m_{00} \equiv \mu$$
$$\mu_{10} = 0$$
$$\mu_{01} = 0$$

$$\mu_{20} = m_{20} - \mu\bar{x}^2$$
$$\mu_{11} = m_{11} - \mu\bar{x}\bar{y}$$
$$\mu_{02} = m_{02} - \mu\bar{y}^2$$
$$\mu_{30} = m_{30} - 3m_{20}\bar{x} + 2\mu\bar{x}^3$$
$$\mu_{21} = m_{21} - m_{20}\bar{y} - 2m_{11}\bar{x} + 2\mu\bar{x}^2\bar{y}$$
$$\mu_{12} = m_{12} - m_{02}\bar{x} - 2m_{11}\bar{y} + 2\mu\bar{x}\bar{y}^2$$
$$\mu_{03} = m_{03} - 3m_{02}\bar{y} + 2\mu\bar{y}^3$$

C. Moments invariant to translation, scale, and orientation

For the second and third order moments, we have the following seven translation, scale, and orientation moment invariants:

$$\nu_1 = \mu_{20} + \mu_{02}$$
$$\nu_2 = (\mu_{20} - \mu_{02})^2 + 4\mu_{11}^2$$
$$\nu_3 = (\mu_{30} - 3\mu_{12})^2 + (3\mu_{21} - \mu_{03})^2$$
$$\nu_4 = (\mu_{30} + \mu_{12})^2 + (\mu_{21} + \mu_{03})^2$$
$$\nu_5 = (\mu_{30} - 3\mu_{12})(\mu_{30} + \mu_{12})[(\mu_{30} + \mu_{12})^2 - 3(\mu_{21} + \mu_{03})^2]$$
$$+(3\mu_{21} - \mu_{03})(\mu_{21} + \mu_{03})$$
$$\cdot[3(\mu_{30} + \mu_{12})^2 - (\mu_{21} + \mu_{03})^2]$$
$$\nu_6 = (\mu_{20} - \mu_{02})[(\mu_{30} + \mu_{12})^2 - (\mu_{21} + \mu_{03})^2]$$
$$+4\mu_{11}(\mu_{30} + \mu_{12})(\mu_{21} + \mu_{03})$$
$$\nu_7 = (3\mu_{21} - \mu_{03})(\mu_{30} + \mu_{12})[(\mu_{30} + \mu_{12})^2 - 3(\mu_{21} + \mu_{03})^2]$$
$$-(\mu_{30} - 3\mu_{12})(\mu_{21} + \mu_{03})[3(\mu_{30} + \mu_{12})^2 - (\mu_{21} + \mu_{03})^2]$$

These moments can used for pattern identification not only independently of position, size, and orientation but also independently of parallel projection.

D. Moments invariant to translation and scale

Under the scale transformation for moment invariants we have

$$\mu'_{pq} = \alpha^{p+q+2}\mu_{pq}. \tag{3}$$

By eliminating α between the zeroth order relation,

$$\mu' = \alpha^2\mu \tag{4}$$

and the remaining ones, we have the following absolute translation and scale moment invariants:

$$\frac{\mu'_{pq}}{(\mu')^{(p+q)/2+1}} = \frac{\mu_{pq}}{\mu^{(p+q)/2+1}} , \tag{5}$$

for $p + q = 2, 3, \cdots$ and $\mu'_{10} = \mu'_{01} \equiv 0$.

HUMAN ACTIVITY RECOGNITION

NIGEL H. GODDARD

Pittsburgh Supercomputing Center
4400 Fifth Avenue Pittsburgh, PA 15213

1. Introduction

A fundamental goal of work in recognition is to discover easily-computed visual features which are efficient indices of members of the class which is to be recognized. The hypothesis behind work in motion-based recognition is that features describing motion in the input can be efficient indices for large classes of objects and activities of interest to the computer vision and biological vision communities. Motion-based recognition encompasses the recognition of objects, object movements, situations, etc., when motion information is used as the primary cue for recognition. For example, optic flow can be used to recognize an imminent collision situation, without any prior recognition of objects or how they are moving. Similarly, objects could be recognized using characteristic motion parameters without prior determination of shape, texture, etc.

In the work described in this chapter I have applied the motion-based approach to a category of recognition problems for which it seems particularly appropriate: recognition of structured movement. Structured movement is any movement which is extended in, and undergoes systematic changes over, time and space. Examples include many movements generated biologically (e.g., animal gait) or mechanistically (e.g., the pistons driving a steam engine's wheels) or both (e.g., legs pedaling a bicycle). Animal gait is a form of motion which captures the essence of structured movement. For this and other reasons outlined below, human gait has been the focus of this work on movement recognition.

Recognition typically includes the following two processes, each of which have individually been referred to by the term "recognition": 1) indexing quickly a large database of models to yield a much smaller ordered list of candidates, followed by 2) search of the "candidate space" by a detailed process which matches candidates with the input, often in 3D. In systems

M. Shah and R. Jain (eds.), Motion-Based Recognition, 147–170.

targeting a highly restricted domain of models, the candidate space may
be sufficiently small that the indexing process can be ignored. In more
general systems, fast indexing is required to reduce the candidate space to a
tractable size. In this chapter we focus on the indexing problem, envisioning
the everyday human visual environment.

It is important to distinguish recognition of movement, on the one hand,
and motion-based indexing on the other. Recognition of animal gait can
proceed from analysis of the sequence of static qualities of each image in an
image sequence, for example the sequence of poses [23]; this is movement-
recognition but not motion-based indexing. Conversely, a tree could be
recognized in an image sequence from the texture of motion of its leaves
fluttering in a breeze without an explicit model of each leaf fluttering or of
the spatial structure of a tree; this would be motion-based recognition of an
object. While this chapter is concerned with structured movement recognition *and* the motion-based approach, the reader should remain clear that
these are *not* the same thing. However, as will be seen later, the use of the
motion-based approach to achieve movement recognition affords unusual
opportunities in processes and cues for attention.

We use the methodology of Structured Connectionist Modeling [5]. This
is an approach to computer vision which rests on a two-part hypothesis: significant advances in vision algorithms come from studying extant biological
vision systems; and understanding these systems is greatly facilitated by
adopting a model of computation based on what is known about biological
sensory systems. The work described in this chapter is an example of structured connectionist modeling which provides support for this hypothesis. It
starts by examining the psychophysical data leading to severe constraints
on the computational model. Next an analysis of the computational tasks
involved in movement recognition leads to a framework for the proposed solution. An overview of the architecture of the model is followed by detailed
analysis of the function of the major components, and a description of their
implementation in a massively parallel network. Simulations demonstrate
the ability of the implemented model to recognize human gaits under various scene conditions. Finally I discuss potential future developments and
related issues.

2. Psychophysical Data

Extensive psychophysical testing in humans has demonstrated the range
of human ability in recognizing human gait and some related movements,
The classic illustration of this ability is Johansson's Moving Light Display
(MLD) [11, 12]. Reflective pads were placed at the joints of an actor dressed
in black, and the actor illuminated. Films were taken of the actor walking,

jumping and making various other movements against a black backdrop. When these films were shown to subjects, they all recognized the display to be of a person walking, jumping, *etc.*, but reported single frames to be meaningless patterns of dots. A presentation time of no more than 200 msec of movement was sufficient for all subjects to correctly identify an MLD of a person walking. In forced choice experiments, all subjects accurately identified 6 human and 3 puppet generated gait patterns with a presentation time of 400 msec. Further experiments [14, 4, 2, 20, 29, 6, 30, 24, 28, 22, 3] demonstrated the generality, sensitivity and early development of this faculty. The focus of the work here is in modeling rapid recognition, in people, of human gait, as demonstrated in Johannson's experiments. Hence the chapter title should be read both ways: human activity-recognition, and human-activity recognition.

3. Computational Tasks

In computational modeling of perception one of the hardest tasks is to delineate a suitable portion of the observed phenomenon to be modeled. It must simultaneously be sufficiently compact and well-defined as to admit to modeling given current circumstances; and sufficiently broad as to be of relevance to biological and machine vision in the real world. Choosing suitable boundaries requires a task framework in which the modeled subsystem is embedded. The major sub-tasks subsumed under the visual motion recognition task, roughly in order from low to high level, are: low-level feature extraction, correspondence, segmentation, tracking, structure from motion, intermediate-level feature composition, indexing, viewpoint recovery, attention, verification, planning, gestalt perception. Although we can neatly list these problems as separate tasks, in fact they are tightly coupled and overlapping in a functioning visual system. For example, planning, attending and verification are obviously related; segmentation and feature extraction are aided by the contextual information provided by the results of the indexing and gestalt processes; correspondence is aided by feature extraction; tracking requires some pre-conscious attention mechanism at least. All of these processes are involved even in laboratory interpretation of Moving Light Displays. Our strategy is to ground the model at the low level in established computational work, and bound it at the high level by choosing a task that is highly unlikely to involve the highest level processes. In the work described in this chapter, we limit the scope of the model to human recognition of activity in MLDs. This requires on the order of a half second of presentation, insufficient time for a significant role for verification or planning, or even external attention shifts (saccades) which require 150 msec to initiate[15]. Correspondence, segmentation and tracking have received a good deal of attention in the literature e.g., [19, 13, 26]. Our focus is on composition of complex motion features, indexing, attention and gestalt formation.

Many attempts to interpret changes in dynamic scenes performed static analysis on each frame (perhaps using motion as a cue, e.g., for segmentation) with integration of frames occurring at a high level (e.g., [1, 23]). Recently there has been some work which shares with the present approach the idea of using motion itself as the basis for interpretation as evidenced by other chapters in this book. Most of that work has focused on recovering complex motion descriptions (features) of the changing scene, with the idea that these descriptions could be the basis for recognition. The present work concentrates on using such motion descriptions to construct models of gait and to index these models. In this context it is interesting to note that recent work in neurophysiology has begun to reveal neurons tuned specifically for complex motion patterns e.g., [7].

Recognition of a movement requires a model of the movement. In contrast to prior work on representation of movement [16, 10, 23], the scenario representation of movement is not based on a 3D model augmented by some motion parameters. The scenario represents a movement as a sequence of motion *events* linked by time *intervals*. The events are characterized by changes in the expected motion configuration rather than the more common use of changes in the expected pose (e.g., [23]). Spatial information is incorporated implicitly in the specification of events.

If a representation of movement and an associated matching process are to be considered candidates for a general purpose vision system, they must meet two basic constraints on the marginal cost of adding a new movement to the system's repertoire as the repertoire's size increases: 1) the additional cost in processing time should tend to zero (or less) and 2) the additional cost in memory should decrease. The representation and matching process presented here meet these constraints. Constraint (1), motivated by the observation that one can recognize many more movements at age 20 than age 2 but with no increase in task time, is particularly hard to satisfy. It implies a parallel algorithm. To ensure the maximum degree of parallel operation, the present system is conceived of and implemented as a massively parallel network of simple computing elements. There is *no learning* involved and considerably more structure than typically used in neural networks.

4. Model Architecture

MARS (Multiple Action Recognition System - Figure 1a) consists of three modules: 1) a visual feature hierarchy that analyzes motion feature maps –

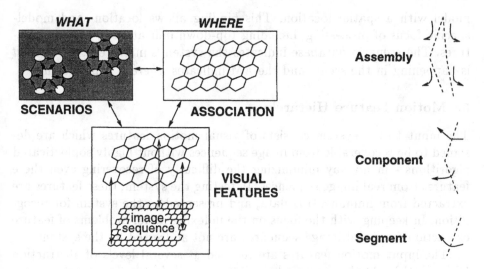

Figure 1. (a) MARS Architecture (b) motion features

which are assumed to have been derived from image sequences – for visual motion parameters, 2) a high level representation of actions, the scenario hierarchy, and 3) an association map (AM) which associates actions with spatial locations. Consider presentation of an image sequence of a person walking. The image sequence can be analyzed to recover moving line segments corresponding to the limbs in motion [23]. These transformations and further combinations in the feature hierarchy produce as output spatially localized complex visual features representing uninterpreted features in the scene, for example pairs of connected line segments moving relative to each other. The visual features index a hierarchical database of scenarios. Scenarios represent named actions that constitute a gait or other complex movement, for example "biped-walking". Recognition involves incrementally matching the scenario with the action in the scene.

Scenarios are conceptualized as *active* memory structures which incrementally analyze the visual features that are derived from the scene as they occur. A scenario is active in the sense that it has continuously varying measures which not only indicates the current belief that the movement represented by the scenario is occurring, but which also provides specific predictions of what will occur next, and when it will occur. These predictions provide a model-based component to the analysis in the visual feature hierarchy.

The scenario representation of movement does not encode location information, which is instead represented in the Association Map described below. The AM uses a time-dependent measure of the match between a movement model and current visual features to bind each partially matched model with a spatial location. This binding allows location- and model-specific focus of processing, including top-down indication of expected features. The scenario database indicates the system's interpretation of *what* is happening in the scene, and the AM indicates *where* it is happening.

5. Motion Feature Hierarchy

The input to the system consists of visual motion features which are designed to be recoverable from image sequences without highly sophisticated algorithms - in no way minimizing the difficulty of recovering even these features from real image sequences. In testing the system, these features are extracted from human MLD data, and presented to the system for recognition. In keeping with the focus on the indexing issue, problems of feature extraction from real image sequences are not addressed by the system.

The input motion features are recoded at several levels of abstraction (Figure 1(b)). At the segment level (input), each feature represents a moving line segment, such as those forming a stick figure idealization of a human. Features are encoded by coarse orientation and coarse angular speed in the image plane. The next level in the motion feature hierarchy, the component level, represents pairs of segment-level features situated in a particular spatial relationship and moving in a particular way. Components are independent of any particular movement represented in the movement database. The top level in the feature hierarchy, assemblies, represents pairs of connected components subject to particular spatial and motion constraints. Conceptually these features are independent of any particular modeled movement

5.1. SPATIAL LOCATION AND COMPOSITION

The visual feature hierarchy is a standard multi-resolution hierarchy: the spatial extent of a feature increases higher in the hierarchy while the spatial resolution decreases. Extent is implicit, spatial location is explicit in the representation of a feature. Each level defines its own coordinate system. These are increasingly coarse with each upward step in the hierarchy.

An $n+1$-feature (i.e., a feature at level $n+1$) is composed of a number of n-features. The locations of the n-features must conform to the set of spatial relationships, specified in the level n coordinate system, which define the spatial structure of the composed feature. These spatial constraints are enforced by the explicit spatial indexing of the features; only those n-features in appropriate spatial locations provide evidence for the $n+1$-feature at a particular location. Composition is enforced by applying a graded AND function to the evidence for each of the constituent features. The graded form works with evidential data and allows for missing features.

First we develop the theory of how evidence for features is combined, and then discuss the implementation and parameter values used.

5.2. THEORY

The spatial structure of an $n+1$-feature θ is defined by a set $\{f_\gamma, r_\gamma\}$ where f_γ is a constituent n-feature and r_γ is its relative offset from the location of θ. If there are p constituents in the set $\{f_\gamma, r_\gamma\}$, then a suitable evidence combination rule is to take the pth root of the product of the weighted evidence for θ of each constituents, a form of graded AND function. Thus we compute the evidence for θ at location l, $F_{\theta,l}$ set of constituents according to:

$$F_{\theta,l} = \sqrt[p]{\prod_{\gamma=1}^{\gamma=p} \left[w_{\gamma,\theta} \left(F_{\gamma,l+r_\gamma}(1 - \alpha^p) + \alpha^p \right) \right]} \tag{1}$$

where $w_{\gamma,\theta}$ is the *a-priori* evidence for θ given γ and α is a parameter controlling the false-positive/false-negative trade off (tolerance of missing constituents and susceptibility to false targets vary with α). Since each $n+1$-location encompasses several n-locations, we take the *max* of this computation over the set of n-locations encompassed by l to be the overall evidence for θ at location l.

5.3. IMPLEMENTATION

It is straightforward to map the feature hierarchy to a parallel network [9]. Each feature at each location is represented by a computing element, whose activation value represents the evidence for the feature. Evidence is communicated by labeled links in the network. Spatial constraints are represented by the pattern of links (i.e., only those n-features of the appropriate type and location send links to a $n+1$-feature). Evidence is combined at the elements according to equation (1). In the absence of psychophysical data indicating otherwise, we set all the weights to unity. Simulations showed that a value of 0.1 for the parameter α in equation (1) gave good results. For the two-constituent features used here, this gives a 90% reduction in activation if one constituent is missing.

A 20x30 input window was used, this being large enough to accommodate two moving stick figures side by side, giving 600 locations at the segment level [9]. This gave rise to approximately 150 locations at the component level and 54 at the assembly level. Segment level features were coded by orientation (quadrant in the image plane) and angular speed (0-100, 100-200 and 200+ degrees/second) in both directions (anti/clockwise), giving 24 different features in all (translational motion in the image plane was not represented). The quantization in orientation and speed was designed

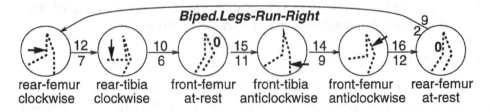

Figure 2. The sequence of events in "Legs-Running"

to be recoverable from image sequences without highly sophisticated algorithms, as discussed earlier. At the component and assembly levels the system need only include those features that are necessary for the discrimination of modeled movements e.g., a walking leg. Although there are potentially very many component and assembly features, only 115 component and 1086 assembly features were needed to represent the 32 collected gait samples.

6. Modeling Movement

Movements are represented by a schema-like structure called the scenario. In its most general form, the scenario is a structure for representing a discrete sequence of events separated in time by specific intervals or ranges of intervals. Thus it is a suitable representation for many spatio-temporal series. A scenario is defined by a set of events, a complete temporal ordering on those events, and a specification of the intervals between consecutive events. An event is characterized by one or more *enabling* conditions which specify a change in the visual features present in the scene that indicate the event has occurred. Enabling conditions are required to locate the time of occurrence of an event explicitly, which is crucial for implementing precise temporal constraints. A time interval is characterized by the range of times to be expected between one event and the next. It is anticipated that the interval will be parameterized by a tempo factor, but this was not used in the implementation. Figure 2 shows, for one scenario, the events (circular icons) with the enabling condition labeled below the icon and the range of times (in frames) between consecutive events. The enabling conditions for the events are indicated in the circular icons. and labeled below the icon. For example, the left most event occurs when the rear-femur begins to move clockwise.

A scenario imposes two types of soft constraints on the matching of scene kinematics to events. Feature constraints specify which visual motion features match which events. Temporal constraints specify the time intervals which are expected to occur between pairs of events (thus temporal constraints include constraints on sequence). Matching is performed incrementally as the scene changes, and the application of constraints adapts dynamically in two ways. First, while evidence for the scenario is low (e.g., initially) feature constraints are weighted much more heavily than temporal constraints, but as matching proceeds and evidence for the scenario accumulates temporal constraints assume more significance. Second, while evidence for the scenario is low temporal constraints are lax about time intervals and strict about sequence, but as evidence for the scenario accumulates expectation about time intervals is applied more strictly. This adaption in the application and nature of constraints allows the matching process to focus on instantaneous scene-derived visual motion features initially. As matches between features and events occur, temporal constraints, which operate across several frames, are applied to check sequence and, increasingly, rhythm.

6.1. COMPOSITION

An important property of any representation scheme is whether or not it exhibits compositionality. For example, it is desirable that a representation for objects be able to compose two object models to form a model for a more complex object. The same is true for representations of movements. In the present case, the lowest level scenarios are used to represent, for example, the movement of a pair of legs during running or the movement of a pair of arms during skipping. It is important to be able to compose two models at this level to form a model of a full-body movement.

The scenario structure described above was easily extended to exhibit this type of compositionality. (Another type of compositionality, concatenation, was not investigated but it would not seem difficult to include it). A composed scenario uses the events of its constituent scenarios to define enabling conditions, instead of using visual features. In addition to the temporal constraints defined over the composed scenario's events, there are temporal constraints specified between events in constituent scenarios (constituent events) and events in the composed scenarios. Figure 3 illustrates how the model of "biped-walking" is composed from "legs-walking" (the "arms-walking" constituent is omitted for clarity). Events are circular and temporal constraints elliptical. The numbers inside the icons represent, for events, the time at which the event takes place, and for temporal constraints, the time interval represented by the constraint (for simplicity, we collapse the time range to a single value in this diagram). The thick solid links show the bottom-up flow of enabling information (delayed by the time indicated) to the the higher-level scenario from the lower-level scenario, and the dashed links show the top-down temporal constraints. These specify a

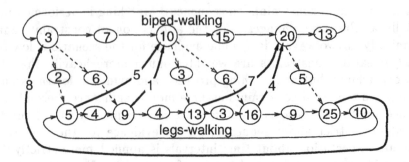

Figure 3. Composed "Biped-Walking" scenario

time interval from a higher-level event to one of its constituent events and provide a top-down augmentation to the bottom-up matching from visual features. This type of compositionality can be repeated for multiple levels although only two levels of scenarios were used in the implementation (the "pair-of-limbs" and "full-body" levels).

The composition of scenarios described above has one major weakness: it cannot enforce spatial constraints on the composition of lower-level scenarios. The "pair-of-arms-walking" and "pair-of-legs-walking" scenarios combine to form the "biped-walking" scenario, without any constraint on their relative spatial locations. The AM represents the location at which scenarios are occurring, and this information is used to enforce the appropriate spatial constraints (e.g., arms above legs) as described in [9].

6.2. THEORY

Matching of scene kinematics to events is a process involving the instantaneous visual motion features in the scene, prior events matched for the scenario, and the temporal constraints on sequence and time intervals. We want the system to be data-driven in the absence of interpretation but increasingly model-driven as confidence builds in an interpretation. Thus an adaptive rule which combines the evidence for visual features with accumulated evidence for prior events is appropriate. Denoting the accumulated evidence for the entire scenario by $0 \leq S(t) \leq 1.0$, the instantaneous evidence for enabling features by $F(t)$, and the evidence provided via temporal constraints from previously matched events by $P(t)$, the instantaneous evidence for an event is:

$$E(t) = \alpha_1 \sqrt{P(t)F(t)}\lambda_1(t) + (1 - \alpha_1)\frac{F(t)}{\lambda_1(t)} \tag{2}$$

where $\lambda_1(t) = \frac{\alpha_2}{1-(1-\alpha_2)S(t)}$, and the $0 < \alpha_i < 1.0$ are parameters controlling, respectively, the contribution of evidence sources and how strictly the

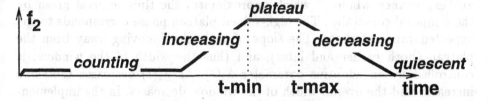

Figure 4. Expectancy: temporal delay and smoothing

constraints should be applied. The first term requires evidence from prior events as well as evidence from visual features. The second term requires evidence provided purely from the detection of the appropriate visual features. $\lambda_1(t)$, the strictness factor, operates in the range $[\alpha_2, 1]$ and increases as confidence in the scenario matching, $S(t)$, increases. It scales the first term so that the requirement for prior evidence is weighted more heavily (at most by a factor $\frac{1}{\alpha_2}$) as the scenario is matched. It scales the second term so that the need for appropriate visual features decreases (at most by a factor α_2) as the scenario is matched. Using this adaptive algorithm, the system self- regulates between being data-driven and model-driven depending on its confidence in the current interpretation.

The component $P(t)$ in equation (2) is the evidence due to the matching of prior events in the scenario modulated by the temporal constraints. If a temporal constraint between events I and J (such that I precedes J) is contained in the scenario, then the component of $P(t)$ due to this constraint is computed by passing the evidence for I through a two-stage filter. The first stage f_1 detects time of occurrence of I (rapid increase of evidence that the event I has been matched), producing value $E_I(t)$ whenever I is detected and zero otherwise. The second stage f_2 convolves the most recent non-zero value produced by the first stage with the parameterized temporal delay-and-smoothing function illustrated in Figure 4. The first stage detection function is computed in the following way:

$$f_1(t) = \begin{cases} E_i(t) & \text{if } f_2(t-1) = 0, \frac{E_i(t)-E_i(t-1)}{E_i(t-1)} > \alpha_3 \\ & \text{or } f_2(t-1) > 0, \frac{E_i(t)-E_i(t-1)}{E_i(t-1)} > \alpha_4 \\ 0 & \text{otherwise} \end{cases} \tag{3}$$

where the parameters $0 < \alpha_3 < \alpha_4 < 1.0$ control the rate of increase of evidence for the event that is required to consider the event as having occurred. The initial detection (first) condition enables the filter to respond to relatively weak indication of occurrence while the redetection (second) condition allows stronger indication later to cause another response.

The second stage convolves the confidence level at the time a match is detected with the function shown in Figure 4. This function describes an "expectancy window" which approximates the time interval given by the temporal constraint. The high-valued plateau phase corresponds to the expected range of times. The slope of the roll-off moving away from the plateau (both earlier and later), and thus the width of the window, is controlled by an adaptive parameter $\lambda_2(t)$. As $\lambda_2(t)$ decreases the slope increases and the overall width of the window decreases. In the implementation the expectancy window is computed in the following piecewise-linear fashion (first match applies):

$$
f_2(t) = \begin{cases}
0 & \text{if } t < t_{min}(1 - \lambda_2(t)) \\
& \text{or } t > t_{max}(1 + \lambda_2(t)) \\
f_1(t)\left(1 - \frac{t_{min}-t}{\lambda_2(t)t_{min}}\right) & \text{if } t < t_{min} \\
f_1(t) & \text{if } t < t_{max} \\
f_1(t)\left(1 - \frac{t-t_{max}}{\lambda_2(t)t_{max}}\right) & \text{if } t > t_{max}
\end{cases}
\tag{4}
$$

where $\lambda_2(t) = \alpha_5 + \alpha_6 S(t)$ and the parameters α_5, α_6 control the minimum and maximum values for the absolute slope of roll-on and roll-off phases of the expectancy window. This adaptive algorithm allows considerable deviation from expected time intervals when confidence in the scenario matching, $S(t)$, is low, but tightens up the window as confidence increases. The effect is that increasing weight is placed on correct rhythm as matching proceeds.

The evidence for the entire scenario is computed by combining the evidence for each of its events. The evidence for an event is taken to be the evidence at the time of its occurrence, modulated by a two-stage filter which is very similar to the temporal constraint filter described in (3) and (4) above. The difference is that the second stage, f_2', is computed by:

$$
f_2'(t) = \begin{cases}
f_1(t) & \text{if } t < t_{max} \\
0 & \text{if } t > t_{max}(1 + \lambda_2(t)) \\
f_1(t)\left(1 - \frac{t-t_{max}}{\lambda_2(t)t_{max}}\right) & \text{if } t > t_{max}
\end{cases}
\tag{5}
$$

which implements an expectancy window similar to Figure 4 with the difference that the plateau phase extends from $t = 0$ to $t = t_{max}$. The instantaneous evidence for the entire scenario, $S(t)$, is taken to be the maximum of the evidence produced by this filter for all the events, modulated by a damping factor α_7:

$$
S(t) = (1 - \alpha_7)S(t - 1) + \alpha_7 \max_{events}(f_2'(t))
\tag{6}
$$

6.3. IMPLEMENTATION

The scenario is easily implemented in a parallel network [9]. Briefly, we assign one computing element to each event, one to each temporal constraint,

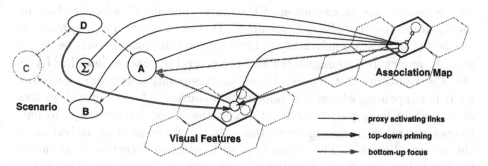

Figure 5. Association Map: Structure and Function

and one to the scenario itself. An event element computes the evidence for the event as in equation (2). The interval element performs the two-stage filter function described in equations (3) and (4). The scenario element computes the overall confidence that the scenario has been matched as in equations (3), (5) and (6). Directed links between these elements pass the evidence values each needs.

Parameter values for the implementation were as follows: $\alpha_1 = 0.8$, $\alpha_2 = 0.5$, $\alpha_3 = 0.2$, $\alpha_4 = 0.75$, $\alpha_5 = 0.2$, $\alpha_6 = 0.3$ and $\alpha_7 = 0.04$. These were experimentally determined to provide perfect discrimination for the non-noisy inputs from which the models were generated (see below). Simulations showed that the system has no great sensitivity to any of these parameter values, a sign of a robust architecture. There were 5 timesteps per input frame at a frame rate of 60 frames/second, thus each timestep corresponded to 3.33 msec of real time. It is plausible that parallel hardware could be built to operate at this speed, which would give real-time performance for an arbitrarily large database of movement models.

The Association Mechanism

As high-level memory structures, scenarios are position-independent representations. In connectionist vision systems, as in biological systems, low level feature detectors are duplicated, each copy having a limited receptive field. This allows parallel processing across the visual field and is an explicit representation of space. For position-independent recognition activation must be integrated across the visual field. Position-independent high-level representations of objects and actions perform this integration. The alternative, duplication of complex object and action representations across the visual field, would require copious hardware (units), and the problem of learning new objects and actions would be made even harder.

In terms of the "what" and "where" distinction in the visual system [17], I postulate that scenarios are analyzed primarily in the "what" path-

way. Scenarios are an extension of the static concept of "what" to include
the dynamic notion "what's happening". Despaced scenario representations
are devoid of location information, but location information must be represented somewhere *and associated with the central representations.* In MARS
the AM performs this representation and association function, encoding
what is happening where. AM information is used to focus bottom-up flow
of activation from particular locations in the visual feature maps to particular central scenario representations, and to focus priming activation in
the reverse direction. The connection with selective attention is discussed
further below, but first I describe the structure and function of the AM.

STRUCTURE

The AM is organized as a spatiotopic map encoding location and action. It
is based on the understanding that the wave of activation flowing around
an active scenario network is a mirror of the changes that are occurring
in the scene. Therefore it is possible to use temporal synchrony between
the visual changes at a location and the internal changes in the scenario
network as the cue for determining an association between the location and
the action (see [8] for an early version of this idea).

At each location in the AM there is a set of *proxy* units (Figure 5
shows two proxies at one location). Each proxy unit represents a scenario.
Its activation indicates the degree of belief that the action is occurring at
that particular location (activation flowing around the scenario network
indicates the belief that the action is occurring *somewhere* in the scene).
Proxy dynamics are described below. The proxies at a given location in
the spatiotopic map inhibit each other (Figure 5), thereby competing for
activation from feature detectors at the location.

FUNCTION

Bottom-up Focus:
A proxy modulates the significance to its scenario of visual features at its location. Proxies have activation levels in the interval $[-\alpha, 1.0]$, $0 \leq \alpha \leq 1.0$,
where polarity indicates evidence for (positive) or against (negative) the
action occurring at the location and magnitude indicates the degree of
belief, so that the neutral or "resting" level is zero[1]. Positive activation
increases the significance to the scenario of the visual features at the location. Negative activation, which occurs through inhibition from other

---

[1] Proxy activations are passed through a scaling function $f(x) = \frac{x+\alpha}{2}$ for transmission
to other units, and its inverse upon reception, which ensures that values passed between
units are always in the interval $[0, 1]$. The "resting" activation level is then $\frac{\alpha}{2}$. This
(de)scaling is ignored here for simplicity.

proxies, decreases the significance. α determines the degree of inhibition between proxies, as described below. Modulation is achieved with a link from the proxy to each of the event units in the scenario (shaded link in Figure 5). The link is labeled with the proxy's location, as are the links from the visual feature units to the event unit (shaded link). The activation $P_{S,L}$ from the proxy for scenario S at location L multiplicatively modulates the activity from the feature units located at L by the factor $(1 + P_{S,L})$ which is in the range $[1 - \alpha, 2]$. In the simulations of gait recognition, α was set at 0.5, so that the modulation factor was always in the interval $[0.5, 2]$.

The modulation factor causes the scenario to "attend" more to locations where its proxy has an activation above resting level and to "neglect" locations where its proxy is below resting level. In a scene containing a small number of actions, the effect is that each location where there is action tends to be "owned" by one scenario, and other scenarios actively ignore that location. Simulations showed that this is important in reducing interference between actions that are occurring in different spatial locations and thus increasing the ability to recognize concurrent actions.

Top-down Priming:

Recall that the scenario network contains clocked interval units that pass on priming activation to the succeeding event unit. This priming is sent when the visual change which the event codes for is expected to occur. This information is also used to enhance the response of the visual feature units representing the change that is expected, *prior* to the change occurring in the input (hollow link from D in Figure 5). The importance of the AM in this process is that the association that has been set up between an action and a location is used to direct the priming to the region in which the action is occurring (hollow link from proxy in Figure 5). This predictive priming acts as a multiplier on the unprimed response of the feature unit. The multiplicative factor at time t for a feature at location L that is selected by one or more events in scenario S is

$$1 + \beta_1 \sqrt{\beta_2 P_{S,L}(t) I_S(t) + (1 - \beta_2) I_S(t)}$$

where I_S is the maximum level of priming from interval units in scenario S that predict the feature, β_1 is a parameter controlling the magnitude of the priming effect, and β_2 is a parameter that controls the modulating of priming by the proxy activation. In the simulations of gait recognition β_1 and β_2 were set to 0.25 and 0.7 respectively. The simulations showed that the predictive priming significantly increased the speed with which the correct scenario was activated.

AM DYNAMICS

A proxy unit has a set of receptive *sites*[2], one for each event its scenario. A site receives input from the event unit and succeeding interval unit in the scenario and from the visual feature units in the feature map that the event unit is selective for (solid links in Figure 5). However, unlike the event unit, the proxy receives visual feature input only from the location it represents. The site compares activity of the event unit with activity of the feature units. It assigns a value which is dependent on simultaneous transients in both event unit and feature units and on the magnitude of those transients. When simultaneous transients are detected, the site value is the geometric mean of the two magnitudes. The site maintains this value during subsequent simulation cycles until the subsequent event has been primed, as indicated by the activation arriving along the link from the interval unit (e.g., unit B in Figure 5), at which time the site value decays to zero. This mechanism allows the proxy to set its activation from the relatively infrequent event transients but for the proxy activation to subside if the predicted events do not occur when expected. The values computed by the sites are combined in a scenario-dependent way to produce the synchrony cue $T(t)$ for the proxy. In the implementation that modeled the gait recognition data [9] the two highest site values are averaged to produce $T(t)$.

$T(t)$ is combined with an overall estimate $S(t)$ of the scenario activity provided by a link from the scenario summator unit. The proxy activation function is given by:

$$P_{S,L}(t+1) = (1 - \gamma_1)P_{S,L}(t) +$$
$$\gamma_1 \left[T(t)(\gamma_2 + (1-\gamma_2)S(t)) - \alpha \max_{i \neq S} P_{i,L}(t) \right]$$

where γ_1 is a parameter controlling the attack and decay rates of the unit and γ_2 is a parameter controlling the extent to which the scenario activation modulates the synchrony cue. The latter is motivated by the observation that it makes no sense for $P_{S,L}(t)$ (S is happening at L) to be higher than $S(t)$ (S is happening somewhere). The final term in the activation function is the mutual inhibition between proxies at each location, controlled by the parameter α introduced above. In the simulations of gait recognition, γ_1 and γ_2 were set to 0.1 and 0.3 respectively and recall that α was set to 0.5.

---

[2] A unit with sites can be thought of as representing a small network of cells, or a single cell and dendritic tree.

7. Experimental Results

Data were obtained for approximately 3 samples of walking, running and skipping, roughly parallel to the camera image plane, from each of two males and two females using a gait analysis system [25], resulting in 12 walking, 12 running and 8 skipping samples in all. These samples were normalized to a 1 second period (to prevent discrimination based on gait period) and were used to develop the scenario models of the three gaits, and their mirror image (e.g., "walk-left" and "walk-right"). The normalized data were sampled at 60 frames per gait period and used to create the moving line segment features that are the input to the system (and which are assumed to be extractable from real image sequences). These data and several distorted forms were introduced to test discrimination under perfect and various adverse conditions.

Five results are reported below (more appear in [9]). The time (number of frames presented at a nominal rate of 60 frames/sec) to achieve discrimination was defined as the time at which the target model had achieved and continued to maintain a confidence level of at least 64% of maximum and none of the non-targets subsequently achieved a confidence level greater than 40% of the target's confidence level. Success by frame 120 was the cutoff for discrimination.

7.1. DISCRIMINATION OF GAIT SAMPLES

Table 1 summarizes the ability of the system to discriminate the gait samples under three conditions. The three lines show the % of samples correctly discriminated by frame 30, 45, 60, 75 and 90 when 1) the original data was presented unmodified and with varied initial phase, 2) self occlusion was simulated by removing the hidden segments from the input when they would normally be occluded by other body parts, and 3) clutter was added in the form of 12 randomly placed segments (the limbs have 8 segments) rotating at randomly chosen angular velocities in the same range as those of the limb segments. In summary, all four conditions produced similar results, namely 100% discrimination within 60-90 frames.

7.2. REJECTING NON-MODELED MOVEMENT

It is extremely important for a recognition system to be able to reject input which is similar to models but not the same as any of them (i.e., to keep the rate of false-positives low). In this experiment the system was presented with a biped figure moving in a manner similar to a real gait but sufficiently different that it should be rejected. The movement was derived by replacing the movement of the distant arm/leg by the mirror image of

experiment	frames presented				
	30	45	60	75	90
Original data	9	91	100		
Occlusion	3	78	99	100	
Clutter	3	66	81	94	100

TABLE 1. % of samples correctly discriminated

	% of max. confidence reached						
	10	20	30	40	50	60	70
% trials	79	13	4	2	1	1	0

TABLE 2. Confidence levels for nonsense movements

the movement of the nearer arm/leg. In these nonsense samples, one half of the data in each frame match one of the scenarios (e.g., "skip-left") and one half match the mirror-scenario ("skip-right"). Table 2 shows the percentage of trials in which any scenario reached a given % of maximum confidence. In only 1% of trials was the 50% confidence level reached and even then the discrimination criteria were not satisfied.

7.3. DYNAMICALLY VARYING MOVEMENT

An experiment was conducted to check the transition from one gait to another. For one of the actors, one walking and one running sample were selected at random. A "common" frame was found in both samples. The walking sample was presented for 90 frames (terminating with the common frame) followed by 90 frames from the running sample (commencing with the common frame), thus simulating breaking into a run from a walk. Figure 6 shows the results. The graph shows time along the horizontal axis and confidence level (maximum value 125) along the vertical axis. Confidence levels for the six scenarios (three gaits x 2 directions) are plotted. Within 45 frames "walk-right" is recognized (thick line 0-90 frames). No more than 45 frames after the transition from walking to running in the input, "run-right" is discriminated (thick line 90-180 frames). This demonstrates the systems ability to cope with dynamically varying movement.

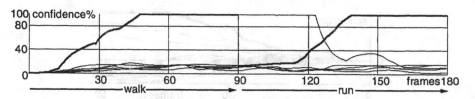

Figure 6. Transition from Walking to Running

Figure 7. Interpolation Coding of Location

7.4. LOCATION CODING IN THE AM:

Simulations in which single actions (e.g., "legs-walking-right") were presented showed that the AM codes location information more finely than the resolution of a single cell. In Figure 7, shading illustrates the activation levels of the proxy units for "legs-walking-right" when the legs were in the location shown (leg actions were modeled to be "located" at the hip). Figure 8 shows a trace of the activation of all the proxies at all locations over time (60 frames/sec simulated). The thick line in Figure 8 indicates the proxy activation for "leg-walking-right" at the heavily shaded location in Figure 7. The next highest activation trace (0.75) is at the moderately shaded location and the third highest (0.5) is at the two lightly-shaded locations. The location of the action can be recovered by interpolating the active locations using their activation as a weighting. Other proxies end up at or below resting level (0).

7.5. WHAT-WHERE INTERACTION:

Figure 9 plots the time-course of activation in the scenarios (summator outputs are shown). Note that by frame 30 in Figures 8 and 9 the leading trace shows high levels of activation and the two continue to rise together. This demonstrates the location binding in the association mechanism ("where") and the activation of scenarios ("what") occurring in parallel. The two

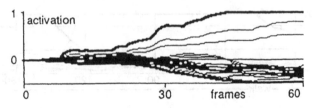

Figure 8. Proxies' Activation

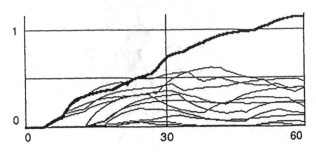

Figure 9. Scenarios' Activation

cesses act cooperatively to settle on a consistent solution.

7.6. PLASTICITY AND PHASE INSENSITIVITY:

A simulation was run to show that the binding in the AM is plastic. 60 frames of "legs-walking-right" were presented, recognized, and an association formed in the AM (Figure 8). Without resetting the network, 60 frames of "legs-running-right" were presented (Figure 10). The previous association dies away after about 30 frames (0.5 sec) of the new action, and the correct new association to "legs-running-right" (thick trace) is formed soon thereafter. The AM is a plastic mechanism.

The scenarios are capable of aligning themselves with the input, independent of initial phase of an action, as described in [9]. The AM receives all its timing expectations from the scenarios and is is therefore also insensitive to phase,

7.7. MULTIPLE ACTIONS:

Two actions were were presented simultaneously. When the two were presented in approximately the same location, there was usually too much cross talk for the AM to establish any scenario/location association. When the actions were spatially separated, the AM formed the correct association with each location (Figure 11). The thick and thin lines that asymptote at

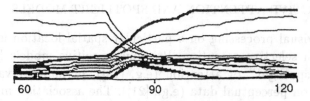

Figure 10. Plasticity

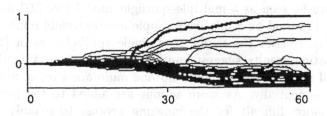

Figure 11. Multiple Actions

1.0 are the activation of the proxies of the two actions at the closest AM location. The other pairs at about 0.75 and 0.5 are the corresponding proxies at the other AM locations used in the interpolation-coding. It takes the AM longer to establish the associations when two actions are presented simultaneously. I presented three spatially-separated actions, and the AM took much longer to establish the associations. The AM is capable of *maintaining* multiple associations, but it shows capacity limitations in *establishing* multiple associations in parallel. This suggests a role for sequential attention.

8. Discussion

Discrimination of articulated movement is a central problem in perception. I have described a model-based system which can discriminate articulated movements and have demonstrated its capabilities in distinguishing three human gaits. While this is, I believe, the first system to demonstrate, in the context of general purpose vision, the ability to *discriminate* between articulated movements, there are a number of deficiencies. Prominent is the inability to deal with significantly varying tempo. This is a topic of current research and some ideas are outlined in [9]. Another desirable line of work would be to incorporate the low-level image-processing capabilities which would allow us to build an end-to-end system that can operate on real image sequences, and thus fully validate the system.

8.1. SELECTIVE ATTENTION AND SPOTLIGHT MODELS

Focusing visual processing on a particular spatial location is usually associated with selective attention. Previous cognitive models have proposed a single "spotlight" of attention (e.g, [18]), perhaps of variable diameter, based on perceptual data (e.g., [21]). The association mechanism outlined here is capable of forming, in parallel, multiple associations between simultaneously-presented spatially-separated actions and their locations. Thus it can be seen as a multiple-spotlight model (see [27] for a review of the data). It would be a relatively simple matter to add inhibition between locations to restrict the model to a single spotlight, as in [18]. However the simulation results suggest another interpretation. A mechanism such as the AM may be used to build up *and maintain* a set of action/location bindings sequentially. As more actions are added to the presentation, it becomes more difficult for the indexing process to reliably activate any scenario model due to crosstalk. The association mechanism cannot focus processing on a particular location until a scenario is at least partially active. If a separate attentional spotlight were added, it would be possible for the AM to make associations between location and action one pair at a time using a sequential spotlight cued by motion or other parameters.

The AM forms a crucial part of MARS, the first program to model the Johansson biological motion data. Modeling arm- and leg-movements separately, I found that the AM was required to enable recognition of full-body human gait. Using as a cue temporal synchrony between scene-action and internal active memory structures, it associates *what's happening* with *where* it is happening. The AM displays an ability to maintain multiple associations in parallel but cannot necessarily form those associations in parallel, suggesting a role complementary to that of sequential attention.

8.2. STRUCTURED CONNECTIONIST MODELING

Critical aspects of the system arise from the structured connectionist approach, which is a natural choice given the need for a parallel algorithm with constant time complexity and the biological limit of order 100 sequential steps for recognition [5]. Movement recognition without recovery of structure is motivated by that same limit. Separation of "what" and "where" processing is motivated by the huge representation space needed in a connectionist system if they are not separated. Incremental recognition over time is motivated by the complexities involved in retaining uninterpreted perceptual data in a connectionist system. The adaptive nature of scenario-based recognition is a natural development in a connectionist system. In sum, the work is a significant demonstration of the effectiveness of the structured connectionist approach.

References

1. Norman I. Badler. *Temporal Scene Analysis: Conceptual Descriptions of Object Movements.* PhD thesis, Department of Computer Science, University of Toronto, 1975.

2. John N. Bassili. Facial motion in the perception of faces and of emotional expression. *Journal of Experimental Psychology: Human Perception and Performance,* 4(3):373–379, 1978.

3. James E. Cutting, Cassandra Moore, and Roger Morrison. Masking the motions of human gait. *Perception and Psychophysics,* 44:339–347, 1988.

4. J.E. Cutting and L.T. Kozlowski. Recognizing friends by their walk: Gait perception without familiarity cues. *Bull. Psychonometric Soc.,* 9(5):353–356, 1977.

5. Jerome A. Feldman, Mark A. Fanty, Nigel H. Goddard, and Kenton Lynne. Computing with structured connectionist networks. In Steven F. Zornetzer, Joel L. Davis, and Clifford Lau, editors, *An Introduction to Neural and Electronic Networks,* pages 433–454. Academic Press, 1990.

6. Robert Fox and Cynthia McDaniel. The perception of biological mottion by human infants. *Science,* 218:486–487, October 1982.

7. Bard J. Geesaman and Richard A. Andersen. The analysis of complex motion patterns by form/cue invariant MSTd neurons. *The Journal of Neuroscience,* 16(15):4716–4732, 1996.

8. Nigel H. Goddard. Representation and recognition of biological motion. In *Program of the Tenth Annual Conference Cognitive Science Society,* pages 230–236, Hillsdale, NJ, August 1988. Lawrence Erlbaum.

9. Nigel H. Goddard. *The Perception of Articulated Motion: Recognizing Moving Light Displays.* PhD thesis, Dept. of Computer Science, University of Rochester, 1992. also Technical Report 405, Computer Science Department.

10. David Crossland Hogg. *Interpreting Images of a Known Moving Object.* PhD thesis, Department of Computer Science, University of Sussex, 1984.

11. Gunnar Johansson. Visual perception of biological motion and a model for its analysis. *Perception and Psychophysics,* 14:201–211, 1973.

12. Gunnar Johansson. Spatio-temporal differentiation and integration in visual motion perception. *Psychological Research,* 38:379–393, 1976.

13. Christof Koch, Jose Marroquin, and Alan Yuille. Analog "neuronal" networks in early vision. *Proc. National Acadamy of Science of the USA,* 83:4263–4267, June 1986.

14. L.T. Kozlowski and J.E. Cutting. Recognizing the sex of walker from dynamic point-light displays. *Perception and Psychophysics,* 21(6):575–580, 1977.

15. Stephen G. Lisberger, E.J. Morris, and L. Tychsen. Visual motion processing and sensory-motor integrationfor smooth-pursuit eye movements. *Annual Review Neuroscience,* 10:97–129, 1987.

16. D. Marr and Lucia Vaina. Representation and recognition of the movements of shapes. *Proceedings of the Royal Society of London B,* 214:501–524, 1982.

17. M. Mishkin, L. G. Ungerleider, and K.A. Macko. Object vision and spatial vision: Two cortical pathways. *Trends in Neuroscience,* 6:414–417, 1983.

18. Michael C. Mozer. *The Perception of Multiple Objects: a Connectionist Approach.* MIT Press, Cambridge, MA, 1991.

19. Thomas J. Olson. *An Architectural Model of Visual Motion Understanding.* PhD thesis, University of Rochester, 1989. also Technical Report 305, Computer Science Department.

20. H. Poizner, U. Bellugi, and V. Lutes-Driscoll. Perception of american sign language in dynamic point light displays. *Journal of Experimental Psychology: Human Perception and Performance,* 7:430–440, 1981.

21. M. I. Posner, C. R. R. Snyder, and B. J. Davidson. Attention and the detection of signals. *Journal of Experimental Psychology: General,* 109:160–174, 1980.

22. Dennis R. Proffitt. Recovering connectivity from moving point-light displays. In Martin and Aggarwal, editors, *Motion Understanding: Robot and Human Vision*, chapter 9, pages 297–327. Boston: Kluwer, 1988.

23. K. Rohr. Towards model-based recognition of human movements in image sequences. *Computer Vision, Graphics and Image Processing: Image Understanding*, 59:94–115, 1994.

24. Sverker Runeson and Gunilla Frykholm. Kinematic specification of dynamics as an informational basis for person-an-action perception: expectation, gender recognition, and deceptive intention. *Journal of Experimental Psychology: General*, 112(4):585–615, 1983.

25. John P. Scholz. Reliability and validity of the WATSMART three-dimensional optoelectronic motion analysis system. *Physical Therapy*, 69(8):679–689, August 1989.

26. M. Seibert and A. Waxman. Spreading activation layers, visual saccades, and invariant representations for neural pattern recognition systems. *Neural Networks*, 2:9–27, 1989.

27. Richard M. Shiffrin. Attention. In R. C. Atkinson, R. J. Herrnstein, G. Lindzey, and R. D. Luce, editors, *Steven's Handbook of Experimental Psychology, Volume 2: Learning and Cognition*, chapter 11, pages 739–811. John Wiley & Sons, New York, 1988.

28. Shigemasa Sumi. Upside-down presentation of the johansson moving light-spot pattern. *Perception*, 13:283–286, 1984.

29. V.C. Tartter and K. C. Knowlton. Perceiving sign language from an array of 27 moving spots. *Nature*, 289:676–678, 1981.

30. James T. Todd. Perception of gait. *Journal of Experimental Psychology: Human Perception and Performance*, 9(1):31–42, 1983.

HUMAN MOVEMENT ANALYSIS BASED ON EXPLICIT MOTION MODELS

K. ROHR
Arbeitsbereich Kognitive Systeme
Fachbereich Informatik, Universität Hamburg
Vogt-Kölln-Str. 30, D-22527 Hamburg, Germany

1. Introduction

Within the field of computer vision the automatic interpretation of human movements is one of the most challenging tasks. A central problem in analyzing such movements is due to the fact that the human body consists of body parts linked to each other at joints which allows different movements of the parts. Therefore, the human body generally has to be treated as a nonrigid or more precisely as an articulated body. In addition, for general camera positions always some of the body parts are occluded. Although occlusions can provide important cues in a recognition task, the automatic interpretation is more difficult. Another problem that has to be dealt with is the clothing which can have a large influence on the appearence of a person (wide or tight trousers, different jackets, etc.). Clothing can also cause complex illumination phenomena that, in addition, change during movement (compare with efforts in the field of computer graphics to simulate cloth objects, e.g., [83]).

Because of these difficulties most existing approaches for analyzing human movements assume the joints of the human body to be marked (e.g., [66],[33],[81],[3],[27],[18],[65],[74]), or they are applied to synthetic images only (e.g., [61],[78],[76]). When using real-world images often special gymnastic movements are analyzed but not locomotion. In this case, the interpretation is generally less difficult because the effect of self-occlusion is diminished (e.g., [1],[47],[62]). Other approaches use stereo-images (e.g., [19]), restrict their analysis to certain parts of the body (e.g., [79],[85],[41],[25]), or analyze image sequences with more or less homogeneous background to diminish the segmentation problem (e.g., [49],[45],[28]). Often, image analysis is not carried out on an incremental basis but uses the entire sequence

M. Shah and R. Jain (eds.), Motion-Based Recognition, 171–198.

(e.g., [60]). For references, see also [40] and the recent survey on motion-based recognition in [17].

In this chapter, we describe a motion-based approach for analyzing human movements in monocular real-world image sequences. We explicitly represent the human body as well as its movement and use this knowledge to estimate 3D positions and postures of persons from images. Central to our approach is the use of an explicit motion model which is based on analytically given motion curves for the body parts. These motion curves exploit data from medical motion studies and represent an average over a relatively large number of test persons. Additionally, we use a Kalman filter to incrementally estimate the model parameters from consecutive images. Since estimates from previous images are taken into account we obtain a smooth and robust result. Moreover, an initialization phase is performed to automatically estimate the initial model parameters. Our algorithm is designed for analyzing the movement of human walking which is the most frequent type of locomotion of persons. However, a generalization to other movement types such as, for example, running is straightforward.

Hogg [34],[35] in his pioneering work also has introduced a model-based approach for determining the 3D positions and postures of walking persons from real-world images. However, the motion model he uses does not exploit data from medical motion studies but has been acquired interactively from one prototype image sequence. Also, the initial model parameters are not provided completely automatically. Moreover, the estimates of the model parameters for the current image do not take into account previous estimates by using a Kalman filter scheme.

Automatic analysis of human movements has a wide spectrum of potential applications. For example, in street-traffic scenes it is important to early recognize situations which might lead to accidents. In the medical area it is important to analyze abnormal movement patterns.

The organization of this chapter is as follows. First, we give an overview of our motion-based approach. Then, we describe in more detail the human body model and the motion model of walking. After that, we show how this knowledge can be used to incrementally estimate the 3D positions and postures of persons from image data. The applicability of our approach will be demonstrated for real-world images. To indicate a possible extension of our work we finally describe a model of a cyclist.

2. Overview of our Motion-Based Approach

In our approach to the recognition of human movements from images we represent the shape of the human body by a volume model which is build of cylinders connected by joints ([51],[52]). Our motion model consists of a set of analytically given motion curves which represent the postures of a person, i.e., the relative positions of all body parts. A nice property is that only one parameter, named *pose*, is needed to fully specify all postures during walking. Together with the 3D space coordinates $\vec{X} = (X, Y, Z)$ of the center of the torso (the origin of the person's coordinate system) and assuming the person to move parallel to the image plane we generally have to estimate four parameters from consecutive images. This is done using a Kalman filter where the system description represents the overall movement of the person, i.e., the movement of the center of the torso. The system description together with the motion curves of the body parts constitute our motion model. The model is explicitly given in an analytic form.

Our algorithm can be subdivided into two phases: initialization and incremental estimation. Whereas in the first phase the images are evaluated in a batch type manner, in the second phase processing is done on an incremental basis. The main parts can be summarized as follows.

1. *Initialization*

   Independent evaluation of about 10-15 images:

   - Detection of image regions corresponding to moving persons (using a change detection algorithm and binary image operations)
   - Estimation of the movement states, i.e., 3D positions and postures (using a calibration matrix for central projection and matching contours of the 3D model with grey-value edges)
   - Determination of starting values for the Kalman filter (using linear regression and the estimates from above)

2. *Incremental estimation*

   After initialization a Kalman filter scheme is applied to each image:

   - Prediction of the movement state (using estimation results from previous images)
   - Determination of measurements (using matching results of the 3D model to the current image)
   - Estimation of the current movement state (using the predicted movement state and the measurements)

The part of our work which deals with modelling the human body and its movement is a central subject within the fields of computer graphics, computer animation, or image synthesis. Thus, our approach for image sequence analysis utilizes methods from image synthesis, and therefore we can speak of an analysis-by-synthesis approach. The general relation between image analysis and image synthesis has been depicted in Fig. 1 (see also,

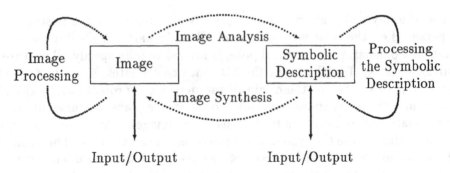

Figure 1. Relation between image analysis and image synthesis

e.g., [80],[46],[64],[15]). Whereas the aim of image analysis consists in deriving a symbolic description from real images, it is the aim of image synthesis to produce realistic images from a symbolic description. By using certain operations images can be transformed into images (image processing) or symbolic descriptions may be transformed into new symbolic descriptions.

3. Model of the Human Body

To represent static 3D models for the purpose of efficient recognition Marr and Nishihara [51] proposed to use volumetric primitives, an object-centered coordinate system, and a modular hierarchical organization on the description. As an example, they considered a 3D model of the human body consisting of cylinders.

In our approach we use this representation and describe the human body by 14 cylinders with elliptic cross-sections (head, torso and three primitives for each arm and leg) which are connected by joints (see also [23],[8],[6]). Cylinders are a good compromise between the number of parameters and the quality of representation of the human body. Each cylinder is described by three parameters: one for the length and two for the sizes of the semi-axes. The coordinate systems for the body parts are aligned with the natural axes. The origin of the coordinate system of the whole body is at the center of the torso (see Fig. 2). Transformations between different coordinate systems are described by homogeneous coordinates $\vec{X} = (X, Y, Z, 1)^T$:

$$\vec{X}' = \underline{A}\,\vec{X}, \qquad\qquad \underline{A} = \left(\begin{array}{cc} \underline{R} & \vec{T} \\ \vec{0}^T & 1 \end{array} \right), \qquad\qquad (1)$$

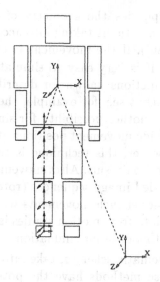

Figure 2. Human body representation

Figure 3. Rendered 3D model of the human body

where $\underline{R}$ is the 3×3 rotation matrix and $\vec{T}$ the translation vector. The inverse of $\underline{A}$ is given by

$$\underline{A}^{-1} = \begin{pmatrix} \underline{R}^T & -\underline{R}^T \vec{T} \\ \vec{0}^T & 1 \end{pmatrix} \quad (2)$$

If several transformations are applied in series then we can multiply the corresponding matrices $\underline{A}_i$, for example,

$$\underline{A}_i \underline{A}_{i+1} = \begin{pmatrix} \underline{R}_i \underline{R}_{i+1} & \underline{R}_i \vec{T}_{i+1} + \vec{T}_i \\ \vec{0}^T & 1 \end{pmatrix} \quad (3)$$

In our model of the human body we use absolute sizes of the body parts. Since persons are "in general" dressed, and since clothing can strongly influence the appearance of a person, the usefulness of existing catalogues of body measurements of unclothed persons (e.g., [21]) is limited. Therefore, we use sizes of the human body parts obtained by direct measurements of a normal person with average clothing. A visualization of our 3D model is displayed in Fig. 3 (see also [46]).

4. Motion Model of Walking

Models of human movements are studied within the fields of biomechanics, computer graphics, and robotics. Generally, one can distinguish between kinematic and dynamic methods for representing movements (e.g., [5],[77]).

A *kinematic* description explicitly specifies the geometry of objects, i.e., position, orientation, and deformation without taking into account the cause of the movement (e.g., [30],[14],[86]). If the movement is explicitly given by time-dependent functions then it is very easy to simulate movements. However, there are hardly any functions known for describing the complex movements of the human body (but see, for example, the laws of motion in [82] published early in 1836). Another possibility for simulating human movements is to interactively provide movement positions at certain time instants (keyframe-technique). However, this technique is time consuming and often does not lead to the desired result. Also, movements can be reconstructed interactively from recorded image sequences (rotoscopie). By this approach it is only possible to model such movements which have previously been performed. If, however, data from motion studies is already available then it is advantageous to use this data for simulation.

Dynamic methods, in contrast to kinematic schemes, take into account forces and torques (e.g., [84],[13]). These methods have the potential to produce realistic motions, however, they are computationally expensive and specifying forces and torques can be difficult. In addition, the resulting movements are not always satisfying and sometimes they are improved by kinematic adjustment (see also [44]).

In our approach to the recognition of human movements in real-world image sequences the agreement of the model with actual movements is important. Also, the number of parameters needed for specifying the model should be kept small in order to facilitate their estimation from images. Therefore, we decided to use a kinematic approach which exploits data from human motion studies. Studies of the human motion have a long tradition (see [2],[82]). At the end of the 19th century, photographic methods have been developed ([56],[50],[11]). For medical purposes Murray et al. [55],[54] have analyzed the movements of sixty normal men ranging in age from twenty to sixty-five years to obtain the basic elements of walking. In this study it is demonstrated that the motion curves of the body parts for different persons are very similar. Note, however, that this similarity is very astonishing if one imagines that it is often possible to identify persons by their gait. The fact that the motion curves are very similar opens us the possibility to use this data as knowledge source.

To represent the movement of walking in our motion model we have taken the average data from [55],[54]. For each of the joints at the shoulder, elbow, hip, and knee within one walking cycle we have taken function values which measure the relative angles between connected body parts, e.g., for the knee-joint the angle between the thigh and the lower leg. Additionally, we have used the values for the vertical displacement of the whole body which describe the periodic ups and downs of the center of the torso during

walking (see also [70]). Note, that with our motion model the overall movement is simplified in such a way that we have only modelled the movement
of four joints and only considered joint rotations within the sagittal plane
(which is spanned by the walking direction and the vertical direction). Since
the joint movements of one side of the human body agree with those on
the other side but are displaced about half of the walking cycle, we only
need four (plus one) 1D motion curves for describing the overall movement
of walking.

The motion curves of the joints at the shoulder, elbow, hip, and knee
as well as for the vertical displacement of the whole body are based on
the function values y_ν at certain points of time $t_\nu, \nu = 0, .., N - 1$. Since
walking is a periodic movement these values have been interpolated by
periodic cubic splines $B(t)$ (see [73] and also [35]). Each interval $[t_\nu, t_{\nu+1}]$
is described by

$$B_\nu(t) = b_{1\nu}(t - t_\nu)^3 + b_{2\nu}(t - t_\nu)^2 + b_{3\nu}(t - t_\nu) + b_{4\nu}. \qquad (4)$$

For periodic splines there is $t_N = t_0 + T$ (with T being the period). At the
boundaries the function values as well as the first and the second derivatives
agree, i.e., $y_0 = y_N$, $y_0' = y_N'$ and $y_0'' = y_N''$. In the resulting system of N
linear equations with the N unknowns $\vec{y}'' = (y_0'', .., y_{N-1}'')$:

$$\underline{H}\,\vec{y}'' + \vec{d} = \vec{0}, \qquad (5)$$

the matrix $\underline{H}$ represents the intervals $h_\nu = t_{\nu+1} - t_\nu$ between consecutive
points of time t_ν, and the vector $\vec{d}$ represents the function values y_ν in
conjunction with the h_ν. $\underline{H}$ is symmetric, diagonal dominant and positive
definite. We solve (5) by Cholesky decomposition. With $\vec{y}''$ we can compute
$b_{1\nu}, b_{2\nu}, b_{3\nu}, b_{4\nu}$ and therefore $B_\nu(t)$ for each interval. In our case we took
$N = 10$ function values at equidistant points of time and standardized the
walking cycle to $T = 1$. The interval between two consecutive points of
time therefore is $h_\nu = 0.1$.

The resulting motion curves of the joints and the vertical displacement
are shown in Figs. 4 and 5, respectively. We use the parameter *pose* as
some kind of time parameter which specifies the relative positions of all
body parts within one walking cycle. A nice property is that we only need
one parameter for this specification. Movement states for half of the walking
cycle are shown in Fig. 6 (*pose* $= 0, 0.1, 0.2, 0.3, 0.4, 0.5$). Depicted are the
contours of the cylinders under central projection. Hidden contours have
been removed (see, e.g., [59],[29]). Fast reproduction of these motion states
on a screen reveals that our motion model appears to be fairly realistic.

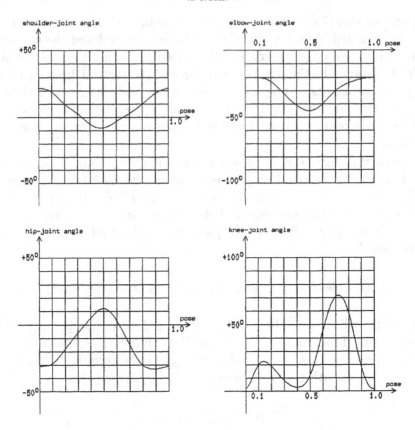

Figure 4. Motion curves of the shoulder, elbow, hip and knee joints

5. Incremental Estimation of the Model Parameters

In this section, we describe how our model of the human body and its motion is used to incrementally estimate the 3D positions and postures of persons from image sequences. Before applying a Kalman filter scheme we first perform an initialization phase. The main parts of this phase are change detection, 3D position estimation, and contour matching.

5.1. INITIALIZATION

5.1.1. *Change detection*
We assume the image sequences to be recorded with a stationary camera. To segment the images into regions corresponding to moving and nonmoving objects we apply a change detection algorithm. In each image point the intensities are approximated locally by a polynomial of second order and

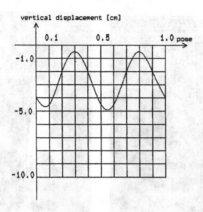

Figure 5. Vertical displacement of the whole body

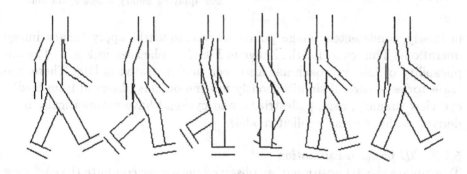

Figure 6.   Movement states of walking

these polynomial fits are compared for consecutive images (see [10],[9]). Let
$\underline{X}$ be the matrix representing the image coordinates within a 5 × 5 window
and $\vec{\beta}$ the parameters of the fit. The n grey values within the window
are given by $\vec{g}_k = \underline{X}\vec{\beta}_k$ and we can estimate $\hat{\vec{\beta}}_k = \underline{X}^{\#}\vec{g}_k$ by using the
generalized inverse $\underline{X}^{\#} = (\underline{X}^T\underline{X})^{-1}\underline{X}^T$. We consider a point of an image
k to represent systematic changes, if

$$\frac{1}{\sqrt{n}}\|\underline{X}\hat{\vec{\beta}}_k - \underline{X}\hat{\vec{\beta}}_j\|_2 = \frac{1}{\sqrt{n}}\|\underline{X}\,\underline{X}^{\#}(\vec{g}_k - \vec{g}_j)\|_2 > T \tag{6}$$

holds for the preceding $(j = k - 1)$ or for the following $(j = k + 1)$ image,
where $\|\cdot\|_2$ denotes the Euclidean norm and T is a threshold. Experi-
mentally we have found that this procedure yields better results than the
approach in [36]. To remove false detections due to noise as well as to fill

Figure 7. Image of a walking person

Figure 8. Result after change detection and applying binary image operations

in falsely nondetected image points we subsequently apply binary image operations. The result for the image in Fig. 7 can be seen in Fig. 8. The applicability of this approach has also been demonstrated in [10]. There, the trajectories of several simultaneously moving objects observed from bird's-eye view (namely, cars, pedestrians, and cyclists) have automatically been derived without using explicit models.

5.1.2. *3D position estimation*

To estimate the 3D position of an observed person we compute the enclosing rectangle of the detected object candidate (see Fig. 8) and make an assumption about the absolute height of the person. The camera is supposed to be upright w.r.t. gravity and a 4×4 calibration matrix $\underline{T}$ is assumed to be available (e.g., [69],[24],[67]). Using homogeneous coordinates the projection of a 3D point $\vec{X} = (X, Y, Z, 1)^T$ onto the image plane is then given by

$$h \, \vec{x} = \underline{T} \, \vec{X}, \tag{7}$$

where $\vec{x} = (x, y, *, 1)$, $*$ denotes an arbitrary value, and h being nonzero (Fig. 9). The midpoints of the bottom and the top edges of the detected rectangle $\vec{x}_u$ and $\vec{x}_o$ represent the sole and the top of the human body, respectively. The connection line between the corresponding 3D positions is supposed to lie perpendicular to the plane upon which the person moves. With H denoting the assumed height of the person and $\vec{H} = (0, H, 0, 0)^T$ we can write

$$h_u \vec{x}_u = \underline{T} \vec{X}_u \tag{8}$$

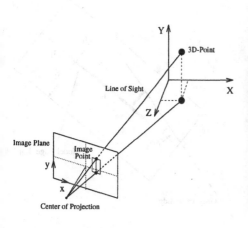

Figure 9. Coordinate systems

Figure 10. Enclosing rectangle and extracted grey-value edge lines

$$h_o \vec{x}_o = \underline{T}(\vec{X}_u + \vec{H}). \tag{9}$$

Based on (8) and (9) we can derive a linear system of equations which is solved to obtain the 3D position of the person (for details, see [72]).

5.1.3. Contour matching

After change detection and 3D position estimation we then apply a contour matching algorithm to estimate the posture of the person. Since the contours of our model consist of straight lines we decided to compare them with edge lines. In comparison to edge points the number of image features is smaller and the influence of noise is reduced. Since we compute the edges only within the detected rectangle the search space for matching has been considerably reduced. Edge detection is done using the approach in [43] which is similar to that in [16]. After that, we use an edge linking procedure followed by an Eigenvector line fitting as described in [22] (see Fig. 10).

For matching the model with the detected grey-value edges we compute a search window for each visible model contour. The size of the search window depends on the length of the model contour (Figs. 11, 12). If a grey-value edge overlaps a window we first cut this edge to the portion inside the window. Then, we compute a measure of similarity between the grey-value edge and the model contour. This measure takes into account the length l_i as the projection of the grey-value edge onto the model contour l_{Mi}, second, the distance d_i between the midpoint of the grey-value edge and its corresponding projection onto the model contour, and third, the angle between the two edges $\Delta\varphi_i$. The larger l_i and the smaller $(l_{Mi} - l_i)$, d_i and

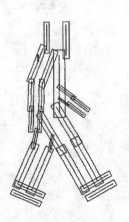

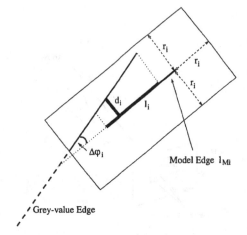

Figure 11. Search windows for the model

Figure 12. Comparison between grey-value edge and model contour

$\Delta\varphi_i$ the more similar are the two edges. We use the following measure of similarity:

$$s_i = l_i \, e^{-\frac{1}{2}\left(\frac{(l_{Mi} - l_i)^2}{\sigma_{li}^2} + \frac{d_i^2}{\sigma_{di}^2} + \frac{\Delta\varphi_i^2}{\sigma_{\Delta\varphi}^2}\right)} \tag{10}$$

With

$$\underline{\Sigma}_i = \begin{pmatrix} \sigma_{li}^2 & 0 & 0 \\ 0 & \sigma_{di}^2 & 0 \\ 0 & 0 & \sigma_{\Delta\varphi}^2 \end{pmatrix}, \qquad \vec{k}_i = \begin{pmatrix} l_{Mi} - l_i \\ d_i \\ \Delta\varphi_i \end{pmatrix} \tag{11}$$

the measure can also be written as

$$s_i = l_i \, e^{-\frac{1}{2}\vec{k}_i^T \underline{\Sigma}_i^{-1} \vec{k}_i}. \tag{12}$$

The parameters σ_{li} and σ_{di} of the measure are determined in dependence of the length of the model contour: $\sigma_{li} = c_l \, l_{Mi}$ and $\sigma_{di} = c_d \, r_i = c_d \, c_r \, l_{Mi}$, where c_l, c_d and c_r are constant for all model contours; $\sigma_{\Delta\varphi}$ is constant, too. Therefore, the exponent in (12) is independent of scaling the model. Since we want grey-value edges with larger values l_i to have a larger influence on the overall similarity we weight the exponential function by this value. Alternatively, we could weight by l_{Mi}.

Many approaches for line matching compare the midpoints or the starting and endpoints of model contours with the grey-value edge lines supposing that the two edges are similar in length (e.g., [48],[53],[75],[7]). In our application the grey-value edges of the lower and upper part of the arms and the legs often are connected to one grey-value edge. Then, a comparison between midpoints or starting and endpoints would lead to large discrepancies. Therefore, we first cut the grey-value edges and then use the quantities of similarity as described above. The use of the exponential function in (12) has the advantage that the resulting expression is defined for all values and decreases fast to zero for less similar edges. If several greyvalue edges overlap the search window, then we take the one with highest similarity value. The overall measure of similarity $s(\vec{p})$ in dependence of the model parameters $\vec{p}$ is the sum of the values s_i for all visible contours of the human body model normalized by the sum of their lengths l_{Mi}:

$$s(\vec{p}) = \frac{\sum\limits_{i=1}^{n} w_i \, s_i}{\sum\limits_{i=1}^{n} l_{Mi}} \longrightarrow max. \tag{13}$$

The parameters w_i can be used to weight the values s_i by other information, e.g., by the magnitude of the grey-value gradient at the matched grey-value edge or by values derived from prior knowledge about the recognition significance of the body parts. With (13) we search those parameters $\vec{p}$ which maximize $s(\vec{p})$. To obtain a more stable recognition result we remove hidden model contours. However, one disadvantage is that no analytic relation between $s(\vec{p})$ and $\vec{p}$ can be determined. Therefore, for maximizing $s(\vec{p})$ we use a grid search method with equally spaced points and take those parameters which yield the highest similarity value.

The procedure described above has been tested on single images. We have fixed the 3D position of the model and have varied the posture parameter *pose* within the whole walking cycle (see [72]). It turned out, that the estimated postures generally agree with the observation. However, the similarity curve has several secondary maxima and therefore it is no good choice to use a downhill optimization procedure. Also, we obtained relatively large similarity values for values of *pose* displaced about half of the walking cycle from the correct posture. However, this is what we expect if we imagine that walking is a symmetric movement w.r.t. to time. Even for a human observer it is generally hard to decide which of the two legs is in front of the other when seeing a walking person in some distance. In either case, to make the estimation result more robust it seems to be important to take into account information derived from several consecutive images.

5.1.4. *Starting values for the Kalman filter*
In the initialization phase of the Kalman filter we apply the procedures described above for each image independently. We use a number of 10-15

images which represent about half of a walking cycle. In comparison to that, a human observer only needs $0.2s$ (less than a quarter of a walking cycle) to recognize a walking human represented by moving light displays ([38],[63]). Starting values for the model parameters are obtained by applying linear regression to the estimated values from the initialization phase. Since the estimates of *pose* for real images often are displaced by about half of the walking cycle, some preprocessing is necessary for this parameter. The values of *pose* in dependence of the image number approximately lie on two lines. Therefore, to exclude systematic errors, we first group those values together which approximately lie on the two lines, and then apply linear regression. The selection of one of the two lines is then done by taking into account the overall number of estimates as well as the mean value of similarity. In our experiments, it was possible to automatically estimate the correct initial posture with this procedure. However, it should be noted, that generally this is a very hard problem which should be investigated further.

5.2. INCREMENTAL ESTIMATION

After initialization we apply a Kalman filter scheme ([39],[26]) to incrementally estimate the model parameters in consecutive images. In the field of computer vision, Kalman filter approaches have been introduced, for example, in [12] and [20]. In [12] an approach is described for incrementally estimating the model parameters of rigid objects from measured image points, whereas in [20] grey-value edge lines are tracked in the image plane. Our aim is to analyze walking persons. Since for general camera positions the body parts during walking often are occluded by one another (self-occlusion) tracking of single grey-value edge lines in general would lead to severe problems. Therefore, to cope with the problem of self-occlusion, in our case we fit the model as a whole and use a hidden-line algorithm to remove occluded model contours.

In the following, we assume the observed person to walk with constant velocity. By interpreting the parameter values for the best fit of the model contours with the grey-value edges as measurements for the corresponding time instant, the system description as well as the measurement model of the Kalman filter approach can be expressed by a linear relation. Therefore, we use a discrete linear Kalman filter.

The general discrete linear model is given by:

$$\vec{p}_k = \underline{\Phi}_{k,k-1}\,\vec{p}_{k-1} + \underline{\Gamma}_{k-1}\,\vec{w}_{k-1} \tag{14}$$
$$\vec{z}_k = \underline{H}_k\,\vec{p}_k + \vec{v}_k, \tag{15}$$

where the searched parameters are represented by the state vector $\vec{p}_k$ at the time instant k, and $\underline{\Phi}_{k,k-1}$ is the transition matrix. $\underline{\Gamma}_k \vec{w}_k$ represents modeling errors, where $\underline{\Gamma}_k$ often is chosen to be the unity matrix and $\vec{w}_k$ is assumed to be Gaussian distributed with expected value $E\{\vec{w}_k\} = \vec{0}$ and covariance matrix $E\{\vec{w}_k \vec{w}_k^T\} = \underline{Q}_k$, i.e., $\vec{w}_k \sim N(\vec{0}, \underline{Q}_k)$. The current measurements are represented by $\vec{z}_k$. $\underline{H}_k$ denotes the measurement matrix and $\vec{v}_k \sim N(\vec{0}, \underline{R}_k)$ the measurement errors. Often it is reasonable to assume that the errors $\vec{w}_k$ and $\vec{v}_j$ are uncorrelated, i.e., $E\{\vec{w}_k \vec{v}_j^T\} = \underline{0}$ for all j, k. The prediction of the parameters and the covariance matrix is given by:

$$\vec{p}_k^* = \underline{\Phi}_{k,k-1} \hat{\vec{p}}_{k-1} \tag{16}$$

$$\underline{P}_k^* = \underline{\Phi}_{k,k-1} \hat{\underline{P}}_{k-1} \underline{\Phi}_{k,k-1}^T + \underline{Q}_{k-1} \tag{17}$$

With these predictions and the current measurement $\vec{z}_k$ the estimates $\hat{\vec{p}}_k$ and $\hat{\underline{P}}_k$ in the current image can be computed by

$$\hat{\vec{p}}_k = \vec{p}_k^* + \hat{\underline{K}}_k (\vec{z}_k - \underline{H}_k \vec{p}_k^*) \tag{18}$$

$$\hat{\underline{P}}_k = (\underline{I} - \hat{\underline{K}}_k \underline{H}_k) \underline{P}_k^* \tag{19}$$

$$\hat{\underline{K}}_k = \underline{P}_k^* \underline{H}_k^T (\underline{H}_k \underline{P}_k^* \underline{H}_k^T + \underline{R}_k)^{-1}, \tag{20}$$

where $\underline{I}$ is the unity matrix.

In our application, we want to estimate the 3D position $\vec{X} = (X, Y, Z)$ and the posture *pose* of walking persons. The state vector is $\vec{p}_k = (X_k, \dot{X}_k, Y_k, \dot{Y}_k, Z_k, \dot{Z}_k, pose_k, \dot{pose}_k)^T$, where $\dot{X}_k, \dot{Y}_k, \dot{Z}_k$, and $\dot{pose}_k$ are the velocities. The time difference between two successive images is denoted by Δt. Supposing constant velocities we have

$$\underline{\Phi}_{k,k-1} = \begin{pmatrix} 1 & \Delta t & 0 & 0 & 0 & 0 & 0 & 0 \\ 0 & 1 & 0 & 0 & 0 & 0 & 0 & 0 \\ 0 & 0 & 1 & \Delta t & 0 & 0 & 0 & 0 \\ 0 & 0 & 0 & 1 & 0 & 0 & 0 & 0 \\ 0 & 0 & 0 & 0 & 1 & \Delta t & 0 & 0 \\ 0 & 0 & 0 & 0 & 0 & 1 & 0 & 0 \\ 0 & 0 & 0 & 0 & 0 & 0 & 1 & \Delta t \\ 0 & 0 & 0 & 0 & 0 & 0 & 0 & 1 \end{pmatrix} \tag{21}$$

Errors for the velocities are taken into account by the covariance matrix $\underline{Q}_k$. At the beginning, cross-correlations between the single parameters are set to zero.

$$Q_k = \begin{pmatrix} 0 & 0 & 0 & 0 & 0 & 0 & 0 & 0 \\ 0 & \sigma^2_{Q\dot{X}} & 0 & 0 & 0 & 0 & 0 & 0 \\ 0 & 0 & 0 & 0 & 0 & 0 & 0 & 0 \\ 0 & 0 & 0 & \sigma^2_{Q\dot{Y}} & 0 & 0 & 0 & 0 \\ 0 & 0 & 0 & 0 & 0 & 0 & 0 & 0 \\ 0 & 0 & 0 & 0 & 0 & \sigma^2_{Q\dot{Z}} & 0 & 0 \\ 0 & 0 & 0 & 0 & 0 & 0 & 0 & 0 \\ 0 & 0 & 0 & 0 & 0 & 0 & 0 & \sigma^2_{Qpose} \end{pmatrix} \tag{22}$$

In our case, we have measurements for $\vec{X} = (X, Y, Z)$ and *pose*. The measurement matrix is given by

$$H_k = \begin{pmatrix} 1 & 0 & 0 & 0 & 0 & 0 & 0 & 0 \\ 0 & 0 & 1 & 0 & 0 & 0 & 0 & 0 \\ 0 & 0 & 0 & 0 & 1 & 0 & 0 & 0 \\ 0 & 0 & 0 & 0 & 0 & 0 & 1 & 0 \end{pmatrix} \tag{23}$$

and uncertainties of the measurements are characterized by

$$R_k = \begin{pmatrix} \sigma^2_{RX} & 0 & 0 & 0 \\ 0 & \sigma^2_{RY} & 0 & 0 \\ 0 & 0 & \sigma^2_{RZ} & 0 \\ 0 & 0 & 0 & \sigma^2_{Rpose} \end{pmatrix} \tag{24}$$

6. Experimental Results

Our approach has been tested on synthetic as well as on real-world image sequences. To reduce the complexity of the recognition task we assume the observed pedestrian to move parallel to the image plane. Therefore, the scene coordinate Z describing the depth of the walking person is supposed to remain constant and we take over the value from the initialization phase. Incrementally we estimate the height Y of the person above the plane of movement taking into account the vertical displacements of the whole body as represented by the motion curve in Fig. 5. Note, that since we normalize the values for Y to the height of a standing person we can assume that the velocity for this parameter is zero and also the velocity's uncertainty can be set to zero. In addition to Y, we estimate the coordinate X in the direction of the movement as well as the posture *pose*. The initial uncertainties for the velocities of these parameters are derived from prior known average velocities of pedestrians. We assume $1s$ time duration for a walking cycle and

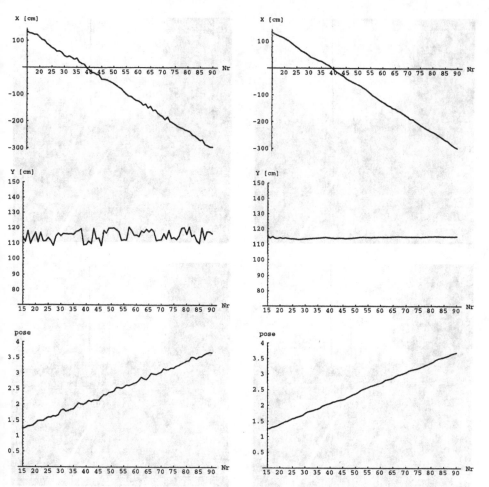

Figure 13. Estimation result of X, Y and *pose*; left: measurements; right: Kalman filter estimates

the covered distance to be $1.6m$ (e.g., [55],[54],[37]). The other elements for the covariance matrices have been chosen heuristically in such a way that we get reasonable values for automatically controlling the search spaces for the parameters. Note, that for incrementally estimating the model parameters we only exploit the matching results of the model contours with the grey-value edges. In this phase, we do not use the change detection and 3D position estimation procedures as described in Section 5.1. Note also, that based on the predictions of the Kalman filter the search space for matching the model contours with grey-value edges is considerably be reduced in comparison to the initialization phase.

Application of our approach to a real-world image sequence consisting of

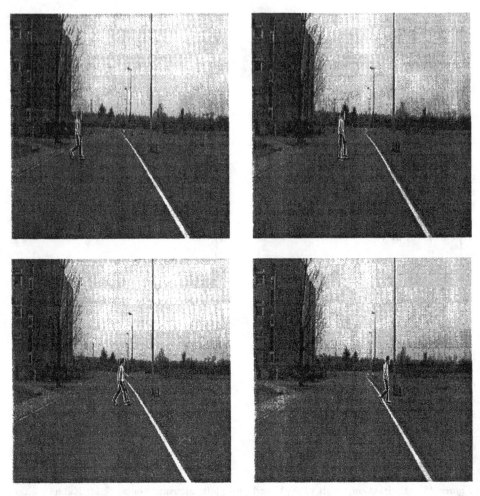

Figure 14. Estimated movement states superimposed onto the original images (images number 20, 40, 60 and 80)

80 images yields the estimation results shown in Fig. 13. For the similarity measure in (13) we have used equal weights $w_i = 1, i = 1, ..., n$. On the left side are the measurements for X, Y and *pose* due to contour matching, where X and Y specify the center of the person's torso. On the right side are the Kalman filter estimates. Especially for Y the smoothing properties of the Kalman filter can clearly be seen. Although the image portion covered by the pedestrian is relatively small, the person has been tracked over the whole sequence. Estimated movement states superimposed onto some of the original images are shown in Fig. 14. Generally, the agreement is fairly well. However, in some images we can observe slight deviations.

To analyze these deviations more precisely we have investigated the re-

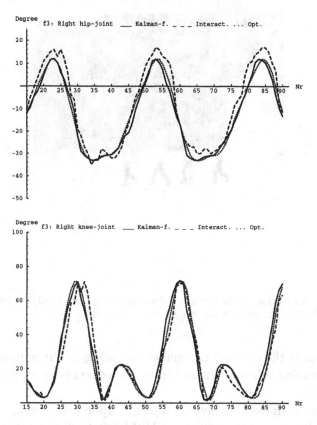

Figure 15. Motion curves for hip and knee joints on the right side of the human body

sult of our approach for another real-world image sequence where the image portion covered by the pedestrian is larger and therefore the deviations are easier to recognize. We have used the same parameter setting as before (see also [72]). For this image sequence the estimated motion curves of the hip and knee joint for the right side of the human body can be seen in Fig. 15 (solid lines). Since we have no ground truth for these images we have interactively determined reference values which we regard as ground truth. This has been done by manually adjusting the model for each image in such a way that for a human observer the agreement is best (see the dashed lines in Fig. 15). Note, that such an adjustment is not very easy since even small changes of the angle at the hip-joint (e.g., $3^o - 5^o$) lead to relatively large posture changes of the whole model. Nevertheless, from Fig. 15 we see that the estimated curves generally agree with the interactively obtained curves. However, at the hip-joint especially for positive angles there are larger deviations which result in the overall difference between the model and the image data. For a further comparison we have also drawn those curves

Figure 16. Visualization of the geometrical scene description (GSD) for images number 20, 40, 60 and 80 of the image sequence shown in Fig. 14

which represent the optimal movement according to our motion model for the corresponding velocity. These curves also generally agree with the other motion curves.

Recently, we have also investigated our approach in conjunction with the natural language access system VITRA [32] to derive natural language descriptions such as "The pedestrian walks across the street" (see [31]). As intermediate representation between our computer vision system and the natural language access system we have used the 'geometrical scene description' (GSD) as proposed in [58],[57]. Generally, the GSD represents all available information about the visible objects and their locations over time. In our case, the GSD consists of the 3D positions and postures of the pedestrian for each image. Additionally, we have coarsely represented the stationary background. The GSD corresponding to the real image sequence in Fig. 14 has been visualized in Fig. 16.

7. Model of a Cyclist

In this section, we indicate a possible extension of our work dealing with articulated movements in street-traffic scenes. We consider cyclists which, besides pedestrians and cars, belong to the most often occurring class of objects in this kind of domain.

To model a cyclist we combine the representation of the human body together with a geometric description of a bicycle. As with our motion model

Figure 17. Movement states of a cyclist

of walking the movement of the cyclist is modeled by using a kinematic method. However, for the cyclist we do not need any data from motion studies. This is possible because the geometric relations of the person and the bicycle fully constrain the movement if we consider "normal" driving straight ahead. With the movement of walking there are more degrees of freedom. For the cyclist we assume that the person sits on the saddle and that the upper part of the body remains unchanged during the movement. Note, however, that this model is a strong simplification in comparison to the complex movements that can be observed, for example, in a finish of a bicycle race. Anyway, the relative positions of the human legs are generally determined by the positions of the pedals of the bicycle. If we assume that the ankle-joint does not move, then the positions of the legs can be computed by intersecting two circles. Since only one solution is physically possible we thus obtain a unique position for the legs. This is because for the knee-joint only positive angles can occur (compare with the motion curves of walking in Fig. 4). The overall movement of the cyclist is periodic and symmetric w.r.t. the different body sides. Therefore, we can apply the same kinematic scheme as we have used before for modeling human walking. For the motion model of the cyclist we have evaluated ten positions of the pedals within one cycle to obtain function values for the joints. These values have been interpolated by periodic cubic splines as described in Section 4 above. Movement states of the cyclist for about half of the cycle are shown in Fig. 17. In Fig. 18 some images from an animated sequence of the cyclist together with a walking person can be seen.

Analogously to our approach for recognizing walking persons, the model of the cyclist, i.e., the geometric and the motion model, could be used to analyze the movements of cyclists in image sequences. To efficiently recognize the wheels of the bicycle a matching algorithm for circles (or ellipses) seems to be necessary.

Figure 18. Moving cyclist and pedestrian

8. Summary and Future Work

The movement of articulated bodies such as the human body is enabled by the coordinated movement of its rigid body parts. The body parts are connected by joints and, in general, move differently. For interpreting these movements as the movement of one single body, it seems to be necessary to incorporate knowledge in the image analysis process.

In our approach for analyzing human movements in real-world image sequences we exploit knowledge about the human body as well as its movement. Central to this approach is the use of an explicit motion model which is based on analytically given motion curves for the body parts. These motion curves represent data from medical motion studies. The fact that the

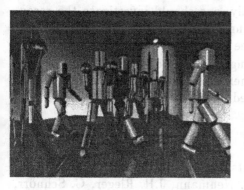

Figure 19. Several walking persons   *Figure 20.* Crossing pedestrian and car

motion curves are very similar for different persons opens us the possibility to use this data as knowledge source. A nice property is that only one parameter is needed to fully specify the relative positions of all body part. To incrementally estimate the 3D positions and postures of a walking person in consecutive images we have applied a Kalman filter scheme yielding smooth and robust results. We have assumed that the person moves parallel to the image plane with constant velocity. Starting values for the Kalman filter have automatically been determined through an initialization phase. Additionally, we have described a model of a cyclist to give an idea of extending our work for more complex street-traffic scenes. Estimation on an incremental basis rather than using the entire sequence is important, for example, in street-traffic scenes for the purpose of early recognition of situations which might lead to accidents, followed by giving a warning to the involved driver.

Future work on the recognition of walking persons should address the problem of reducing the deviations between the observed individual movements and those of our motion model which represents an average over a relatively large number of test persons walking at about the same velocity. An important point is that the motion curves of the body parts in general depend on the speed of movement. One possibility for improvement is to start the algorithm with the predefined motion curves and then try to adapt those curves to the observed movement. In this way, an incremental update of the model curves would be used for further evaluation. Also, to make the estimation result more robust, additional features could be taken into account, e.g., one could compare the model velocity fields with image velocity fields. A more efficient match of the model to the image data could possibly be achieved by using an aspect graph (view graph) for the model (e.g., [42],[68]).

In further investigations also other human movement types, such as running, should be analyzed. Additionally, one should investigate the recognition of several simultaneously moving bodies as well as the classification between different movement types (see Figs. 18, 19, and 20 for simulations of such scenes). Another future challenge is the automatic natural language description of observed human movements. Recently, investigations towards this long-term goal have been reported in [31].

Acknowledgment

For discussions and critical comments I thank D. Bister, C. Bregler, K. Daniilidis, G. Herzog, W. Leister, B. Neumann, J.H. Rieger, C. Schnörr, H.S. Stiehl, J. Weisbrod, and G. Zimmermann. This work has been supported by the Deutsche Forschungsgemeinschaft (DFG), Sonderforschungsbereich 314 "Künstliche Intelligenz – Wissensbasierte Systeme".

References

1. Akita, K. Image sequence analysis of real world human motion, *Pattern Recognition* 17 (1984) 73-83
2. Aristoteles, *Über die Bewegung der Lebewesen; Über die Fortbewegung der Lebewesen*, Datierung um 330 v.Chr., Teil II und III, J. Kollesch, Aristoteles Werke in deutscher Übersetzung, Band 17, Zoologische Schriften II, E. Grumach, H. Flashar (Hrsg.), Wissenschaftliche Buchgesellschaft, Darmstadt 1985
3. Attwood, C.I., Sullivan, G.D. and Baker, K.D. Model-based Recognition of Human Posture Using Single Synthetic Images, *Proc. Fifth Alvey Vision Conf.*, Univ. of Reading, Reading/UK, 25-28 Sept. 1989, 25-30
4. Badler, N.I. Temporal scene analysis: conceptual descriptions of object movements, Tech. Rep. No. 80, Dept. Computer Science, Univ. Toronto, Feb. 1975
5. Badler, N.I. and Manoochehri, K.H. and Walters, G. Articulated Figure Positioning by Multiple Constraints, *IEEE Computer Graphics & Appl.* 7:6 (June 1987) 28-38
6. Baum, L.F. and Denslow, W.W. *The Wonderful Wizard of Oz*, Geo. M. Hill Co. Chicago New York 1900, Justin Knowles Publishing Group London 1987
7. Beveridge, J.R., Weiss, R. and Riseman, E.M. Optimization of 2-Dimensional Model Matching, *Proc. Image Understanding Workshop*, Palo Alto, California, May 23-26, 1989, 815-830
8. Binford, T.O. Visual perception by computer, *IEEE Conf. on Systems and Control*, Dec. 1971
9. Bister, D. Bestimmung der Trajektorien von zeitweise verdeckten Objekten aus einer Bildfolge, Diplomarbeit, Institut für Algorithmen und Kognitive Systeme, Fakultät für Informatik der Universität Karlsruhe (TH), April 1991
10. Bister, D., Rohr, K. and Schnörr, C. Automatische Bestimmung der Trajektorien von sich bewegenden Objekten aus einer Bildfolge, *12. DAGM - Symposium Mustererkennung*, 24.-26. Sept. 1990, Oberkochen-Aalen, *Informatik-Fachberichte 254*, R.E. Großkopf (Hrsg.), Springer-Verlag Berlin Heidelberg 1990, 44-51
11. Braune, W. and Fischer, O. *Der Gang des Menschen, 1. Theil: Versuche am unbelasteten und belasteten Menschen*, Abhandlungen der Mathematisch-Physischen Classe der königlich sächsischen Gesellschaft der Wissenschaften, Einundzwanzigster Band, S. Hirzel Leipzig 1895
12. Broida, T.J. and Chellappa, R. Estimation of Object Motion Parameters from Noisy Images, *IEEE Trans. on Pattern Anal. and Machine Intell.* 8 (1986) 90-99

13. Bruderlin, A. and Calvert, T.W. Goal-Directed, Dynamic Animation of Human Walking, *Computer Graphics* 23:3 (July 1989) 233-242
14. Calvert, T.W. and Chapman, J. Aspects of the Kinematic Simulation of Human Movement, *IEEE Computer Graphics & Appl.* 2:9 (Nov. 1982) 41-49
15. Calvert, T.W. and Chapman, A.E. Analysis and synthesis of human movement, in T.Y. Young (Ed.), Handbook of Pattern Recognition and Image Processing (Vol. 2): Computer Vision, Academic Press, SanDiego, CA, 1994, 431-474
16. Canny, F. "A computational approach to edge detection", *IEEE Trans. on Pattern Anal. and Machine Intell.* 8 (1986) 679-698
17. Cédras, C. and Shah, M. Motion-based recognition: a survey, *Image and Vision Computing* 13:2 (1995) 129-155
18. Chen, Z. and Lee, H.-J. Knowledge-Guided Visual Perception of 3-D Human Gait from a Single Image Sequence, *IEEE Trans. on Systems, Man and Cyb.* 22:2 (1992) 336-342
19. Cipolla, R. and Yamamoto, M. Stereoscopic Tracking of Bodies in Motion, *Proc. Fifth Alvey Vision Conf.*, Univ. of Reading, Reading/UK, 25-28 Sept. 1989, 109-114
20. Deriche, R. and Faugeras, O. Tracking line segments, *Image and Vision Computing* 8:4 (Nov. 1990) 261-270
21. DIN 33402, Deutsche Normen, *Körpermaße des Menschen*, Beuth Verlag Berlin 1987
22. Duda, R.O. and Hart, P.E. *Pattern classification and scene analysis*, Wiley New York 1973
23. Fischer, O. *Theoretische Grundlagen für eine Mechanik der lebenden Körper mit speziellen Anwendungen auf den Menschen, sowie auf einige Bewegungsvorgänge an Menschen*, B.G. Teubner Leipzig Berlin 1906
24. Ganapathy, S. Decomposition of transformation matrices for robot vision, *Pattern Recognition Letters* 2:6 (1984) 401-412
25. Gavrila, D.M. and Davis, L.S. 3-D model-based tracking of human upper body movement: a multi-view approach, *Proc. Intern. Symposium Computer Vision (ISCV'95)*, Los Alamitos, 21-23 Nov. 1995, IEEE Computer Society Press 1995, 253-258
26. Gelb, A. *Applied Optimal Estimation*, MIT Press, Cambridge/MA, 1974
27. Goddard, N.H. The Interpretation of Visual Motion: Recognizing Moving Light Displays, *Proc. Workshop on Visual Motion*, Irvine, CA, March 20-22, 1989, 212-220
28. Guo, Y., Xu, G. and Tsuji, S. Tracking human body motion based on a stick figure model, *J. of Visual Communication and Image Representation* 5:1 (1994) 1-9
29. Hartmann, E. *Computerunterstützte Darstellende Geometrie*, B.G. Teubner Stuttgart 1988
30. Hartrum, T.C. Computer implementation of a parametric model for biped locomotion kinematics, Ph.D. Thesis, School of the Ohio State University, Columbus, Ohio, 1973
31. Herzog, G. and Rohr, K. "Integrating Vision and Language: Towards Automatic Description of Human Movements", *Proc. 19th Conf. on Artificial Intelligence, KI-95: Advances in Artificial Intelligence*, Sept. 1995, Bielefeld/Germany, *Lecture Notes in Artificial Intelligence* 981, I. Wachsmuth, C.-R. Rollinger, and W. Brauer (Eds.), Springer-Verlag Berlin Heidelberg 1995, 259-268
32. Herzog, G. and Wazinski, P. VIsual TRAnslator: Linking Perceptions and Natural Language Descriptions, *Artificial Intelligence Review* 8:2/3 (1994) 175-187
33. Hoffman, D.D. and Flinchbaugh, B.E. Interpretation of Biological Motion, *Biol. Cybern.* 42 (1982) 195-204
34. Hogg, D. Model based vision: a program to see a walking person, *Image and Vision Computing* 1:1 (1983) 5-20
35. Hogg, D. Interpreting Images of a Known Moving Object, PhD dissertation, University of Sussex, Brighton/UK 1984

36. Hsu, Y.Z., Nagel, H.-H. and Rekers, G. New Likelihood Test Methods for Change Detection in Image Sequences, *Computer Vision, Graphics, and Image Processing* 26 (1984) 73-106

37. Inman, H.V.T., Ralston, H.J. and Todd, F. *Human walking*, Williams & Wilkins Baltimore/London 1980

38. Johansson, G. Spatio-temporal differentiation and integration in visual motion perception, *Psychological Research* 38 (1976) 379-396

39. Kalman, R.E. A New Approach to Linear Filtering and Prediction Problems, *Trans. ASME J. Basic Eng.*, Series 82D (March 1960) 35-45

40. Kambhamettu, C., Goldgof, D.B., Terzopoulos, D. and Huang, T.S. Nonrigid motion analysis, in T.Y. Young (Ed.), Handbook of Pattern Recognition and Image Processing (Vol. 2): Computer Vision, Academic Press, SanDiego, CA, 1994, 405-430.

41. Kinzel, W. and Dickmanns, E.D. Moving Humans Recognition using Spatio-Temporal Models, *Proc. ISPRS'92*, Washington, D.C., *Internat. Archives for Photogrammetry and Remote Sensing*, Vol. XXIX, Part B5, L.W. Fritz and J.R. Lucas (Eds.), 1992, 885-892

42. Koenderink, J.J. and van Doorn, A.J. The Internal Representation of Solid Shape with Respect to Vision, *Biol. Cybernetics* 32 (1979) 211-216

43. Korn, A.F. Toward a Symbolic Representation of Intensity Changes in Images, *IEEE Trans. on Pattern Analysis and Machine Intelligence* 10 (1988) 610-625

44. Kunii, T.L. and Sun, L. Dynamic Analysis-Based Human Animation, *CG International'90*, T.S. Chua, T.L. Kunii (Eds.), Springer-Verlag Tokyo Berlin Heidelberg 1990, 3-15

45. Kurakake, S. and Nevatia, R. Description and Tracking of Moving Articulated Objects, *Proc. 11th Intern. Conf. on Pattern Recogn.*, The Hague, The Netherlands, Aug. 30 - Sept. 3, 1992, Vol. I, 491-495

46. Leister, W. and Rohr, K. Voruntersuchungen von Bildsynthesesmethoden zur Analyse von Bildfolgen, Techn. Report Nr. 25/90, Universität Karlsruhe (TH), Fakultät für Informatik, Sept. 1990

47. Leung, M.K. and Yang, Y.H. A region based approach for human body motion analysis, *Pattern Recognition* 20:3 (1987) 321-339

48. Lowe, D.G. The Viewpoint Consistency Constraint, *Internat. J. of Computer Vision* 1 (1987) 57-72

49. Luo, Y., Perales, F.J. and Villanueva, J.J. An Automatic Rotoscopy System for Human Motion Based on a Biomechanic Graphical Model, *Comput. & Graphics* 16:4 (1992) 355-362

50. Marey, É.-J. *Movement*, William Heine London 1895

51. Marr, D. and Nishihara, H.K. Representation and recognition of the spatial organization of three-dimensional shapes, *Proc. R. Soc. Lond.* B 200 (1978) 269-294

52. Marr, D. and Vaina, L. Representation and recognition of the movements of shapes, *Proc. R. Soc. Lond.* B 214 (1982) 501-524

53. McIntosh, J.H. and Mutch, K.M. Matching straight Lines, *Computer Vision, Graphics, and Image Processing* 43 (1988) 386-408

54. Murray, M.P. Gait as a total pattern of movement, *American J. of Physical Med.* 46:1 (1967) 290-332

55. Murray, M.P., Drought, A.B. and Kory, R.C. Walking Patterns of Normal Men, *The J. of Bone and Joint Surgery* 46-A:2 (1964) 335-360

56. Muybridge, E. *Muybridge's complete Human and Animal Locomotion. All 781 Plates from the 1887 'Animal Locomotion'*, Vol. 1, Dover Publications, Inc., New York 1979

57. Neumann, B. Natural Language Description of Time-Varying Scenes, in *Semantic Structures, Advances in Natural Language Processing*, D.L. Waltz (Ed.), Lawrence Erlbaum, Hillsdale/NJ, 1989, 167-206

58. Neumann, B. and Novak, H.-J. Event models for recognition and natural language

description of events in real-world sequences, *Proc. Internat. Joint Conf. on Artificial Intell. (IJCAI'83)*, 1983, 724-726

59. Newman, W.M. and Sproull, R.F. *Grundzüge der interaktiven Computergrafik*, McGraw-Hill Book Company Hamburg 1986

60. Niyogi, S.A. and Adelson, E.H. Analyzing and Recognizing Walking Figures in XYT, *Proc. IEEE Conf. on Computer Vision & Pattern Recognition*, Seattle, WA, June 21-23, 1994, 469-474

61. O'Rourke, J. and Badler, N.I. Model-based image analysis of human motion using constraint propagation, *IEEE Trans. on Pattern Anal. and Machine Intell.* 2:6 (1980) 522-536

62. Pentland, A. and Horowitz, B. Recovery of Non-Rigid Motion and Structure, *IEEE Trans. on Pattern Anal. and Machine Intell.* 13:7 (1991) 730-742

63. Perrett, D.I., Harries, M.H., Benson, P.J., Chitty, A.J. and Mistlin, A.J. *Retrieval of Structure from Rigid and Biological Motion: An Analysis of the Visual Responses of Neurones in the Macaque Temporal Cortex*, in *AI and the Eye*, A. Blake and T. Troscianko (Eds.), John Wiley & Sons Chichester/UK New York/NY 1990, 181-200

64. Pun, T. and Blake, E. Relationships between Image Synthesis and Analysis: Towards Unification, *Computer Graphics Forum* 9:2 (1990) 149-163

65. Qian, R.J. and Huang, T.S. Motion Analysis of Human Ambulatory Patterns, *Proc. 11th Intern. Conf. on Pattern Recogn.*, The Hague, The Netherlands, Aug. 30 - Sept. 3, 1992, Vol. I, 220-223

66. Rashid, R.F. Towards a System for the Interpretation of Moving Light Displays, *IEEE Trans. on Pattern Anal. and Machine Intell.* 2:6 (Nov. 1980) 574-581

67. Rehfeld, N. Auswertung von Stereobildfolgen mit Kantenmerkmalen, Dissertation, Fakultät für Informatik der Universität Karlsruhe (TH), Juni 1990

68. Rieger, J.H. On the complexity and computation of view graphs of piecewise smooth algebraic surfaces, *Phil. Trans. R. Soc. Lond. A* 354 (1996) 1899-1940

69. Rogers, D.F. and Adams, J.A. *Mathematical Elements for Computer Graphics*, McGraw-Hill Book Company New York 1976

70. Rohr, K. Auf dem Wege zu modellgestütztem Erkennen von bewegten nicht-starren Körpern in Realweltbildfolgen, 11. *DAGM - Symposium Mustererkennung*, 2.-4. Okt. 1989, Hamburg, *Informatik-Fachberichte* 219, H. Burkhardt, K.H. Höhne, B. Neumann (Hrsg.), Springer-Verlag Berlin Heidelberg 1989, 324-328

71. Rohr, K. Incremental Recognition of Pedestrians from Image Sequences, *Proc. IEEE Conf. on Computer Vision & Pattern Recognition*, New York/NY, USA, June 15-17, 1993, 8-13

72. Rohr, K. Towards Model-Based Recognition of Human Movements in Image Sequences, *Computer Vision, Graphics, and Image Processing: Image Understanding* 59:1 (1994) 94-115

73. Schwarz, H.R. *Numerische Mathematik*, B.G. Teubner Stuttgart 1986

74. Seitz, S.M. and Dyer, C.R. Affine Invariant Detection of Periodic Motion, *Proc. IEEE Conf. on Computer Vision & Pattern Recognition*, Seattle, WA, June 21-23, 1994, 970-975

75. Sester, M. and Förstner, W. Object Location Based on Uncertain Models, 11. *DAGM-Symposium Mustererkennung*, Hamburg, 2.-4. Okt. 1989, *Informatik-Fachberichte* 219, H. Burkhardt, K.H. Höhne, and B. Neumann (Hrsg.), Springer-Verlag Berlin Heidelberg 1989, 457-464

76. Shiohara, M., Gotoh, T., Nakagawa, Y.M. and Yoshida, Surface Correspondence Based on Three-dimensional Structure Inference In Animation Images, *10th Intern. Conf. on Pattern Recognition*, 16-21 June 1990, Atlantic City, New Jersey, USA, 194-197

77. Tost, D. and Pueyo, X. Human body animation: a survey, *The Visual Computer* 3 (1988) 254-264

78. Tsuji, S., Osada, M. and Yachida, M. Tracking and Segmentation of Moving Objects

in Dynamic Line Images, *IEEE Trans. on Pattern Anal. and Machine Intell.* 2:6 (1980) 516-522

79. Tsukiyama, V.T. and Shirai, Y. Detection of the movements of persons from a sparse sequence of TV images, *Pattern Recognition* 18 (1985) 207-213

80. Voss, K. *Theoretische Grundlagen der digitalen Bildverarbeitung,* Akademie Verlag Berlin 1988

81. Webb, J.A. and Aggarwal, J.K. Structure from Motion of Rigid and Jointed Objects, *Artificial Intelligence* 19 (1982) 107-130

82. Weber, W. and Weber, E. *Mechanik der menschlichen Gehwerkzeuge,* Dietrichsche Buchhandlung, Göttingen 1836

83. Weil, J. The Synthesis of Cloth Objects, *Computer Graphics* 20:4 (1986) 49-54

84. Wilhelms, J. Using Dynamic Analysis for Realistic Animation of Articulated Bodies, *IEEE Computer Graphics & Appl.* 7:6 (June 1987) 12-27

85. Yamamoto, M. and Koshikawa, K. Human Motion Analysis Based on a Robot Arm Model, *Proc. Computer Vision and Pattern Recogn.,* Lahaina, Maui, Hawaii, June 3-6, 1991, 664-665

86. Zeltzer, D. Motor Control Techniques for Figure Animation, *IEEE Computer Graphics & Appl.* 2:9 (Nov. 1982) 53-59

Part II

Gesture Recognition and Facial Expression Recognition

STATE-BASED RECOGNITION OF GESTURE

AARON F. BOBICK AND ANDREW D. WILSON

MIT Media Laboratory
20 Ames St., Cambridge, MA 02139
E-mail: bobick,drew@media.mit.edu

1. Introduction

A gesture is a motion that has special status in a domain or context. Recent interest in gesture recognition has been spurred by its broad range of applicability in more natural user interface designs. However, the recognition of gestures, especially natural gestures, is difficult because gestures exhibit human variability. We present a technique for quantifying this variability for the purposes of representing and recognizing gesture.

We make the assumption that the useful constraints of the domain or context of a gesture recognition task are captured implicitly by a number of examples of each gesture. Specifically, we require that by observing an adequate set of examples one can (1) determine the important aspects of the gesture by noting components of the motion that are reliably repeated; and (2) learn which aspects are loosely constrained by measuring high variability. Therefore, training consists of summarizing a set of motion trajectories that are smooth in time by representing the variance of the motion at local regions in the space of measurements. These local variances can be translated into a natural symbolic description of the movement which represent gesture as a sequence of measurement or *configuration* states. Recognition is then performed by determining whether a new trajectory is consistent with the required sequence of states.

In this chapter we apply the configuration state representation to a range of gesture-related sensory data: the two-dimensional movements of a mouse input device, the movement of the hand measured by a magnetic spatial position and orientation sensor, and the changing eigenvector

An initial version of this chapter appeared as a chapter in the International Conference on Computer Vision, Cambridge, 1995.

M. Shah and R. Jain (eds.), Motion-Based Recognition, 201–226.
© 1997 *Kluwer Academic Publishers.*

projection coefficients computed from an image sequence. The successful application of the technique to all these domains demonstrates the general utility of the approach.

We begin by describing related work on gesture recognition. We then motivate our particular choice of state-based representation and present a technique for computing it from generic sensor data. This computation requires the development of a novel technique for collapsing an ensemble of time-varying data while preserving the qualitative, topological structure of the trajectories. We develop and test methods for using the state-based representation to concurrently segment and recognize a stream of gesture data. Finally we consider the relationship between the configuration states proposed here and Hidden Markov Models (HMMs).

2. Related Work

Early experiments by Johansson [11] with Moving Light Display (MLD) sequences suggest that body movements, and therefore perhaps gestures as well, may be recognized by motion information alone. Work with very low resolution American Sign Language images by Sperling et al. [23] further supports the notion that in some domains a full geometric reconstruction of the moving object is unnecessary for recognition. For example, Polana and Nelson [17] use low level features of motion to recognize periodic motions such as walking.

A number of researchers have developed novel trajectory representations that are useful for gesture recognition. These works are relevant to ours in that the representation, like ours, is not grounded in a particular type of sensor. Gould and Shah [9] show the analysis of motion trajectories to identify event boundaries. These are recorded in their trajectory primal sketch to be used for motion recognition. Rangarajan et al. [19] demonstrate two-dimensional motion trajectory matching through scale-space. Davis and Shah [7] recognize simple hand gestures by matching image-plane trajectories made by markers on the fingertips. Rohr [20] smoothes a number of example joint angle trajectories to build a representation of one walking cycle parameterized by a pose variable.

Simpler trajectory representations have been used in the real time recognition of gestures made with a mouse input device. Tew and Gray [25] use dynamic programming to match mouse trajectories to prototype trajectories. Lipscomb [13] concentrates on filtering mouse movement data to obtain robust recognition of similarly filtered models. Mardia et al. [14] computes many features of each trajectory and use a learned decision tree for each gesture to best utilizes the features for recognition.

The work presented in this chapter has been applied to the problem of recognizing gestures from image sequences. Other works that have addressed this problem include Darrell and Pentland [6], which uses dynamic time warping to match changing normalized image correlation template scores to learned models. The correlation templates are distributed over duration of a model gesture such that at least one template has a high match score. Cui et al. [5] concatenates a series of static images spanning a sequence and do a Fisher-discriminant reduction of dimensionality to distinguish between gestures.

In work relevant to image-based movement recognition, Murase and Nayar [15] match changing eigenvector projection coefficients taken from the image data to determine the orientation and illumination angle of a previously seen object. Bregler and Omohundro [3] learn a surface representing constraints on the image sequence for the task of nonlinear image interpolation.

The popularity of Hidden Markov Models (HMMs) has lead to its use in the movement and gesture recognition. Part of the appeal of the HMM approach is that recognition and segmentation in time occur simultaneously. Yamato et al. [29], in recognizing tennis strokes, compute a simple region-based statistic from each frame of an image sequence, which is then vector-quantized. Sequences of the discrete symbols are then identified by a trained HMM. HMM use in various kinds of gesture has been explored in [24, 27, 21, 22].

3. Motivation for a Representation of Gesture

If all the constraints on the motion that make up a gesture were known exactly, recognition would simply be a matter of determining if a given movement met a set of known constraints. However, especially in the case of natural gesture, the exact movement seen is almost certainly governed by processes inaccessible to the observer. For example, the motion the gesturer is planning to execute after a gesture will influence the end of the current gesture; this effect is similar to co-articulation in speech. The incomplete knowledge of the constraints manifests itself as variance in the measurements of the movement. A representation for gesture must quantify this variance and how it changes over the course of the gesture.

Secondly, we desire a representation that is invariant to nonuniform changes in the speed of the gesture to be recognized. A global shift of the gesture in time caused by a slight pause early in the gesture should not affect the recognition of most of the gesture. The space of measurements that define each point of an example gesture is its *configuration space*. The goal of time invariance is motivated by the informal observation that the important quality in a gesture is how it traverses configuration space and not exactly when it reaches a certain point in the space. In particular, we would like the representation to be time invariant but order-preserving: e.g. first the hand goes up, then it goes down. Finally, strong time invariance is probably not required: in natural gestures, many kinds of time shifts are simply not physically plausible given the dynamics and kinematics of the human body.

Our basic approach to quantifying the variances in configuration space and simultaneously achieving sufficient temporal invariance is to represents a gesture as a sequence of states in configuration space. Each *configuration state* is intended to capture the degree of variability of the motion when traversing a region of configuration space. Since gestures are smooth movements through configuration space and not a set of naturally defined discrete states, the configuration states $S = \{s_i, 1 \leq i \leq M\}$ should be thought of as being "fuzzy", with fuzziness defined by the variance of the points that fall near it. A point moving smoothly through configuration space will move smoothly among the fuzzy states defined in the space.

Formally, we define a *gesture* as an ordered sequence of fuzzy states $s_i \in S$ in configuration space. This definition contrasts with a *trajectory* which is simply a path through configuration space representing a movement. A d-dimensional point x in configuration space has a membership to state s_i described by the fuzzy membership function $\mu_{s_i}(x) \in [0, 1]$. The states along the gesture should be defined so that all examples of the gesture follow the same sequence of states. We represent a gesture as a sequence of n states, $G_\alpha = \langle \alpha_1 \alpha_2 .. \alpha_n \rangle$.

We can now consider the state membership function of an entire trajectory. Let $T_i(t)$ be the ith trajectory. We need to choose a *combination rule* that defines the state membership of a point x in configuration space with respect to a group of states. For convenience let us choose **max**, which assigns the combined membership of x:

$$M_S(x) = \max_i (\mu_{s_i}(x)).$$

The combined membership value of a trajectory is a function of time while the assigned state of the trajectory at each time instant is the state whose membership is greatest. Thus, a set of configuration states translates a trajectory into a symbolic description, a sequence of states.

Defining gestures in this manner allows an intuitive definition of a *prototype gesture*: the motion trajectory that gives the highest combined membership to the sequence of states that define the gesture. In a training situation in which we have only examples of a gesture, we can first compute a prototype trajectory, and then define states that lie along that curve that capture the relevant variances. In the next section we will develop such

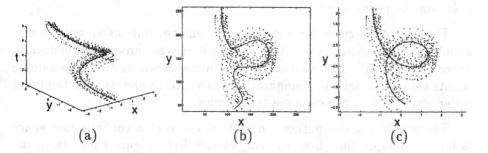

Figure 1. (a) Example trajectories as a function of time. (b) Projection of trajectory points into configuration space. Normal principal curve routines would lose the intersection. (c) Prototype curve recovered using the time-collapsing technique (see Appendix to this chapter).

a method. Afterwards we define a recognition procedure which evaluates whether a test input trajectory successfully passes through the states in the prescribed order.

4. Computing the Representation

The technique presented to compute the configuration state representation proceeds in three steps. First, an ensemble of training gestures is reduced to a prototype curve through configuration space; time is no longer a dimension and the prototype is parameterized only according to arc length. Second, the prototype curve is partitioned in regions according to spatial extent and variance of direction. Third, each region of the prototype, in conjunction with the training examples, is used to define a fuzzy state representing traversal through that segment of the gesture. In this section we detail each of the above procedures.

4.1. COMPUTING THE PROTOTYPE

Each example of a gesture is a trajectory in configuration space defined by a set of discrete samples evenly spaced in time. At first, it is convenient to parameterize the ith trajectory by the time of each sample: $T_i(t) \in \Re^d$, the d-dimensional feature space.

Our definition of a prototype curve of an ensemble of training trajectories $T_i(t)$ is a continuous one-dimensional curve that best fits the sample points in configuration space according to a least squares criterion. For ensembles of space curves in metric spaces there are several well known techniques that compute a "principal curve" [10] that attempts to minimize distance between each point of each of the trajectories and the nearest point on the principal curve.

The prototype curve for a gesture, P, removes time as an axis, and is simply parameterized by arc length λ as it moves through configuration space, $P(\lambda) \in \Re^d$. The goal of the parameterization is to group sample points that are nearby in configuration space and to preserve the temporal order along each of the example trajectories.

The problem of computing a prototype curve P in configuration space is how to collapse time from the trajectories $T_i(t)$. Figure 1 illustrates the difficulty. If the points that make up the trajectories (a) are simply projected into configuration space by removing time (b), there is no clear way to generate a connected curve that preserves the temporal coherence of the path through configuration space. Likewise, if each of the trajectories is projected into configuration space small variations in temporal alignment make it impossible to group corresponding sections of the trajectories without the consideration of time.

The details of our method are presented in an appendix of this chapter but we give the intuition here. Our approach is to begin with the trajectories in a time-augmented configuration space. Since a trajectory is a function of time, we construct a corresponding curve in a space consisting of the dimensions of configuration space plus a time axis. After computing the principal curve in this space, the trajectories and the recovered principal curve are slightly compressed in the time direction. The new principal curve is computed using the previous solution as an initial condition for an iterative technique. One requirement of our prototype construction method is that the training examples be well behaved in time, requiring only linear warping to be placed in temporal alignment; our recognition procedure, however, is insensitive to any order-preserving temporal variation.

By placing constraints on how dramatically the principal curve can change at each time step, the system converges gracefully to a prototype curve in configuration space that minimizes distance between the example trajectories and the prototype, while preserving temporal ordering. Figure 1(c) shows the results of the algorithm. The resulting prototype curve captures the path through configuration space while maintaining temporal ordering.

An important by-product of calculating the prototype is a map of each sample point x_i of a trajectory to an arc length along the prototype curve $\lambda_i = \lambda(x_i)$. This mapping is employed twice: first to cluster the input points with respect to sections of the prototype curve, and then again as the index with which to consider points along an new input gesture. Testing for a proper state traversal proceeds with respect to this index.

4.2. CLUSTERING THE SAMPLE POINTS

To define the fuzzy states s_i, the sample points of the trajectories must be partitioned into coherent groups. Instead of clustering the sample points directly, we cluster unit-length vectors tangent to the prototype curve $P(\lambda)$. These vectors are chosen by evenly re-sampling the curve $P(\lambda)$ and computing the tangent by finite-difference approximation at each of the samples. By clustering tangent vectors along $P(\lambda)$ instead of the original sample points, every point x_i that projects to a certain arc length $\lambda(x_i)$ along the prototype will belong to exactly one cluster. One desirable consequence of this is that the clusters will be ordered one after the other along the prototype. This ordered sequence of states is recorded as $G_\alpha = \langle \alpha_1 \alpha_2 .. \alpha_n \rangle$.

The prototype tangent vectors are clustered by a k-means algorithm [8], in which the distance between two vectors is a weighted sum of the Euclidean distance between the bases of the vectors and a measure of the difference in (unsigned) direction[1] of the vectors. This difference in direction is defined to be at a minimum when two vectors are parallel and at a maximum when perpendicular.

Clustering with this metric groups vectors that are oriented similarly, regardless of the temporal ordering associated with the prototype. If the prototype visits a part of configuration space and then later revisits the same part while moving in nearly the same (unsigned) direction, both sets of vectors from each of the visits will be clustered together. The sample points associated with both sets will then belong to the single state which appears multiply in the sequence G_α. In this way, the clustering leads to a parsimonious allocation of states, and is useful in detecting periodicity in the gesture.

Each cluster found by k-means algorithm corresponds to a fuzzy state. In the results presented in this chapter, the number k is set manually so that there are sufficiently many states to describe the movement. An alternative to this subjective measure is to use cross-validation techniques [2] to select the number of states.

4.3. DETERMINING STATE SHAPES

The center of each of the clusters found by the k-means algorithm is the average location c and average orientation $\vec{v}$ (a unit vector on a hemisphere if using unsigned direction) of the prototype curve tangent vectors belonging to the cluster. The membership function for the state is computed from

---

[1]The choice of whether to use signed or unsigned direction is dependent upon whether it is desirable for states to require motion in a particular direction. For the work discussed here we use unsigned values, implying that a gesture performed backwards can be represented by the same states as the forward gesture, except that the sequence is reversed.

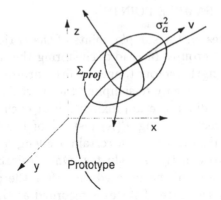

Figure 2. Decoupling the variance in a state into the axial component σ_a^2 in the direction of $\vec{v}$ and the covariance matrix Σ_{proj} in the hyper-plane perpendicular to $\vec{v}$.

these center vectors and the sample points that map to the prototype curve vectors in each cluster.

For a given state s_i, the membership function $\mu_{s_i}(x)$ should be defined so that membership is highest along the prototype curve; this direction is approximated by $\vec{v}$. Membership should also decrease at the boundaries of the cluster to smoothly blend into the membership of neighboring fuzzy states. Call this membership the "axial" membership. The membership in directions perpendicular to the curve determines the degree to which the state generalizes membership to points on perhaps significantly different trajectories. Call this membership the "cross sectional" membership.

Since we define a state as a fuzzy blob with decoupled variances, a single oriented Gaussian is well suited to model the local, smooth membership function. Orienting the Gaussian so that one axis of the Gaussian coincides with the orientation $\vec{v}$ of the center of the state, the axial membership is computed simply as the variance of the sample points in the axial direction. The cross-sectional membership is computed as the variance of the points projected on a hyper-plane normal to the axis. Figure 2 illustrates a Gaussian density where σ_a^2 is variance along the axis in the direction of $\vec{v}$ and Σ_{proj} is meant to represent the remaining variance in the hyper-plane perpendicular to the axis direction.

We now construct a new state covariance matrix Σ_S of the oriented Gaussian which decouples the along-axis variance from the perpendicular variation. Let Σ be the covariance matrix of the state points with the center location subtracted. First, a rotation matrix R is constructed, whose first column is $\vec{v}$, the axial direction, and whose remaining columns are generated by a Gram-Schmidt orthogonalization. Next, R is applied to the covariance matrix Σ:

$$\Sigma_R = R^T \Sigma R = \begin{bmatrix} \sigma_a^2 & \cdots \\ \vdots & [\Sigma_{proj}] \end{bmatrix}$$

where σ_a^2 is the variance of the points along $\vec{v}$, and Σ_{proj} is the covariance of the points projected onto the hyper-plane normal to $\vec{v}$. We can scale each of these variances to adjust the cross sectional and axial variances independently by the scalars k_c and k_a, respectively. The scalar k_c can be chosen to reflect only the training data ($k_c = 1.0$) or to control the amount of generalization independently from the magnitude with which neighboring states blend into one another. In the results reported here, k_a is automatically adjusted so that all training examples are accepted by the parsing algorithm.

Setting the first row and column to zero except for the variance in direction $\vec{v}$:

$$\Sigma_{R'} = \begin{bmatrix} k_a \sigma_a^2 & \cdots 0 \cdots \\ \vdots & \\ 0 & k_c [\Sigma_{proj}] \\ \vdots & \end{bmatrix}$$

decouples the axis direction from the remaining dimensions. This gives $\Sigma_S = R \Sigma_{R'} R^T$, and a new inverse covariance matrix: $\Sigma_S^{-1} = (R \Sigma_{R'} R^T)^{-1}$.

A state s_i is then defined by c, $\vec{v}$, and Σ_S^{-1}, with membership $\mu_{s_i}(x) = e^{-(x-c)\Sigma_S^{-1}(x-c)^T}$. The memberships of a number of states can be combined to find the membership $\mu_{s_i,s_j,..}(x)$ to a group of states $\{s_i, s_j, ..\}$. As mentioned, one combination rule simply returns the maximum membership of the individual state memberships:

$$\mu_{s_i,s_j,..}(x) = \max_{s \in \{s_i, s_j, ..\}} \mu_s(x)$$

The result is a function of position in configuration space whose value is approximately equal to the variance-normalized distance to the prototype trajectory.

5. Recognition

The online recognition of motion trajectories consists of parsing sequences of sample points as they are taken from the movement where the start-time of a gesture is not known. Given a set of trajectory sample points $x_1, x_2, ... x_N$ taken during the previous N time steps, we wish to find a gesture G and a start time t_s, $t_1 \leq t_s \leq t_N$ such that $x_s..x_N$ has an average combined membership above some threshold, and adequately passes through the states required by G.

Given the sequence of sample points at $t_1..t_N$, we can compute t_s and the average combined membership for a gesture $G_\alpha = \langle \alpha_1 \alpha_2 .. \alpha_n \rangle$ by a dynamic programming algorithm. The dynamic programming formulation used is a simplified version of a more general algorithm to compute the minimum cost path between two nodes in a graph [1].

For the dynamic programming solution, each possible state at time t is a node in a graph. The cost of a path between two nodes or states is the sum of the cost assigned to each transition between adjacent nodes in the graph. To detect the performance of a gesture, we enforce a forward progress through the states of the gesture and prefer states with high membership. The cost of a transition between a state α_i at time t and a state α_j at time $t + 1$ is

$$c_t(\alpha_i, \alpha_j) = \begin{cases} \infty & \text{for } j < i \\ 1 - \mu_{\alpha_j}(T(t)) & \text{otherwise} \end{cases}$$

The dynamic programming algorithm uses a partial sum variable, $C_{t_i,t_j}(\alpha_i, \alpha_j)$ to recursively compute a minimal solution. $C_{t_i,t_j}(\alpha_i, \alpha_j)$ is defined to be the minimal cost of a path between state α_i at a time t_i and α_j at a time t_j:

$$C_{t_i,t_j}(\alpha_k, \alpha_m) = \min_{\alpha_l \in G_\alpha} \left\{ c_{t_i}(\alpha_k, \alpha_l) + C_{t_{i+1},t_j}(\alpha_l, \alpha_m) \right\}$$

with the recursion condition

$$C_{t_i,t_i}(\alpha_k, \alpha_m) = 0$$

The total cost associated with explaining all samples by the gesture G_α is then $C_{t_1,t_N}(\alpha_1, \alpha_n)$.

The start of the gesture is not likely to fall exactly at time t_1, but at some later time t_s. Given t_s, we can compute the average combined membership of the match from the total cost to give an overall match score for a match starting at time t_s:

$$\bar{\mu}_{G_\alpha} = 1 - \frac{C_{t_s,t_1}(\alpha_1, \alpha_n)}{(t_N - t_s)}$$

Thus, using $\bar{\mu}_{G_\alpha}$ we can search backward in time looking for t_s. Darrell and Pentland [6] use the same technique for the on-line parsing of gestures.

To be classified as a gesture G_α, the trajectory must have a high match score and pass through all the states in G_α as well. For the latter we can compute the minimum of the maximum membership observed in each state α_i in G_α. This quantity indicates the *completeness* of the trajectory with respect to the model G_α. If the quantity is less than a certain threshold, the match is rejected. The start of a matching gesture is then

$$t_s = \underset{t}{\arg\min}\ C_{t,t_N}(\alpha_1, \alpha_n),\ completeness > threshold$$

Causal segmentation of a stream of samples is performed using the dynamic programming algorithm at successive time steps. At a time step t, the highest match score and the match start time t_s is computed for all samples from t_0 to t. If the match score is greater than a threshold η, and the gesture is judged complete, then all points up to time t are explained by a gesture model and so are removed from the stream of points, giving a new value $t_0 = t$. Otherwise, the points remain in the stream possibly to be explained at a later time step. This is repeated for all time steps t successively.

In general, the complexity of a dynamic programming problem is $O(nk)$ where n is the number of time steps in the path and k is the number of possible states at each step. Simply trying each possible t_s would yield a computation of order $O(n^2k)$. However, the recursive nature of the algorithm allows us to use the computed paths back in time to some $t = t_0$ to generate the solution back one more time step to $t_0 - 1$. Therefore the complexity is still $O(nk)$.

6. Experiments

The configuration-state representation has been tested using motion trajectory measurements taken from three devices that are useful in gesture recognition tasks: a mouse input device, a magnetic spatial position and orientation sensor, and a video camera. In each case we segment by hand a number of training examples from a stream of smooth motion trajectories collected from the device. We demonstrate how the representation characterizes the training examples, recognizes new examples of the same gesture, and is useful in a segmentation and tracking task.

In each case, the measurements collected were uniformly scaled to fall within the interval from -1 to 1. With this scaling, the same initial time scaling s of 4.0 was found to work well in computing the prototype trajectory (see Appendix). The prototype curve was computed in turn for $s = \{4, 3, 2, 1, 0\}$ in each case. Because of the dramatic difference in the input signals considered it was necessary in some cases to change the value of the smoothing kernel parameter h and the values of σ_c and σ_a. Otherwise no adjustment of parameters was made between the experiments.

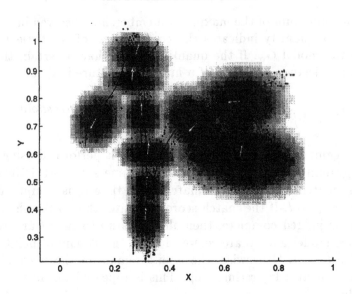

Figure 3. The prototype curves (black) for G_α and G_β are combined to find a set of states to describe both gestures. The clustering of the prototype curve vectors for G_α and G_β gives the clustering of all sample points. Each of the ten clusters depicted here is used to define a state centered about each of the white vectors. The combined membership of the 10 states in G_α and G_β is depicted under the curves; darker regions indicate high membership.

6.1. MOUSE GESTURES

In the first experiment, we demonstrate how the representation permits the combination of multiple gestures, and generalizes to unseen examples for recognition. Furthermore, by considering two dimensional data, the cluster points and the resulting states are easily visualized.

The (x, y) position of a mouse input device was sampled at a rate of 20Hz for two different gestures G_α and G_β. Ten different examples of each gesture were collected; each consisted of about one hundred sample points. Half of the examples were used to compute the prototype curve for each gesture. The smoothing kernel parameter h had a value of 0.2. The vectors along both of the curves were then clustered to find a single set of ten states useful in tracking either gesture. Because the gestures overlap, six of the states are shared by the two gestures. The prototype curves and the assignments of the sample points to the clusters are shown in Figure 3.

The shapes of the states computed from the clustering is shown in Figure 3. The generalization parameter σ_c had a value 3.0. The state sequences G_α and G_β were computed by looking at the sequence of state assignments of the vectors along the prototype. The sequences G_α and G_β reflect the six shared states: $G_\alpha = \langle s_1 s_2 s_3 s_2 s_4 s_5 s_6 s_7 \rangle$, $G_\beta = \langle s_3 s_2 s_4 s_5 s_6 s_7 s_6 s_5 s_8 s_9 s_{10} s_8 s_9 \rangle$

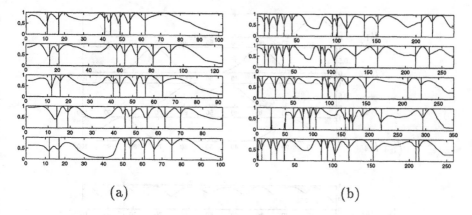

(a)   (b)

Figure 4. The state transition and membership plots for the testing examples for G_α (a) and G_β (b). The state transitions are marked by vertical bars. The transitions are calculated to maximize the average membership while still giving an interpretation that is complete with respect to G_α and G_β. No complete and consistent interpretation could be found for the fourth example of G_β.

The match scores of the hand-segmented[2] test gestures were then computed using the dynamic programming algorithm outlined in Section 5. In computing the maximum match score, the algorithm assigns each point to a state consistent with the sequence G_α. As described, a match is made only if it is considered complete.

The state transitions and the membership values computed for each sample are shown for the new examples of G_α and G_β in Figure 4. The plots show the value of $\max \mu_{s_i}$ — the maximum membership — as a function of time. The state transitions are marked as vertical bars; thus all sample points between two adjacent vertical bars are attributed to the same state. As described in Section 5, a transition to a state is only allowed if it is consistent with G_α. A match was considered complete if the minimum of the maximum memberships between each transition was greater than the empirically established threshold of 0.4. The state transition plots for this experiment and the others graphically depict the time-invariant but order-preserving nature of the representation.

The representation provides a convenient way to specify a gesture as an ordered sequence of states while permitting the combination of states shared by multiple gestures. By using the sequence of states in the recognition

---

[2]In this experiment the test gestures were hand segmented in time to focus only on the state-membership aspect of the system; the next two experiments use the backward searching algorithm described.

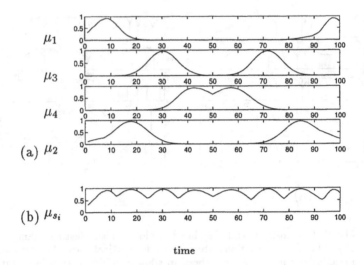

Figure 5.   (a) The membership plot for the prototype curves for G_{wave} show for each of the configuration states how the states lie along the prototype. (b) Combined membership (maximum) of all states at each point along the prototype. The prototype for G_{point} is similar.

task, nearby and possibly overlapping states belonging to another gesture or another part of the same do not confuse the matching process.

6.2. SPATIAL POSITION AND ORIENTATION SENSOR GESTURES

For the second experiment, we compute the representation with somewhat sparse, higher dimensional data. We show its use in the automatic, causal segmentation of two different gestures as if the samples were collected in a real time recognition application.

An Ascension Technology Flock of Birds magnetic position and orientation sensor was worn on the back of the hand and polled at 20Hz. For each sample, the position of the hand and the normal vector out of the palm of the hand was recorded (six dimensions). Ten large wave gestures (about 40 samples each) and ten pointing gestures (about 70 samples each) were collected. To insure that there were enough points available to compute the prototype curve, each example was up-sampled using Catmull-Rom [4] splines so that each wave gesture is about 40 samples and each point gesture about 70 samples.

The prototype curves for each gesture were computed separately. The membership plots for the prototype wave is shown for each state in Figure 5. The gesture sequence G_{wave} is found by analyzing the combined membership function (Figure 5b): $G_{wave} = \langle s_1 s_2 s_3 s_4 s_3 s_2 s_1 \rangle$. Similarly, G_{point} (not shown) is defined by $\langle s_5 s_6 s_7 s_8 s_7 s_6 s_5 \rangle$. Note how both the sequences

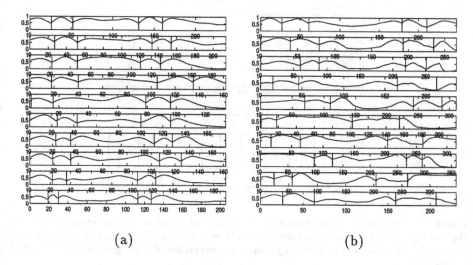

(a)   (b)

Figure 6. The state transition and membership plots for the testing examples for G_{wave} (a) and G_{point} (b). The state transitions are marked by vertical bars.

and the plots capture the similarity between the initiation and retraction phases of the gesture.

The state transition and membership plots as calculated by the dynamic programming algorithm are shown for all the example wave gestures and point gestures in Figure 6. Because the example gestures started at slightly different spatial positions, it was necessary to ignore the first and last states in the calculation to obtain good matches. This situation can also occur, for example, due to gesture co-articulation effects that were not observed during training. The lesson we learn here is that training in isolation will result in an underestimation of the variance of the start and end states caused by previous and following action, respectively.

A stream of samples consisting of an alternating sequence of all the example wave and point gestures was causally segmented to find all wave and point gestures. For a matching threshold of $\eta = 0.5$, all the examples were correctly segmented (Figure 7).

Even with sparse and high dimensional data, the representation is capable of determining segmentation. Additionally, the representation provides a useful way to visualize the tracking process in a high dimensional configuration space.

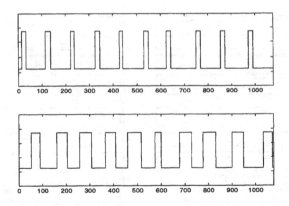

Figure 7.   The causal segmentation of the stream of position and orientation data, for the wave gesture (top) and point gesture (bottom). A high value indicates the detection of a single complete and consistent gesture. All examples of both gesture were detected in this test.

6.3. IMAGE-SPACE GESTURES

As a final experiment, we consider a digitized image sequence of a waving hand (Figure 8). Given a set of smoothly varying measurements taken from the images and a few hand-segmented example waves, the goal is to automatically segment the sequence to recover the remaining waves. This example shows how the measurements do not require a direct physical or geometric interpretation, but should vary smoothly in a meaningful and regular way.

Each frame of the sequence is point in a 4800-dimensional space of pixel values, or image-space. If the motion is smooth and the images are smooth, then the sequence will trace a smooth path in image-space [16]. Rather than approximate trajectories in the 4800-dimensional space, we instead approximate the trajectories of the coefficients of projection onto the first few eigenvectors computed from a part of the sequence.

The first five example waves were used in training. The three eigenvectors with the largest eigenvalues were computed by the Karhunen-Loeve Transform (as in [26, 15]) of the training frames, treating each frame as a column of pixel values. The first three eigenvectors accounted for 71% of the variance of the pixel intensity values of the training frames; the coefficients of those eigenvectors are the features used as input for the experiment.

The training frames were then projected onto the eigenvectors to give the smooth trajectories shown in Figure 9. The membership along the prototype curve ($h = 0.4$) computed from the projection trajectories is shown in Figure 10. The recovered state sequence $G_{wave} = \langle s_1 s_4 s_3 s_2 s_3 s_4 \rangle$ again shows the periodicity of the motion.

Figure 8.   Each row of images depicts a complete wave sequence taken from the larger image sequence of 830 frames, 30 fps, 60 by 80 pixels each. Only 5 frames of each sequence are presented. The variation in appearance between each example (along columns) is typical of the entire sequence.

Figure 11 shows the state transitions and membership values for ten other examples projected onto the same eigenvectors. Again, the first and last states were ignored in the matching process due to variations in the exact beginning and ending of the motion. Included in these examples is the only instance not recognized correctly. Human observers agreed that that particular example looked the quite distinct from the rest.

Lastly, Figure 12 shows the automatic causal segmentation of the entire image sequence. The top trace is the manual annotation of the boundaries of the wave patterns; the bottom trace indicates time points considered to be contained in a complete and consistent gesture. Of the 32 waves in the sequence, all but one (the same incomplete example above) are correctly segmented.

This example demonstrates how the representation may be used in a broad range of tracking and segmentation tasks; namely, those in which

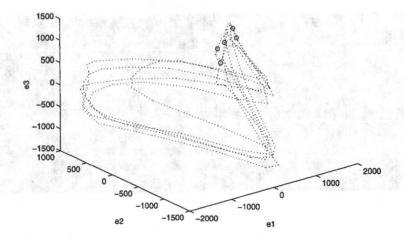

Figure 9. Each of the axes e_1, e_2, and e_3 represents the projection onto each of the three eigenvectors. The image sequences for the five training examples project to the smooth trajectories shown here. A circle represents the end of one example and the start of the next. The trajectory plots show the coherence of the examples as a group, as well as the periodicity of the movement.

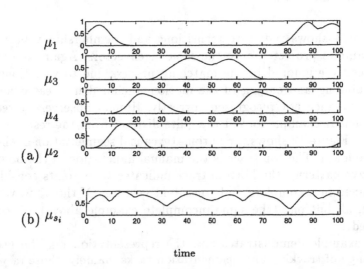

Figure 10. The membership along the prototype curve for each state is shown in the top four plots (a). (b) shows the combined membership of all states along the prototype.

smoothly varying and meaningful measurements can be collected from the example gestures.

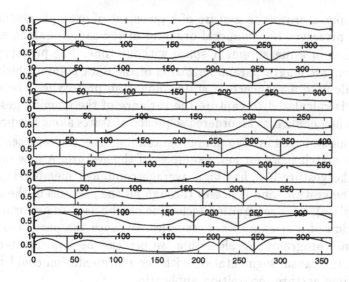

Figure 11. The state transition and membership plots for testing examples 16 through 25, taken from the image sequence of 32 examples. The state transitions are marked by vertical bars. The fifth example was deemed incomplete. On most of the successfully tracked examples, the dynamic programming algorithm chooses to make an immediate transition to the second state; this is caused by an overly high degree of overlap between states, controlled by σ_a.

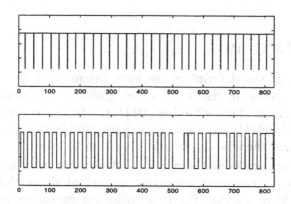

Figure 12. The stream recognition process results in a segmentation of the entire video sequence. A high value indicates a complete and consistent gesture. All but one of the 32 examples of the wave gesture are correctly identified.

7. Conclusion

7.1. SUMMARY

A novel technique for computing a representation for gesture that is time-invariant but order-preserving has been presented. The technique proceeds by computing a prototype gesture of a given set of example gestures. The prototype preserves the temporal ordering of the samples along each gesture, but lies in a measurement space without time. The prototype offers a convenient arc length parameterization of the data points, which is then used to calculate a sequence of states along the prototype. The shape of the states is calculated to capture the variance of the training gestures. A gesture is then defined as an ordered sequence of states along the prototype.

A technique based on dynamic programming uses the representation to compute a match score for new examples of the gesture. A new movement matches the gesture if it has high average combined membership for the states in the gesture, and it passes through all the gestures in the sequence (it is complete). This recognition technique can be used to perform a causal segmentation of a stream of new samples. Because the particular form of the dynamic programming algorithm we use can be implemented to run efficiently, the causal segmentation with the representation could be useful in a real time gesture recognition application.

Three experiments were conducted, each taking data from input devices that are typical of gesture recognition applications. The variety of inputs addressed demonstrates the general utility of the technique. Our intuition is that there are only a few requirements on the measurements for the technique to be useful; we would like to make these requirements more explicit.

7.2. RELATION TO HMMS

At the heart of the description we have proposed is the idea of *states*. Recently, researchers in visual gesture recognition have proposed the incorporation of Hidden Markov Models (HMMs) into the recognition procedure [21, 22, 27, 24]. HMMs have been quite successful in the speech recognition domain [18] and there is the natural extension to vision.

While we acknowledge the relation between our work here and HMMs, we would like to comment upon one important distinction, namely the existence of the prototype. The time collapsing procedure discussed in the appendix yields a reasonable prototypical model even if only one training example is present. The statistical nature of an HMM precludes a rapid training phase: many free parameters need to be computed.

However, we believe that the HMM framework can be constrained to obtain a representation approximately similar to the configuration-state representation. Starting with standard HMM training formulation [18], (1) the Markov state transition matrix must be constrained to be causal, (2) the feature space must be augmented with velocity or orientation (as in a *phase space*), and lastly, (3) a "prototype" could be generated after training by performing a nonlinear interpolation over the states (as in [3] for example). To match the parsing algorithm presented here, the usual (non-hierarchical) HMM parsing algorithm must be augmented to allow state sequence specifications that are not captured by a first-order Markov model (for example, our configuration-state parsing algorithm could easily be extended to accept $G = \langle s_1 s_2 s_3 \rangle$ OR $\langle s_4 s_2 s_5 \rangle$ as a gesture).

Both the method described in this chapter and the constrained HMM formulation outlined above would generate our representation of gesture as a sequence of fuzzy states.

7.3. NATURAL GESTURE

The work presented falls under the traditional paradigm for hand gesture recognition involving the construction of a model for each gesture to be recognized. We believe that this approach is well suited to the recognition of stylized or literal gesture, such as the gestures made by a user navigating aeronautical data in a virtual reality system by contorting their hands. These actions are less gestures than particular literal movements. Others examples are the emblematic gestures substituting for simple linguistic constructs: the ubiquitous OK sign, or the gestures used to signal crane operators.

However, as an approach to *natural gesture* understanding, this methodology seems inappropriate. By "natural gesture" we mean the types of gestures spontaneously generated by a person telling a story. The reasons for this skepticism are clear. First, the particular configurations and motions observed in natural gesture are inherently speaker *dependent*, influenced by cultural, educational, and situational factors [12]. An approach employing fixed, physical descriptions of gesture might find no cross-speaker invariances.

Second, and more important, is that the literal representation of the gesture assumes that the spatial configuration is the most significant aspect of the signal to be extracted. Given that we are observing a sequence, it is plausible that more abstract *temporal* properties are the important elements of a gesture.

Recently [28] we have extended the work presented here to consider the temporal structure — the *gestural phases* — in natural gesture. The basic idea there is that gestural states may not be defined primarily by their spatial configuration, but by their temporal properties. The remainder of the technology described here — variance-based state membership and backward-looking dynamic programming — is still directly applicable.

There seems to be little consensus in the literature on a useful definition of "gesture". Part of the problem in arriving at a concise notion of gesture is the broad applicability of gesture recognition, and the difficulty in reasoning about gesture without respect to a particular domain (e.g., hand gestures). The development of the configuration state technique presented is an attempt to formalize the notion of gesture without limiting its applicability to a particular domain. That is, we wish to find what distinguishes gesture from the larger background of all motion, and incorporate that knowledge into a representation.

References

1. A. V. Aho, J. E. Hopcroft, and J. D. Ullman. *Data structures and algorithms.* Addison-Wesley, Reading, 1983.
2. C. M. Bishop. *Neural networks for pattern recognition.* Clarendon Press, Oxford, 1995.
3. C. Bregler and S. M. Omohundro. Nonlinear image interpolation using surface learning. In G. Tesauro J. D. Cowan and J. Alspector, editors, *Advances in neural information processing systems 6*, pages 43–50, San Fransisco, CA, 1994. Morgan Kaufmann Publishers.
4. E. Catmull and R. Rom. A class of local interpolating splines. In R. Barnhill and R. Riesenfeld, editors, *Computer Aided Geometric Design*, pages 317–326, San Francisco, 1974. Academic Press.
5. Y. Cui and J. Weng. Learning-based hand sign recognition. In *Proc. of the Intl. Workshop on Automatic Face- and Gesture-Recognition*, Zurich, 1995.
6. T.J. Darrell and A.P. Pentland. Space-time gestures. *Proc. Comp. Vis. and Pattern Rec.*, pages 335–340, 1993.
7. J. W. Davis and M. Shah. Gesture recognition. *Proc. European Conf. Comp. Vis.*, pages 331–340, 1994.
8. R. Duda and P. Hart. *Pattern classification and scene analysis.* John Wiley, New York, 1973.
9. K. Gould and M. Shah. The trajectory primal sketch: a multi-scale scheme for representing motion characteristics. *Proc. Comp. Vis. and Pattern Rec.*, pages 79–85, 1989.
10. T. Hastie and W. Stuetzle. Principal curves. *Journal of the American Statistical Association*, 84(406):502–516, 1989.
11. G. Johansson. Visual perception of biological motion and a model for its analysis. *Perception and Psychophysics*, 14(2):201–211, 1973.
12. A. Kendon. How gestures can become like words. In F. Poyatos, editor, *Cross-cultural perspectives in nonverbal communication*, New York, 1988. C.J. Hogrefe.
13. J.S. Lipscomb. A trainable gesture recognizer. *Pattern Recognition*, 24(9):895–907, 1991.
14. K. V. Mardia, N. M. Ghali, M. Howes T. J. Hainsworth, and N. Sheehy. Techniques for online gesture recognition on workstations. *Image and Vision Computing*, 11(5):283–294, 1993.
15. H. Murase and S. Nayar. Learning and recognition of 3d objects from appearance. In *IEEE 2nd Qualitative Vision Workshop*, New York, June 1993.
16. H. Murase and S. Nayar. Visual learning and recognition of 3-D objects from appearance. *Int. J. of Comp. Vis.*, 14:5–24, 1995.
17. R. Polana and R. Nelson. Low level recognition of human motion. In *Proc. of the Workshop on Motion of Non-Rigid and Articulated Objects*, pages 77–82, Austin, Texas, Nov. 1994.
18. L. R. Rabiner and B. H. Juang. *Fundamentals of speech recognition.* Prentice Hall, Englewood Cliffs, 1993.

19. K. Rangarajan, W. Allen, and M. Shah. Matching motion trajectories using scale-space. *Pattern Recognition*, 26(4):595–610, 1993.
20. K. Rohr. Towards model-based recognition of human movements in image sequences. *Comp. Vis., Graph., and Img. Proc.*, 59(1):94–115, 1994.
21. J. Schlenzig, E. Hunter, and R. Jain. Recursive identification of gesture inputs using hidden markov models. *Proc. Second Annual Conference on Applications of Computer Vision*, pages 187–194, December 1994.
22. J. Schlenzig, E. Hunter, and R. Jain. Vision based hand gesture interpretation using recursive estimation. In *Proc. of the Twenty-Eighth Asilomar Conf. on Signals, Systems and Comp.*, October 1994.
23. G. Sperling, M. Landy, Y. Cohen, and M. Pavel. Intelligible encoding of ASL image sequences at extremely low information rates. *Comp. Vis., Graph., and Img. Proc.*, 31:335–391, 1985.
24. T. E. Starner and A. Pentland. Visual recognition of American Sign Language using hidden markov models. In *Proc. of the Intl. Workshop on Automatic Face- and Gesture-Recognition*, Zurich, 1995.
25. A.I. Tew and C.J. Gray. A real-time gesture recognizer based on dynamic programming. *Journal of Biomedical Eng.*, 15:181–187, May 1993.
26. M. Turk and A. Pentland. Eigenfaces for recognition. *Journal of Cognitive Neuroscience*, 3(1):71–86, 1991.
27. A. D. Wilson and A. F. Bobick. Learning visual behavior for gesture analysis. In *Proc. IEEE Int'l. Symp. on Comp. Vis.*, Coral Gables, Florida, November 1995.
28. A. D. Wilson, A. F. Bobick, and J. Cassell. Recovering the temporal structure of natural gesture. In *Proc. of the Intl. Workshop on Automatic Face- and Gesture-Recognition*, Killington, 1996.
29. J. Yamato, J. Ohya, and K. Ishii. Recognizing human action in time-sequential images using hidden markov model. *Proc. Comp. Vis. and Pattern Rec.*, pages 379–385, 1992.

A. Computation of time collapsed prototype curve

For each trajectory $T_i(t)$, we have a $\hat{T}_i(t) = T_i(\frac{t}{s})$, where s is a scalar that maps the time parameterization of $T_i(t)$ to $\hat{T}_i(t)$. The time course of all example trajectories are first normalized to the same time interval $[0, s]$. The smooth approximation of the time-normalized sample points gives a rough starting point in determining which of the sample points correspond to a point on the prototype. These correspondences can be refined by iteratively recomputing the approximation while successively reducing the time scale s. If the prototype curve is not allowed to change drastically from one iteration to the next, a temporally coherent prototype curve in the original configuration space will result.

To compute the prototype curve $P(\lambda)$, we use Hastie and Stuetzle's "principal curves" [10]. Their technique results in a smooth curve which minimizes the sum of perpendicular distances of each sample to the nearest point on the curve. The arc length along the prototype of the nearest point is a useful way to parameterize the samples independently of the time of each sample. That is, for each sample x_i there is a lambda which minimizes

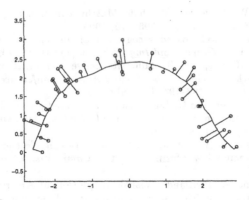

Figure 13.   A principal curve calculated from a number of points randomly scattered about an arc. Only a fraction of the data points are shown; the distances of these points to the closest points on the curve are indicated.

the distance to $P(\lambda)$: $\lambda(x_i) =_{\underset{\lambda}{\arg\min}} \|P(\lambda) - x_i\|$. An example is shown in Figure 13.

The algorithm for finding principal curves is iterative and begins by computing the line along the first principal component of the samples. Each data point is then projected to its nearest point on the curve and the arc length of each projected point is saved. All the points that project to the same arc length along the curve are then averaged in space. These average points define the new curve. This projection and averaging iteration proceeds until the change in approximation error is small.

In practice, only one sample point will project to a particular arc length along the curve. Therefore, a number of points that project to approximately equal arc lengths are averaged. The approach suggested by Hastie and Stuetzle and used here is to compute a weighted least squares line fit of the nearby points, where the weights are derived from a smooth, symmetric and decreasing kernel centered about the target arc length. The weight w for a sample x_i and curve point $p = P(\lambda)$ is given by

$$w = \left(1.0 - (\frac{|\lambda(p) - \lambda(x_i)|}{h})^3 \right)^3$$

where h controls the width of the kernel.

The new location of the curve point is then the point on the fitted line that has the same arc length. For efficiency, if the least squares solution involves many points, a fixed number of the points may be selected randomly to obtain a reliable fit.

By starting with a time scaling s which renders the trajectories slowly varying in the configuration space parameters as a function of arc length, the principal curve algorithm computes a curve which is consistent with the temporal order of the trajectory samples. Then the time scale s can be reduced somewhat and the algorithm run again, starting with the previous curve. In the style of a continuation method, this process of computing the curve and rescaling time repeats until the time scale is zero, and the curve is in the original configuration space. To ensure that points along the prototype do not coincide nor spread too far from one another as the curve assumes its final shape, the principal curve is re-sampled between time scaling iterations so that the distance between adjacent points is constant.

The inductive assumption in the continuation method is that the curve found in the previous iteration is consistent with the temporal order of the trajectory samples. This assumption is maintained in the current iteration by a modification of the local averaging procedure in the principal curves algorithm. When the arc length of each point projected on the curve is computed, its value is checked against the point's arc length computed in the previous iteration. If the new arc length is drastically different from the previously computed arc length ($|\lambda_t(x_i) - \lambda_{t-1}(x_i)| > threshold$), it must be the case that by reducing the time scale some other part of the curve is now closer to the sample point. This sample point to prototype arc length correspondence is temporally inconsistent with the previous iteration, and should be rejected. The next closest point on the curve $P(\lambda)$ is found and checked. This process repeats until a temporally consistent projection of the data point is found.

By repeatedly applying the principal curve algorithm and collapsing time, a temporally consistent prototype $P(\lambda)$ is found in configuration space. Additionally, the arc length associated with each projected point, $\lambda(x_i)$, is a useful time-invariant but order-preserving parameterization of the samples. An 2-dimensional example (such as mouse data) of this time-collapsing process is shown in Figure 14.

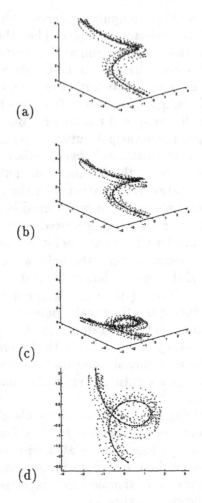

Figure 14. The principal curve is tracked while time is slowly collapsed in this series: (a) $s = 8$, (b) $s = 5$, (c) $s = 0$. In each of these graphs, the vertical axis is time. (d) shows the final, temporally consistent curve.

REAL-TIME AMERICAN SIGN LANGUAGE RECOGNITION FROM VIDEO USING HIDDEN MARKOV MODELS

THAD STARNER AND ALEX PENTLAND
Room E15-383, The Media Laboratory
Massachusetts Institute of Technology
20 Ames Street, Cambridge MA 02139
thad,sandy@media.mit.edu

1. Introduction

While there are many different types of gestures, the most structured sets belong to the sign languages. In sign language, each gesture already has assigned meaning, and strong rules of context and grammar may be applied to make recognition tractable.

To date, most work on sign language recognition has employed expensive wired "datagloves" which the user must wear [19]. In addition, these systems have mostly concentrated on finger signing, in which the user spells each word with finger signs corresponding to the letters of the alphabet [5]. However, most signing does not involve finger spelling but instead, gestures which represent whole words, allowing signed conversations to proceed at about the pace of spoken conversation.

In this chapter, we describe an extensible system which uses one color camera to track hands in real time and interprets American Sign Language (ASL) using Hidden Markov Models (HMM's). The hand tracking stage of the system does not attempt a fine description of hand shape; studies of human sign readers have shown that such detailed information is not necessary for humans to interpret sign language [12, 16]. Instead, the tracking process produces only a coarse description of hand shape, orientation, and trajectory. The hands are tracked by their color: in the first experiment via solidly colored gloves and in the second, via their natural skin tone. In both cases the resultant shape, orientation, and trajectory information is input to a HMM for recognition of the signed words.

Hidden Markov models have intrinsic properties which make them very attractive for sign language recognition. Explicit segmentation on the word

M. Shah and R. Jain (eds.), Motion-Based Recognition, 227–243.
© *1997 Kluwer Academic Publishers.*

level is not necessary for either training or recognition [18]. Language and context models can be applied on several different levels, and much related development of this technology has already been done by the speech recognition community [8]. Consequently, sign language recognition seems an ideal machine vision application of HMM technology, offering the benefits of problem scalability, well defined meanings, a pre-determined language model, a large base of users, and immediate applications for a recognizer.

American Sign Language (ASL) is the language of choice for most deaf people in the United States. ASL's grammar allows more flexibility in word order than English and sometimes uses redundancy for emphasis. Another variant, English Sign Language, has more in common with spoken English but is not in widespread use in America. ASL uses approximately 6000 gestures for common words and finger spelling for communication of obscure words or proper nouns.

Conversants in ASL may describe a person, place, or thing and then point to a place in space to store that object temporarily for later reference [16]. For the purposes of this experiment, this aspect of ASL will be ignored. Furthermore, in ASL the eyebrows are raised for a question, relaxed for a statement, and furrowed for a directive. While we have also built systems that track facial features [6], this source of information will not be used to aid recognition in the task addressed here.

While the scope of this work is not to create a user independent, full lexicon system for recognizing ASL, the system should be extensible toward this goal. Another goal is real-time recognition which allows easier experimentation, demonstrates the possibility of a commercial product in the future, and simplifies archiving of test data. "Continuous" sign language recognition of full sentences is necessary to demonstrate the feasibility of recognizing complicated series of gestures. Of course, a low error rate is also a high priority.

TABLE 1. ASL Test Lexicon

Part of Speech	Vocabulary
Pronoun	I, You, He, We, You(pl), They
Verb	Want, Like, Lose, Dontwant, Dontlike, Love, Pack, Hit, Loan
Noun	Box, Car, Book, Table, Paper, Pants, Bicycle, Bottle, Can, Wristwatch, Umbrella, Coat, Pencil, Shoes, Food, Magazine, Fish, Mouse, Pill, Bowl
Adjective	Red, Brown, Black, Gray, Yellow

For this recognition system, sentences of the form "personal pronoun, verb, noun, adjective, (the same) personal pronoun" are to be recognized. This sentence structure emphasizes the need for a distinct grammar for ASL recognition and allows a large variety of meaningful sentences to be generated randomly using words from each class. Table 1 shows the words chosen for each class. Six personal pronouns, nine verbs, twenty nouns, and five adjectives are included making a total lexicon of forty words. The words were chosen by paging through Humphries *et al.* [9] and selecting those which would generate coherent sentences when chosen randomly for each part of speech.

2. Machine Sign Language Recognition

Attempts at machine sign language recognition have begun to appear in the literature over the past five years. However, these systems have generally concentrated on isolated signs and small training and test sets. Tamura and Kawasaki demonstrate an early image processing system which recognizes 20 Japanese signs based on matching cheremes [20]. Charayaphan and Marble [2] demonstrate a feature set that distinguishes between the 31 isolated ASL signs in their training set (which also acts as the test set). More recently, Cui and Weng [3] have shown an image-based system with 96% accuracy on 28 isolated gestures.

Takahashi and Kishino in [19] discuss a user dependent Dataglove-based system that recognizes 34 of the 46 Japanese kana alphabet gestures using a joint angle and hand orientation coding technique. The test user makes each of the 46 gestures 10 times to provide data for principle component and cluster analysis. A separate test set is created from five iterations of the alphabet by the user, with each gesture well separated in time. Murakami and Taguchi [11] describe a similar Dataglove system using recurrent neural networks. However, in this experiment a 42 static-pose finger alphabet is used, and the system achieves up to 98% recognition for trainers of the system and 77% for users not in the training set. This study also demonstrates a separate 10 word gesture lexicon with user dependent accuracies up to 96% in constrained situations.

3. Use of Hidden Markov Models in Gesture Recognition

While the continuous speech recognition community adopted HMM's many years ago, these techniques are just now accepted by the vision community. An early effort by Yamato *et al.* [22] uses discrete HMM's to recognize image sequences of six different tennis strokes among three subjects. This experiment is significant because it uses a 25x25 pixel quantized subsampled camera image as a feature vector. Even with such low-level information, the model can learn the set of motions and recognize them with respectable accuracy. Darrell and Pentland [4] use dynamic time warping, a technique similar to HMM's, to match the interpolated responses of several learned image templates. Schlenzig *et al.* [15] use hidden Markov models to recognize "hello," "good-bye," and "rotate." While Baum-Welch re-estimation was not implemented, this study shows the continuous gesture recognition capabilities of HMM's by recognizing gesture sequences. Recently, Wilson and Bobick [21] explore incorporating multiple representations in HMM frameworks.

4. Hidden Markov Modeling

While a substantial body of literature exists on HMM technology [1, 8, 13, 23], this section briefly outlines a traditional discussion of the algorithms. After outlining the fundamental theory in training and testing a discrete HMM, this result is then generalized to the continuous density case used in the experiments. For broader discussion of the topic, Huang *et al.* [8] and Starner [17] are recommended.

A time domain process demonstrates a Markov property if the conditional probability density of the current event, given all present and past events, depends only on the jth most recent events. If the current event depends solely on the most recent past event, then the process is a first order Markov process. While the order of words in American Sign Language is not truly a first order Markov process, it is a useful assumption when considering the positions and orientations of the hands of the signer through time.

The initial topology for an HMM can be determined by estimating how many different states are involved in specifying a sign. Fine tuning this topology can be performed empirically. While different topologies can be specified for each sign, a four state HMM with one skip transition was determined to be sufficient for this task (Figure 1).

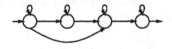

Figure 1. The four state HMM used for recognition

There are three key problems in HMM use: evaluation, estimation, and decoding. The evaluation problem is, given an observation sequence and a model, what is the probability that the observed sequence was generated by the model ($Pr(\mathbf{O}|\lambda)$) (notational style from [8])? If this can be evaluated for all competing models for an observation sequence, then the model with the highest probability can be chosen for recognition.

$Pr(\mathbf{O}|\lambda)$ can be calculated several ways. The naive way is to sum the probability over all the possible state sequences in a model for the observation sequence:

$$Pr(\mathbf{O}|\lambda) = \sum_{allS} \prod_{t=1}^{T} a_{s_{t-1}s_t} b_{s_t}(O_t)$$

However, this method is exponential in time, so the more efficient forward-backward algorithm is used in practice. The following algorithm defines the forward variable c and uses it to generate $Pr(\mathbf{O}|\lambda)$ (π are the initial state probabilities, a are the state transition probabilities, and b are the output probabilities).

- $\alpha_1(i) = \pi_i b_i(O_1)$, for all states i (if $i \epsilon S_I, \pi_i = \frac{1}{n_I}$; otherwise $\pi_i = 0$)
- Calculating $\alpha()$ along the time axis, for $t = 2, ..., T$, and all states j, compute

$$\alpha_t(j) = [\sum_i \alpha_{t-1}(i)a_{ij}]b_j(O_t)$$

- Final probability is given by

$$Pr(\mathbf{O}|\lambda) = \sum_{i \epsilon S_F} \alpha_T(i)$$

The first step initializes the forward variable with the initial probability for all states, while the second step inductively steps the forward variable through time. The final step gives the desired result $Pr(\mathbf{O}|\lambda)$, and it can be shown by constructing a lattice of states and transitions through time that the computation is only order $O(N^2 T)$. The backward algorithm, using a process similar to the above, can also be used to compute $Pr(\mathbf{O}|\lambda)$ and defines the convenience variable β.

The estimation problem concerns how to adjust λ to maximize $Pr(\mathbf{O}|\lambda)$ given an observation sequence $\mathbf{O}$. Given an initial model, which can have flat probabilities, the forward-backward algorithm allows us to evaluate this probability. All that remains is to find a method to improve the initial model. Unfortunately, an analytical solution is not known, but an iterative technique can be employed.

Using the actual evidence from the training data, a new estimate for the respective output probability can be assigned:

$$\bar{b}_j(k) = \frac{\sum_{t \epsilon O_t = v_k} \gamma_t(j)}{\sum_{t=1}^{T} \gamma_t(j)}$$

where $\gamma_t(i)$ is defined as the posterior probability of being in state i at time t given the observation sequence and the model. Similarly, the evidence can be used to develop a new estimate of the probability of a state transition ($\bar{a}_{ij}$) and initial state probabilities ($\bar{\pi}_i$).

Thus all the components of model (λ) can be re-estimated. Since either the forward or backward algorithm can be used to evaluate $Pr(\mathbf{O}|\bar{\lambda})$ versus the previous estimation, the above technique can be used iteratively to converge the model to some limit. While the technique described only handles a single observation sequence, it is easy to extend to a set of observation sequences. A more formal discussion can be found in [1, 8, 23].

While the estimation and evaluation processes described above are sufficient for the development of an HMM system, the Viterbi algorithm provides a quick means of evaluating a set of HMM's in practice as well as providing a solution for the decoding problem. In decoding, the goal is to recover the state sequence given an observation sequence. The Viterbi algorithm can be viewed as a special form of the forward-backward algorithm where only the maximum path at each time step is taken instead of all paths. This optimization reduces computational load and allows the recovery of the most likely state sequence. The steps to the Viterbi are

- Initialization. For all states i,
  $$\delta_1(i) = \pi_i b_i(O_1);$$
  $$\psi_i(i) = 0$$
- Recursion. From $t = 2$ to T and for all states j,
  $$\delta_t(j) = Max_i[\delta_{t-1}(i)a_{ij}]b_j(O_t);$$
  $$\psi_t(j) = argmax_i[\delta_{t-1}(i)a_{ij}]$$
- Termination.
  $$P = Max_{s \in S_F}[\delta_T(s)];$$
  $$s_T = argmax_{s \in S_F}[\delta_T(s)]$$
- Recovering the state sequence. From $t = T - 1$ to 1,
  $$s_t = \psi_{t+1}(s_{t+1})$$

In many HMM system implementations, the Viterbi algorithm is used for evaluation at recognition time. Note that since Viterbi only guarantees the *maximum* of $Pr(\mathbf{O}, S|\lambda)$ over all state sequences S (as a result of the first order Markov assumption) instead of the *sum* over all possible state sequences, the resultant scores are only an approximation. However, Rabiner and Juang [13] show that this is often sufficient.

So far the discussion has assumed some method of quantization of feature vectors into classes. However, instead of using vector quantization, the actual probability densities for the features may be used. Baum-Welch, Viterbi, and the forward-backward algorithms can be modified to handle a variety of characteristic densities [10]. In this context, however, the densities will be assumed to be Gaussian. Specifically,

$$b_j(O_t) = \frac{1}{\sqrt{(2\pi)^n |\sigma_j|}} e^{\frac{1}{2}(O_t - \mu_j)' \sigma_j^{-1}(O_t - \mu_j)}$$

Initial estimations of μ and σ may be calculated by dividing the evidence evenly among the states of the model and calculating the mean and variance in the normal way. Whereas flat densities were used for the initialization step before, the evidence is used here. Now all that is needed is a way to provide new estimates for the output probability. We wish to weight the influence of a particular observation for each state based on the likelihood of that observation occurring in that state. Adapting the solution from the discrete case yields

$$\bar{\mu}_j = \frac{\sum_{t=1}^{T} \gamma_t(j) O_t}{\sum_{t=1}^{T} \gamma_t(j)}$$

and

$$\bar{\sigma}_j = \frac{\sum_{t=1}^{T} \gamma_t(j)(O_t - \bar{\mu}_j)(O_t - \bar{\mu}_j)^t}{\sum_{t=1}^{T} \gamma_t(j)}$$

For convenience, μ_j is used to calculate $\bar{\sigma}_j$ instead of the re-estimated $\bar{\mu}_j$. While this is not strictly proper, the values are approximately equal in contiguous iterations [8] and seem not to make an empirical difference [23]. Since only one stream of data is being used and only one mixture (Gaussian density) is being assumed, the algorithms above can proceed normally, incorporating these changes for the continuous density case.

5. Tracking Hands in Video

Previous systems have shown that, given some constraints, relatively detailed models of the hands can be recovered from video images [5, 14]. However, many of these constraints conflict with recognizing ASL in a natural context, either by requiring simple, unchanging backgrounds (unlike clothing); not allowing occlusion; requiring carefully labeled gloves; or being difficult to run in real time.

In this project we have tried two methods of hand tracking: one, using solidly-colored cloth gloves (thus simplifying the segmentation problem), and two, tracking the hands directly without aid of gloves or markings. Figure 2 shows the view from the camera's perspective in the no-gloves case. In both cases color NTSC composite video is captured and analyzed at 320 by 243 pixel resolution. On a Silicon Graphics Indigo 2 with Galileo video board we can achieve a constant 5 frames per second. Using a Silicon Graphics 200MHz Indy workstation we were able to track the hands at 10 frames per second.

In the first method, the subject wears distinctly colored cloth gloves on each hand (a yellow glove for the right hand and an orange glove for the left) and sits in a chair facing the camera. To find each hand initially, the algorithm scans the image until it finds a pixel of the appropriate color. Given this pixel as a seed, the region is grown by checking the eight nearest neighbors for the appropriate color. Each pixel checked is considered part of the hand. This, in effect, performs a simple morphological dilation upon the resultant image that helps to prevent edge and lighting aberrations. The centroid is calculated as a by-product of the growing step and is stored as the seed for the next frame. Given the resultant bitmap and centroid, second moment analysis is performed as described in the following section.

Figure 2. View from the tracking camera

In the second method, the the hands were tracked based on skin tone. We have found that all human hands have approximately the same hue and saturation, and vary primarily in their brightness. Using this information we can build an *a priori* model of skin color and use this model to track the hands much as was done in the gloved case. Since the hands have the same skin tone, "left" and "right" are simply assigned to whichever hand is leftmost and rightmost. Processing proceeds normally except for simple rules to handle hand and face ambiguity described in the next section.

6. Feature Extraction and Hand Ambiguity

Psychophysical studies of human sign readers have shown that detailed information about hand shape is not necessary for humans to interpret sign language [12, 16]. Consequently, we began by considering only very simple hand shape features, and evolved a more complete feature set as testing progressed.

Since finger spelling is not allowed and there are few ambiguities in the test vocabulary based on individual finger motion, a relatively coarse tracking system may be used. Based on previous work, it was assumed that a system could be designed to separate the hands from the rest of the scene. Traditional vision algorithms could then be applied to the binarized result. Aside from the position of the hands, some concept of the shape of the hand and the angle of the hand relative to horizontal seemed necessary. Thus, an eight element feature vector consisting of each hand's x and y position, angle of axis of least inertia, and eccentricity of bounding ellipse was chosen. The eccentricity of the bounding ellipse was found by determining the ratio of the square roots of the eigenvalues that correspond to the matrix

$$\begin{pmatrix} a & b/2 \\ b/2 & c \end{pmatrix}$$

where a, b, and c are defined as

$$a = \int\int_{I'} (x')^2 dx' dy'$$

$$b = \int\int_{I'} x'y' dx' dy'$$

$$c = \int\int_{I'} (y')^2 dx' dy'$$

(x' and y' are the x and y coordinates normalized to the centroid).

The axis of least inertia is then determined by the major axis of the bounding ellipse, which corresponds to the primary eigenvector of the matrix [7]. Note that this leaves a 180 degree ambiguity in the angle of the ellipses. To address this problem, the angles were only allowed to range from -90 to +90 degrees.

When tracking skin tones, the above analysis helps to model situations of hand ambiguity implicitly. When a hand occludes either the other hand or the face, color tracking alone can not resolve the ambiguity. Since the face remains in the same area of the frame, its position can be determined and discounted. However, the hands move rapidly and occlude each other often. When occlusion occurs, the hands appear to the above system as a single blob of larger than normal mass with significantly different moments than either of the two hands in the previous frame. In this implementation, each of the two hands is assigned the moment and position information of the single blob whenever occlusion occurs. While not as informative as tracking each hand separately, this method still retains a surprising amount of discriminating information. The occlusion event is implicitly modeled, and the combined position and moment information are retained. This method, combined with the time context provided by hidden Markov models, is sufficient to distinguish between many different signs where hand occlusion occurs.

7. Training an HMM network

When using HMM's to recognize strings of data such as continuous speech, cursive handwriting, or ASL sentences, several methods can be used to bring context to bear in training and recognition. A simple context modeling method is embedded training. While initial training of the models might rely on manual segmentation or, in this case, evenly dividing the evidence among the models, embedded training trains the models *in situ* and allows model boundaries to shift through a probabilistic entry into the initial states of each model [23].

Generally, a sign can be affected by both the sign in front of it and the sign behind it. For phonemes in speech, this is called "co-articulation." While this can confuse systems trying to recognize isolated signs, the context information can be used to aid recognition. For example, if two signs are often seen together, recognizing the two signs as one group may be beneficial.

A final use of context is on the word or phrase level. Statistical grammars relating the probability of the co-occurrence of two or more words can be used to weight the recognition process. Grammars that associate two words are called bigrams, whereas grammars that associate three words are called trigrams. Rule-based grammars can also be used to aid recognition.

8. Experimentation

Since we could not exactly recreate the signing conditions between the first and second experiments, direct comparison of the gloved and no-glove experiments is impossible. However, a sense of the increase in error due to removal of the gloves can be obtained since the same vocabulary and sentences were used in both experiments.

8.1. EXPERIMENT 1: GLOVED-HAND TRACKING

The glove-based handtracking system described earlier worked well. Occasionally tracking would be lost (generating error values of 0) due to lighting effects, but recovery was fast enough (within a frame) that this was not a problem. A 5 frame/sec rate was maintained within a tolerance of a few milliseconds. However, frames were deleted where tracking of one or both hands was lost. Thus, a constant data rate was not guaranteed. This hand tracking process produced an eight-element feature vector (each hand's x and y position, angle of axis of least inertia, and eccentricity of bounding ellipse) that was used for subsequent modeling and recognition.

Of the 500 sentences collected, six were eliminated due to subject error or outlier signs. In general, each sign was 1 to 3 seconds long. No intentional pauses were placed between signs within a sentence, but the sentences themselves were distinct.

TABLE 2. Word accuracy of glove-based system

Experiment	Training Set	Independent Test Set
Grammar	99.5% (99.5%)	99.2% (99.2%)
No Grammar	92.0% (97%)	91.3% (96%)
	(D=9, S=67,	(D=1, S=16,
	I=121, N=2470)	I=26, N=495)

Word accuracies are calculated as in the speech recognition literature by the formula $A = \frac{N-D-S-I}{N}$ where N is the total number of words, D is the number of deletions, S is the number of substitutions, and I is the number of insertions. Shown in parentheses is the percent correct, which reflects only substitution errors.

Initial estimates for the means and variances of the output probabilities were provided by iteratively using Viterbi alignment on the training data (after initially dividing the evidence equally among the words in the sentence) and then recomputing the means and variances by pooling the vectors in each segment. Entropic's Hidden Markov Model ToolKit (HTK) is used as a basis for this step and all other HMM modeling and training tasks. The results from the initial alignment program are fed into a Baum-Welch re-estimator, whose estimates are, in turn, refined in embedded training which ignores any initial segmentation. For recognition, HTK's Viterbi recognizer is used both with and without a strong grammar based on the known form of the sentences. Contexts are not used, since a similar effect could be achieved with the strong grammar given this data set. Recognition occurs five times faster than real time.

Word recognition accuracies are shown in Table 2. When testing on training, all 494 sentences were used to train and test the system. For the fair test, the sentences were randomly divided into a set of 395 training sentences and a set of 99 independent test sentences. The 99 test sentences were not used for any portion of the training. Given the strong grammar (pronoun, verb, noun, adjective, pronoun-giving a total of 5 words in that order), insertion and deletion errors are not possible. Thus, with this grammar, all errors are substitution errors (and accuracy is equivalent to percent correct). However, without the grammar, the recognizer can output any of the 40 vocabulary words, in any order, and any number of times. Thus, deletion (D), insertion (I), and substitution (S) errors are possible. Note that, since each type of error is counted against the accuracy rate, it is possible to get large negative accuracies (and corresponding error rates of over 100%). The absolute number of errors of each type are listed in Table 2. It was observed that most insertion errors corresponded to signs with repetitive motion.

8.2. ANALYSIS

The 0.8% error rate of the independent test set shows that the HMM topologies are sound and that the models generalize well. With such low error rates, little can be learned by analyzing the remaining errors.

However, the remaining 8.7% error rate (based on accuracy) of the "no grammar" experiment better indicates where problems may occur when extending the system. Without the grammar, signs with repetitive or long gestures were often inserted twice for each actual occurrence. In fact, insertions caused more errors than substitutions. Thus, the sign "shoes" might be recognized as "shoes shoes," which is a viable hypothesis without a language model. However, a practical solution to this problem is the use of context training and a statistical grammar instead of the rule-based grammar.

Using context modeling as described before may significantly improve recognition accuracy in a more general implementation as shown by the speech and handwriting recognition communities [18]. While a rule-based grammar explicitly constrains the word order, statistical context modeling would have a similar effect while generalizing to allow different sentence structures. In the speech community, such modeling occurs at the "triphone" level, where groups of three phonemes are recognized as one unit. The equivalent in ASL would be to recognize "trisines" (groups of three signs) corresponding to three words, or three letters in the case of finger spelling. In speech recognition, statistics are gathered on word co-occurrence to create "bigram" and "trigram" grammars which can be used to weight the likelihood of a word. In ASL, this might be applied on the phrase level. For example, the random sentence construction used in the experiments allowed "they like pill yellow they," which would probably not occur in natural, everyday conversation. As such, context modeling would tend to suppress this sentence in recognition, perhaps preferring "they like food yellow they," except when the evidence is particularly strong for the previous hypothesis.

Further examination of the errors made without the grammar shows the importance of finger position information. Signs like "pack," "car," and "gray" have very similar motions. In fact, the main difference between "pack" and "car" is that the fingers are pointed down for the former and clenched in the latter. Since this information is not available in the model, confusion occurs. While recovering specific finger positions is difficult with the current testing apparatus, simple palm orientation might be sufficient to resolve these ambiguities.

Since the raw screen coordinates of the hands were used, the system was trained to expect certain gestures in certain locations. When this varied due to subject seating position or arm placement, the system could become confused. A possible solution is to use position deltas in the feature vector, as was done in the second experiment.

8.3. EXPERIMENT 2: NATURAL SKIN TRACKING

The natural hand color tracking method maintained a 10 frame per second rate at 320x240 pixel resolution on a 200MHz SGI Indy. Higher resolution tracking suffered from video interlace effects. While tracking was somewhat noisier due to confusions with similarly-colored background elements, lighting seemed to play a lesser role due to the lower specularity of skin compared to the gloves. Since only one hand "blob" might be expected at a given time, no frames were rejected due to lack of tracking of a hand. Due to the subject's increased familiarity with ASL from the first experiment, the 500 sentences were obtained in a much shorter span of time and in fewer sessions.

During review of this data set and comparison with the earlier set of sentences, it was found that subject error and variability increased. In particular, there was increased variability in imaged hand size (due to changes in depth under perspective) and increased variability in body rotation relative to the camera. Ignoring these unintentional complications, 478 of the sentences were correctly signed; 384 were used for training, and 94 were reserved for testing.

In the first experiment an eight-element feature vector (each hand's x and y position, angle of axis of least inertia, and eccentricity of bounding ellipse) was found to be sufficient. However this feature vector does not include all the information derived from the hand tracking process. In particular, the hand area, the length of the major axis of the first eigenvector, and the change in x and y positions of the hand were not used. In this second experiment these features were added to help resolve the ambiguity when the hands cross. In addition, combinations of these feature elements were also explored to gain an understanding of the their information content. Training and recognition occurred as described previously. The word accuracy results are summarized in Table 3; the percentage of words correctly recognized is shown in parentheses next to the accuracy rates.

TABLE 3. Word accuracy of natural skin system

Experiment	Training Set	Independent Test Set
Original	87.9% (87.9%)	84.7% (84.7%)
+ Area	92.1% (92.1%)	89.2% (89.2%)
Δ + Area	89.6% (89.6%)	87.2% (87.2%)
Full	94.1% (94.1%)	91.9% (91.9%)
Full-No Grammar	81.0% (87%)	74.5% (83%)
	(D=31, S=287,	(D=3, S=76,
	I=137, N=2390)	I=41, N=470)

Word accuracies; percent correct in parentheses. The first test uses the original feature set from the first experiment. The second adds the area of the imaged hand. The change in position of the hands replaces the absolute position in the third test, and the final test uses the full set of features: x, y, Δx, Δy, angle, eccentricity, area, and length of the major eigenvector. All tests use the grammar except for the last result which shows "no grammar" for completeness.

8.4. ANALYSIS

A higher error rate was expected for the gloveless system, and indeed this was the case. The 8 element feature set (x, y, angle, and eccentricity for each hand) was not sufficient for the task due to loss of information when the hands crossed. However, the simple addition of the area of each hand improved the accuracy significantly.

The subject's variability in body rotation and position was known to be a problem with this data set. Thus, signs that are distinguished by the hands' positions in relation to the body were confused since only the absolute positions of the hands in screen coordinates were measured. To minimize this type of error, the absolute positions of the hands can be replaced by their relative motion between frames (Δx and Δy). While this replacement causes the error rate to increase slightly, it demonstrates the feasibility of allowing the subject to vary his location in the room while signing, removing another constraint from the system.

By combining the relative motion and absolute position information with the angle, eccentricity, area, and length of the major eigenvector, the highest fair test accuracy, 91.9%, was reached. Without the context information provided by the grammar, accuracy dropped considerably; reviewing the errors showed considerable insertions and substitutions at points where the hands crossed.

9. Discussion and Conclusion

We have shown an unencumbered, vision-based method of recognizing American Sign Language (ASL). Through use of hidden Markov models, low error rates were achieved on both the training set and an independent test set without invoking complex models of the hands.

With a larger training set and context modeling, lower error rates are expected and generalization to a freer, user independent ASL recognition system should be attainable. To progress toward this goal, the following improvements seem most important:

- Measure hand position relative to each respective shoulder or a fixed point on the body.
- Add finger and palm tracking information. This may be as simple as counting how many fingers are visible along the contour of the hand and whether the palm is facing up or down.
- Use a two camera vision system to help disambiguate the hands in 2D and/or track the hands in 3D.
- Collect appropriate domain or task-oriented data and perform context modeling both on the trisine level as well as the grammar/phrase level.
- Integrate explicit face tracking and facial gestures into the feature set.

These improvements do not address the user independence issue. Just as in speech, making a system which can understand different subjects with their own variations of the language involves collecting data from many subjects. Until such a system is tried, it is hard to estimate the number of subjects and the amount of data that would comprise a suitable training database. Independent recognition often places new requirements on the feature set as well. While the modifications mentioned above may be initially sufficient, the development process is highly empirical.

So far, finger spelling has been ignored. However, incorporating finger spelling into the recognition system is a very interesting problem. Of course, changing the feature vector to address finger information is vital to the problem, but adjusting the context modeling is also of importance. With finger spelling, a closer parallel can be made to speech recognition. Trisine context occurs at the sub-word level while grammar modeling occurs at the word level. However, this is at odds with context across word signs. Can trisine context be used across finger spelling and signing? Is it beneficial to switch to a separate mode for finger spelling recognition? Can natural language techniques be applied, and if so, can they also be used to address the spatial positioning issues in ASL? The answers to these questions may be key to creating an unconstrained sign language recognition system.

10. Acknowledgments

The authors would like to thank Tavenner Hall for her help editing and proofing this document.

References

1. Baum L. (1972) An Inequality and Associated Maximization Technique in Statistical Estimation of Probabilistic Functions of Markov Processes, *Inequalities*, Vol. 3, pp. 1–8.
2. Charayaphan C. and Marble A. (1992) Image Processing System for Interpreting Motion in American Sign Language, *Journal of Biomedical Engineering*, Vol. 14, pp. 419–425.
3. Cui Y. and Weng J. (1995) Learning-based Hand Sign Recognition, *Intl. Work. Auto. Face Gest. Recog. (IWAFGR) '95*, pp. 201–206.
4. Darrell T. and Pentland A. (1993) Space-time Gestures, *CVPR '93*, pp. 335–340.
5. Dorner B. (1993) Hand Shape Identification and Tracking for Sign Language Interpretation, *IJCAI '93 Workshop on Looking at People*.
6. Essa I., Darrell T., and Pentland A. (1994) Tracking Facial Motion, *IEEE Workshop on Nonrigid and Articulated Motion*.
7. Horn B. (1986) *Robot Vision*. MIT Press, NY.
8. Huang X., Ariki Y., and Jack M. (1990) *Hidden Markov Models for Speech Recognition*, Edinburgh University Press, Edinburgh.
9. Humphries T., Padden C., and O'Rourke T. (1980) *A Basic Course in American Sign Language*. T. J. Publishers, Inc., Silver Spring, MD.
10. Juang B. (1985) Maximum Likelihood Estimation for Mixture Multivariate Observations of Markov Chains, *AT&T Tech. J.*, Vol. 64, pp. 1235–1249.
11. Murakami K. and Taguchi H. (1991) Gesture Recognition Using Recurrent Neural Networks, *CHI '91*, pp. 237–241.
12. Poizner H., Bellugi U., and Lutes-Driscoll V. (1981) Perception of American Sign Language in Dynamic Point-light Displays, *J. Exp. Pyschol.: Human Perform.*, Vol. 7, pp. 430–440.
13. Rabiner L. and Juang B. (1986) An Introduction to Hidden Markov Models. *IEEE ASSP Magazine*, January issue, pp. 4–16.
14. Rehg J. and Kanade T. (1993) DigitEyes: Vision-Based Human Hand Tracking, *School of Computer Science Technical Report CMU-CS-93-220*, Carnegie Mellon Univ., Dec. 1993.
15. Schlenzig J., Hunter E., and Jain R. (1994) Recursive Identification of Gesture Inputers Using Hidden Markov Models, *Proc. Second Ann. Conf. on Appl. of Comp. Vision*, pp. 187–194.
16. Sperling G., Landy M., Cohen Y., and Pavel M. (1985) Intelligible Encoding of ASL Image Sequences at Extremely Low Information Rates, *Comp. Vision, Graphics, and Image Proc.*, Vol. 31, pp. 335–391.
17. Starner T. (1995) Visual Recognition of American Sign Language Using Hidden Markov Models, *Master's thesis, MIT Media Laboratory*.
18. Starner T., Makhoul J., Schwartz R., and Chou G. (1994) On-line Cursive Handwriting Recognition Using Speech Recognition Methods, *ICASSP*, Vol. 5, pp. 125–128.
19. Takahashi T. and Kishino F. (1991) Hand Gesture Coding Based on Experiments Using a Hand Gesture Interface Device, *SIGCHI Bul.*, Vol. 23, pp. 67–73.
20. Tamura S. and Kawasaki S. (1988) Recognition of Sign Language Motion Images, *Pattern Recognition*, Vol. 21, pp. 343–353.
21. Wilson A. and Bobick A. (1995) Learning Visual Behavior for Gesture Analysis, *Proc. IEEE Int'l. Symp. on Comp. Vis.*, pp. 229–234.
22. Yamato J., Ohya J., and Ishii K. (1992) Recognizing Human Action in Time-

sequential Images Using Hidden Markov Models, *ICCV*, pp. 379–385.
23. S. Young. (1993) *HTK: Hidden Markov Model Toolkit V1.5*, Cambridge Univ. Eng. Dept. Speech Group and Entropic Research Lab. Inc., Washington DC.

RECOGNIZING HUMAN MOTION USING
PARAMETERIZED MODELS OF OPTICAL FLOW

MICHAEL J. BLACK
Xerox Palo Alto Research Center, Palo Alto, CA 94304, USA

YASER YACOOB
Computer Vision Laboratory, Center for Automation Research
University of Maryland, College Park, MD 20742, USA

AND

SHANON X. JU
Department of Computer Science, University of Toronto
Toronto, Ontario M5S 1A4, Canada

1. Introduction

The tracking and recognition of human motion is a challenging problem with diverse applications in virtual reality, medicine, teleoperations, animation, and human-computer interaction to name a few. The study of human motion has a long history with the use of *images* for analyzing animate motion beginning with the improvements in photography and the development of motion-pictures in the late nineteenth century. Scientists and artists such as Marey [12] and Muybridge [26] were early explorers of human and animal motion in images and image sequences. Today, commercial motion-capture systems can be used to accurately record the 3D movements of an instrumented person, but the motion analysis and motion recognition of an arbitrary person in a video sequence remains an unsolved problem. In this chapter we describe the representation and recognition of human motion using parameterized models of optical flow. A person's limbs, face, and facial features are represented as patches whose motion

Portions of this paper are reprinted, with permission, from the Proceedings of the International Conference on Computer Vision, Boston, Mass., June 1995, pages 374–381; and the Proceedings of the International Conference on Automatic Face and Gesture Recognition, Killington, Vermont, Nov. 1996, pages 38–44; ©1995/1996 IEEE.

M. Shah and R. Jain (eds.), Motion-Based Recognition, 245–269.
© 1997 *Kluwer Academic Publishers.*

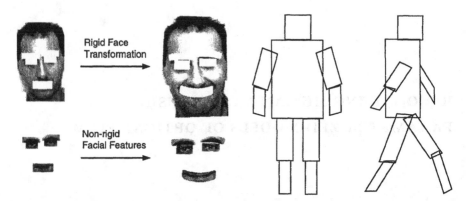

Figure 1. The cardboard person model. The limbs and face of a person are represented by planar patches. The motion of the face is further represented by the relative motions of the features.

in an image sequence can be modeled by low-order polynomials. A robust optical flow estimation technique is used to recover the motion of these patches and the recovered motion parameters provide a rich, yet concise, description of the human motion which can be used to recognize human activities, gestures, and facial expressions.

In our representation we approximate the non-rigid motion of a person using a set of parameterized models of optical flow. This *cardboard-person* model is illustrated in Figure 1. While parameterized flow models (for example affine flow) have been used for representing image motion in rigid scenes, Black and Yacoob [7] observed that simple parameterized models could well approximate more complex motions if localized in space and time. Moreover, they showed that the motion of one body region (for example the face region) could be used to stabilize that body part in a warped image sequence. This allowed the image motions of related body parts (the eyes, mouth, and eyebrows) to be estimated *relative* to the motion of the face. Isolating the motions of these features from the motion of the face is critical for recognizing facial expressions using motion.

These parameterized motion models can be extended to model the articulated motion of the human limbs [18]. Limb segments can be approximated by planes and the motion of these planes can be recovered using a simple eight-parameter optical flow model. Constraints can be added to the optical flow estimation problem to model the articulation of the limbs and the relative image motions of the limbs can be used for recognition.

An example of tracking and recognizing facial motion is provided in Figure 2. Regions corresponding to parts of the face are located in the first image of the sequence. Then between pairs of frames, the image motion within the regions is computed robustly using a parameterized optical flow

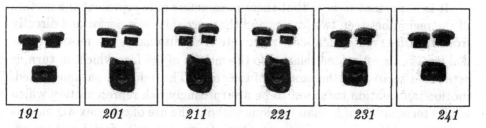

| 191 | 201 | 211 | 221 | 231 | 241 |

Figure 2. Surprise Experiment: facial expression tracking. Features every 10 frames.

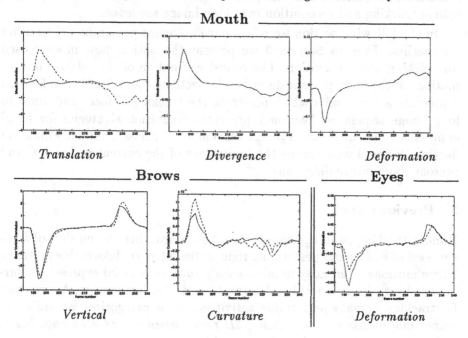

Mouth

Translation   *Divergence*   *Deformation*

Brows   ## Eyes

Vertical   *Curvature*   *Deformation*

Figure 3. Motion parameters. For the mouth translation, the solid line indicates horizontal motion while the dashed line indicates vertical motion. For the eye and brows, the solid and dashed lines indicate left and right respectively.

model. These models capture how the regions move and deform and the motion information is used to track the regions through the image sequence.

The motion of the regions between frames is described by a small set of parameters which can be used for recognition. Some of the information contained in these parameters is shown for the surprise expression in Figure 3. During the initiation of the expression the mouth translates down, diverges, and deforms significantly. Simultaneously, the brows and eyes move upwards, the brows arch, and the eyes deform as they widen. The ending phase in this example shows a more gradual reversal of these parameters returning the face to the resting position.

It is important to note that these parameters only represent the motion of the region between two frames and that recognition is performed directly from the observed optical flow parameters. Additionally, the motion of facial features is estimated relative to the motion of the face which, in turn, is estimated relative to the motion of the torso. This relative, parameterized, motion information turns out to be a surprisingly rich representation which allows recognition of human motions without the use of complex 3D models or the matching of image features. Our experiments with facial and articulated motions indicate that the parameterized motion models can provide robust tracking and recognition over long image sequences.

In the following section we relate our cardboard-person model to other approaches. Then in Section 3 we present the optical flow models used for tracking and recognition. The robust estimation of the relative images motions of the body parts is described in Sections 4 and 5 and then Section 6 presents a few results which illustrate the tracking of faces and limbs in long image sequences. Section 7 presents recognition strategies for facial expressions, gestures, and cyclical movements such as walking. Finally in Sections 8 and 9 we examine the limitations of the current techniques and present some future directions.

2. Previous work

Human tracking in image sequences involves tracking the motion of a diverse set of body parts performing rigid, articulated, or deformable motions. These motions often exist simultaneously such as in facial expressions during head rotations or clothing deformations during locomotion. Approaches for tracking humans performing activities can be categorized according to being: *motion-based* versus *static*, *3D model-based* versus *2D image-based*, and *region-based* versus *boundary-based*.

Motion-based approaches consider the tracking of body parts as involving the recovery of the motion parameters between consecutive images [2, 14, 21, 28, 38]. These motion parameters can either be recovered directly from the spatio-temporal derivatives of the image sequence or from a dense optical flow field. Static approaches view body part tracking as the localization of a body part in a single image or pose estimation in 3D space [15, 16, 31, 40].

3D model-based approaches employ significant structural information about the body parts to recover their 3D pose in space [14, 15, 21, 28, 30] while 2D image-based approaches focus on the intensity (or color) distribution in the images to track the body parts possibly through employing 2D models of body part shape or motion [3, 7, 24, 31].

Region tracking integrates information over areas of the image to track the body part motion [4, 7, 24, 27, 37] while boundary tracking concentrates on the discontinuities in the projected image of the body part in motion [3, 8, 15, 16, 19, 31, 36, 39].

The limitations of each of the above categories are well known, but these limitations are exacerbated in the context of human motions. 3D motion recovery algorithms require a priori structure information (at least a coarse model) in the case of articulated objects or point correspondences for rigid objects [2]. On the other hand, image motions are not easily related to the multiple activities of body parts and their projection on the image plane. Recovering 3D models of the human body or its parts is generally difficult to achieve in an unconstrained environment due to the significant variability of human appearances (clothing, make up, hair styles, etc.). Additionally, 2D image-based tracking is complicated by the articulation and deformation of body parts and the dependence on the observation point of the activity. Boundary tracking allows focusing on information-rich parts of the image, these boundaries can occasionally be ambiguous, small in number, dependent on imaging conditions and may not sufficiently constrain the recovery of certain motions (e.g., rotation of a roughly cylindrical body part, such as a forearm, around its major axis). Region tracking employs considerably more data from images and thus can be more robust to ambiguous data, however, when regions are uniform multiple solutions may exist.

Generally, research on recognition of human activities has focused on one type of human body part motion and has assumed no coincidence of other motions. With the exception of [7], work on facial deformations (facial expressions and lip reading) has assumed that little or no rigid head or body motions are coincident [14, 13, 21, 24, 32, 36, 39, 40]. Articulated motion tracking work has assumed that non-rigid deformations are not coincident (e.g. clothing deformations during locomotion) [4, 3, 15, 16, 27, 28, 30, 31, 38]. Rigid motion recovery approaches such as [2] do not account for deformable and articulated motions (e.g., facial expressions and speech).

Recognition of human motion critically depends on the recovered representations of the body parts' activities. Most recognition work has employed well known techniques such as eigenspaces [22] dynamic time warping [15], hidden Markov models [25, 35], phase space [9, 23, 27], rule-based techniques [7, 39], and neural networks [32] (for a detailed overview see [10]).

In this chapter, we propose a 2D model-based framework for human part tracking using parametrized motion models of these parts. This framework shifts the focus of tracking from edges to the intensity pattern created by each body part in the image plane. It employs a 2D model-based viewer-centered approach to analyzing the data in image sequences. The approach enforces inter-part motion constraints for recovery which results in simplifying the tracking. We show that our paradigm provides a reasonable

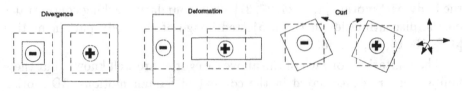

Figure 4. The figure illustrates the motion captured by the various parameters used to represent the motion of the regions. The solid lines indicate the deformed image region and the "-" and "+" indicate the sign of the quantity.

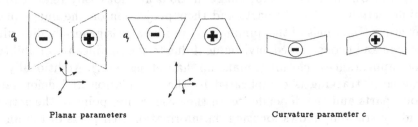

Planar parameters   Curvature parameter c

Figure 5. Additional parameters for planar motion and curvature.

model of motion types prevalent in human activity. We further discuss viewer-centered motion recognition approaches of human activity that involve deformable motions (e.g., facial expressions), rigid motions (e.g., head gestures) and articulated motion (e.g., locomotion).

3. Models of Image Motion

Parameterized models of image motion make explicit assumptions about the spatial variation of the optical flow within a region. Modeling the motion of a human body part involves making simplifying assumptions that approximate the image motion of that part. For example we assume that the limbs of the body and the face (excluding the face features) can be modeled as rigid planes. The image motion of a rigid planar patch of the scene can be described by the following eight-parameter model:

$$u(x, y) = a_0 + a_1 x + a_2 y + a_6 x^2 + a_7 xy, \tag{1}$$
$$v(x, y) = a_3 + a_4 x + a_5 y + a_6 xy + a_7 y^2, \tag{2}$$

where $\mathbf{a} = [a_0, a_1, a_2, a_3, a_4, a_5, a_6, a_7]$ denotes the vector of parameters to be estimated, and $\mathbf{u}(\mathbf{x}, \mathbf{a}) = [u(x, y), v(x, y)]^T$ are the horizontal and vertical components of the flow at image point $\mathbf{x} = (x, y)$. The coordinates (x, y) are defined with respect to a particular point. Here this is taken to be the center of the region but could be taken to be at a point of articulation.

The parameters a_i have qualitative interpretations in terms of image motion. For example, a_0 and a_3 represent horizontal and vertical translation respectively. Additionally, we can approximately express *divergence* (isotropic expansion), *curl* (rotation about the viewing direction), and *deformation* (squashing or stretching) as combinations of the a_i (cf. [11, 20]). We define these quantities as

$$\text{divergence} \overset{\text{def}}{=} a_1 + a_5, \tag{3}$$

$$\text{curl} \overset{\text{def}}{=} -a_2 + a_4, \tag{4}$$

$$\text{deformation} \overset{\text{def}}{=} a_1 - a_5. \tag{5}$$

Note that these terms give qualitative measures that can be used to interprate the motion of a region. Translation, along with divergence, curl, and deformation, will prove to be useful for describing facial expressions and are illustrated in Figure 4. For example, eye blinking can be detected as rapid deformation, divergence, and vertical translation in the eye region.

The parameters a_6 and a_7 roughly represent "yaw" and "pitch" deformations in the image plane respectively and are illustrated in Figure 5. While we have experimented with more complex models of rigid face motion, and Wang *et al.* [38] have used cylindrical models of limbs, we have found that the planar motion approximation is both simple and expressive enough to robustly represent qualitative rigid face and limb motions in a variety of situations.

To recognize the motion of the head using these parameters we need to know the head motion *relative* to the motion of the torso. Similarly, to recognize facial expressions from motion we need to know the motion of the eyes, mouth, and eyebrows relative to the motion of the face. In addition to isolating the motions of interest for recognition, this relative motion estimation allows us to estimate the motion of body parts that occupy small regions of the image; for example, facial features or fingers. The absolute motion of these regions in an image sequence may be very large with respect to their size making motion estimation difficult. The problem of estimating the motion of small parts like fingers is simplified if we know the motion of the torso, arm and hand.

For small regions of the image such as eyes and fingers, the planar model may not be necessary and the motion of these regions can be approximated by the simpler affine model in which the terms a_6 and a_7 are zero. The non-rigid motions of facial features such as the eyebrows and mouth, however, are not well captured by the rigid affine or planar models so we augment the affine model to account for the primary form of curvature seen in mouths and eyebrows. We add a new parameter c to the affine model

$$u(x, y) = a_0 + a_1 x + a_2 y \tag{6}$$

$$v(x, y) = a_3 + a_4 x + a_5 y + cx^2 \tag{7}$$

where c encodes curvature and is illustrated in Figure 5. This curvature parameter must be estimated in the coordinate frame of the face as described in [7]. As the experiments will demonstrate, this seven parameter model captures the essential image motion of the mouth and eyebrows.

4. Parametric Motion Estimation

To estimate the motion parameters, $\mathbf{a}$, for a given patch we make the assumption that the brightness pattern within the patch remains constant while the patch may deform as specified by the model. This brightness constancy assumption gives rise to the optical flow constraint equation

$$\nabla I \cdot \mathbf{u}(\mathbf{x}, \mathbf{a}_s) + I_t = 0, \ \forall \mathbf{x} \in \mathcal{R}_s \tag{8}$$

where $\mathbf{a}_s$ denotes the planar model for patch s, $\mathcal{R}_s$ denotes the points in patch s, I is the image brightness function and t represents time. $\nabla I = [I_x, I_y]$, and the subscripts indicates partial derivatives of image brightness with respect to the spatial dimensions and time at the point $\mathbf{x}$.

Note that the brightness constancy assumption used to estimate the image motion is often violated in practice due to changes in lighting, specular reflections, occlusion boundaries, etc. It may also be violated because the motion model is only a rough approximation to the true motion; for example we model the face as a plane although it is not really rigid or planar.

Robust regression has been shown to provide accurate motion estimates in a variety of situations in which the brightness constancy assumption in violated [5]. To estimate the parameters $\mathbf{a}_s$ robustly we minimize

$$\sum_{\mathbf{x} \in \mathcal{R}_s} \rho(\nabla I \cdot \mathbf{u}(\mathbf{x}, \mathbf{a}_s) + I_t, \sigma), \tag{9}$$

for some robust error norm ρ where σ is a scale parameter. Violations of the brightness constancy assumption can be viewed as "outliers" [17] and we need to choose the function ρ such that it is insensitive to these gross errors.

For the experiments in this chapter we take ρ to be

$$\rho(x, \sigma) = \frac{x^2}{\sigma + x^2} \tag{10}$$

which is the robust error norm used in [5, 7, 18] and is shown in Figure 6. As the magnitudes of residuals $\nabla I \cdot \mathbf{u}(\mathbf{x}, \mathbf{a}_s) + I_t$ grow beyond a point their influence on the solution begins to decrease and the value of $\rho(\cdot)$ approaches a constant. The function $\psi(x, \sigma)$ shown in Figure 6 is the derivative of ρ

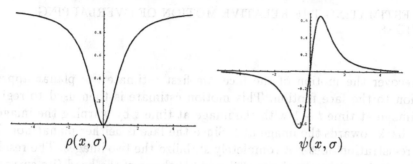

$\rho(x, \sigma)$ $\qquad$ $\psi(x, \sigma)$

Figure 6. Robust error norm (ρ) and influence function (ψ).

and characterizes the influence of the residuals [17]. The value σ is a scale parameter that effects the point at which the influence of outliers begins to decrease.

Equation (9) is minimized using a simple coordinate descent scheme with a continuation method that begins with a high value for σ and lowers it during the minimization (see [5, 7, 18] for details). The effect of this procedure is that initially no data are rejected as outliers then gradually the influence of outliers is reduced. To cope with large motions a coarse-to-fine strategy is used in which the motion is estimated at a coarse level then, at the next finer level, the image at time $t + 1$ is warped towards the image at time t using the current motion estimate. The motion parameters are refined at this level and the process continues until the finest level.

In the remainder of this chapter we use this robust estimation scheme for estimating face motion as described in [7] and for articulated motion as described in [18].

5. Estimating Relative Body Part Motion

The parametric motions of human body parts are inter-related as either *linked* or *overlapping* parts. Linked body parts, such as the "thigh" and "calf," share a joint in common and must satisfy an articulation constraint on their motion. The overlapping relation describes the relationship between regions such as the face and mouth in which the motion of the mouth is estimated relative to the motion of the face but is not constrained by it. These relationships lead to different treatments in terms of how the inter-part motions are estimated. These relations and the associated motion estimation techniques are described in this section and are illustrated with examples of facial motion estimation and articulated leg motion.

5.1. ESTIMATING THE RELATIVE MOTION OF OVERLAPPING REGIONS

To recover the motion of the face, we first estimate the planar approximation to the face motion. This motion estimate is then used to register the image at time $t+1$ with the image at time t by warping the image at $t+1$ back towards the image at t. Since the face is neither planar nor rigid this registration does not completely stabilize the two images. The residual motion is due either to the non-planar 3D shape of the head (its curvature and the nose for example) or the non-rigid motion of the facial features (cf. work on plane+parallax models of motion in rigid scenes [33]). We have observed that the planar model does a very good job of removing the rigid motion of the face and that the dominant residual motion is due to the motion of the facial features. The residual motion in the stabilized sequence is estimated using the appropriate motion model for that feature (i.e., affine or affine+curvature). Thus stablizing the face with respect to the planar approximation of its motion between two images allows the relative motions of the facial features to be estimated.

The estimated parametric motion of the face and facial features estimated between two frames is used to predict the location of the features in the next frame. The face and the eyes are simple quadrilaterals which are represented by the image locations of their four corners. We update the location x of each of the four corners of the face and eyes by applying the planar motion parameters a_f of the face to get $x' = x + u(x, a_f)$. Then the relative motion of the eyes locations is accounted for and the corners become $x' + u(x', a_{le})$ and $x' + u(x', a_{re})$ where a_{le} and a_{re} are the parameters estimated for the motions of the left and right eyes respectively. In updating the eye region we do not use the full affine model since when the eye blinks this would cause the tracked region to deform to the point where the eye region could no longer be tracked. Instead only the horizontal and vertical translation of the eye region is used to update its location relative to the face motion.

The curvature of the mouth and brows means that the simple updating of the corners is not sufficient for tracking. In our current implementation we use image masks to represent the regions of the image corresponding to the brows and the mouth. These masks are updated by warping them first by the planar face motion a_f and then by the motion of the individual features a_m, a_{lb} and a_{rb} which correspond the mouth and the left and right brows respectively. This simple updating scheme works well in practice.

5.2. ESTIMATING ARTICULATED MOTION

For an articulated object, we assume that each patch is connected to only one preceding patch and one following patch; that is, the patches construct a chain structure (see Figure 7). For example, a "thigh" patch may be connected to a preceding "torso" patch and a following "calf" patch. Each patch is represented by its four corners. Our approach is to simultaneously estimate the motions, $\mathbf{a}_s$, of all the patches. We minimize the total energy of the following equation to estimate the motions of each patch (from 0 to n)

$$E = \sum_{s=0}^{n} E_s = \sum_{s=0}^{n} \sum_{\mathbf{x} \in \mathcal{R}_s} \rho(\nabla I \cdot \mathbf{u}(\mathbf{x}, \mathbf{a}_s) + I_t, \sigma) \tag{11}$$

where ρ is the robust error norm defined above.

Equation (11) may be ill-conditioned due to the lack of sufficient brightness variation within the patch. The articulated nature of the patches provides an additional constraint on the solution. This articulation constraint is added to Equation (11) as follows

$$E = \sum_{s=0}^{n} (\frac{1}{|\mathcal{R}_s|} E_s + \lambda \sum_{\mathbf{x} \in \mathcal{A}_s} \|\mathbf{u}(\mathbf{x}, \mathbf{a}_s) - \mathbf{u}(\mathbf{x}, \mathbf{a}')\|^2), \tag{12}$$

where $|\mathcal{R}_s|$ is the number of pixels in patch s, λ controls relative importance of the two terms, $\mathcal{A}_s$ is the set of articulated points for patch s, $\mathbf{a}'$ is the planar motion of the patch which is connected to patch s at the articulated point $\mathbf{x}$, and $\| \cdot \|$ is the standard norm. The use of a quadratic function for the articulation constraint reflects that the assumption that no "outliers" are allowed.

Instead of using a constraint on the image velocity at the articulation points, we can make use of the distance between a pair of points. Assuming $\mathbf{x}'$ is the corresponding image point of the articulated point $\mathbf{x}$, and $\mathbf{x}'$ belongs to the patch connected to patch s at point $\mathbf{x}$ (see Figure 7), Equation (12) can be modified as

$$E = \sum_{s=0}^{n} (\frac{1}{|\mathcal{R}_s|} E_s + \lambda \sum_{\mathbf{x} \in \mathcal{A}_s} \|\mathbf{x} + \mathbf{u}(\mathbf{x}, \mathbf{a}_s) - \mathbf{x}' - \mathbf{u}(\mathbf{x}', \mathbf{a}')\|^2) \tag{13}$$

This formulation has the advantage that the pair of articulated points, $\mathbf{x}$ and $\mathbf{x}'$, will always be close to each other at any time. The second energy term (the "smoothness" term) in Equation (13) can also be considered as a spring force energy term between two points (Figure 7).

The planar motions estimated from the Equation (13) are absolute motions. In order to recognize articulated motion, we need to recover the

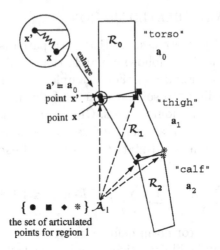

Figure 7. The "chain" structure of a three-segment articulated object.

motions of limbs which are relative to their preceding (parent) patches. We define

$$u(x + u(x, a_{s-1}), a_s^r) = u(x, a_s) - u(x, a_{s-1}), \tag{14}$$

where a_s^r is the relative motion of patch s, $u(x, a_s) - u(x, a_{s-1})$ is the relative displacement at point x, and $x + u(x, a_{s-1})$ is the new location of point x under motion a_{s-1}. A planar motion has eight parameters, therefore four different points of patch s are sufficient to solve a_s^r given the linear equations (14). In our experiments we use the four corners of the patches.

6. Motion Estimation Results

In this section we illustrate the performance of the parameterized flow models on articulated, rigid and deformable body parts. Head motion and facial expressions are used in illustrating the rigid and deformable motions, respectively. For articulated motion we focus on "walking" (on a treadmill, for simplicity) and provide the recovered motion parameters for two leg parts during this cyclic activity.

The image sequence in Figure 8 illustrates facial expressions ("smiling" and "surprise") in conjunction with rigid head motion (in this case looming). The figure plots the regions corresponding to the face and the facial features tracked across the image sequence. The parameters describing the planar motion of the face are plotted in Figure 9 where the divergence due to the looming motion of the head is clearly visible in the plot of divergence. Notice that the rigid motions of the face are not visible in the plotted motions of the facial features in Figure 10. This indicates that the motion of the face has been factored out and that the feature motions are

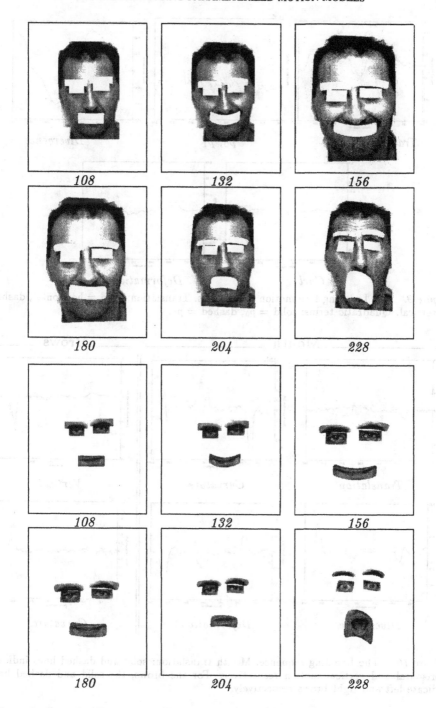

Figure 8. Looming Experiment. Facial expression tracking with rigid head motion (every 24 frames).

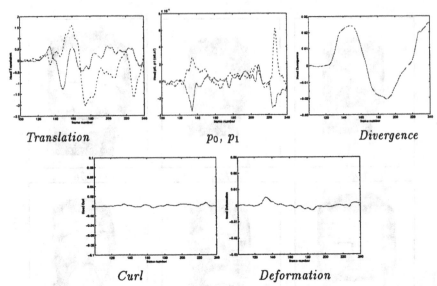

Figure 9. The Looming face motion parameters. Translation: solid = horizontal, dashed = vertical. Quadratic terms: solid = p_0, dashed = p_1.

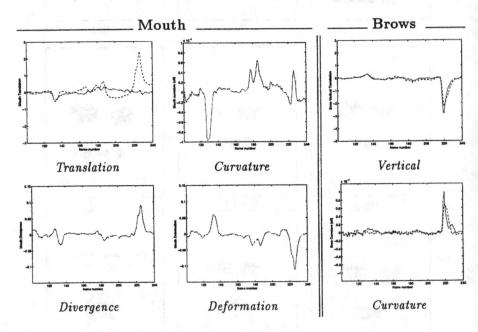

Figure 10. The Looming sequence. Mouth translation: solid and dashed lines indicate horizontal and vertical motion respectively. For the brows, the solid and dashed lines indicate left and right brows respectively.

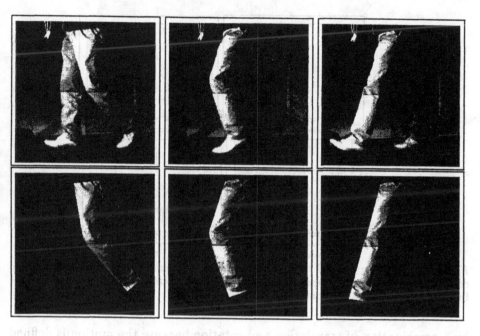

Figure 11. Walking parallel to the imaging plane.

truly relative to the face motion. Analyzing the plots of the facial features reveals that a "smile" expression begins around frame 125 with an increase in mouth curvature followed by a deformation of the mouth. The curvature decreases between frames 175 and 185 and then a "surprise" expression begins around frame 220 with vertical eyebrow motion, brow arching, and mouth deformation.

Figures 11-14 demonstrate two "walking" sequences taken from different view-points. Each sequence contains 500 images and roughly three cycles of the activity. In Figures 11 and 13 the upper row shows three images from the sequences and the second row shows the tracking of two parts (the "thigh" and "calf"). Figures 12 and 14 show relevant recovered motions parameters over the entire 500 frame image sequences. The first rows in these figures show the motion of the thigh while the second rows show the motion of the calf. These graphs are only meant to demonstrate the effectiveness of our tracking model and its ability to capture meaningful parameters of the body movement.

In Figure 12 it is clear that the horizontal translation and "curl" parameters capture quite well the cyclic motion of the two parts of the leg. The translation of the "calf" is relative to that of the "thigh" and therefore it is significantly smaller. On the other hand, the rotation (i.e., "curl") is more significant at the "calf". Note that the motion of these regions is described

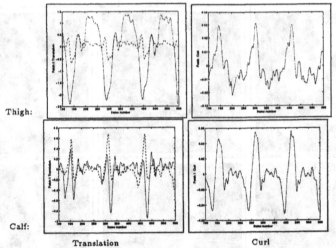

Figure 12. Walking parallel to the imaging plane; motion parameters (translation and curl) over 500 frames. For translation, the dashed line indicates vertical translation while the solid line indicates horizontal translation.

by a combination of translation and rotation because the motion is defined with respect to the center of the regions. A different choice of region center would result in different plots.

When a person is walking away from the camera as shown in Figure 14 the significant parameters which capture the cyclic walking motion are deformation, divergence, and "image pitch." Notice that the "image pitch" measured at the two parts is always reversed since when the "thigh" rotates in one direction the "calf" is viewed to be rotating in an opposite way.

In summary, the reported experiments show that the image motion models are capable of tracking rigid, deformable and articulated motion quite accurately over long sequences and recovering a meaningful set of parameters that can be exploited by a recognition system.

7. Recognition of Movement

Recognizing human movement requires answering:

- When does the activity begin and end?
- What class does the observed activity most closely resemble?
- What is the period (if cyclic) of the activity?

The answers to these questions involve spatio-temporal reasoning about a large parameter space. Our choice of parameterized motion models for tracking the diverse motions of human body parts dictates that recognition be considered as analysis of the spatio-temporal curves and surfaces that are created by the parameter values of each part. These curves are based

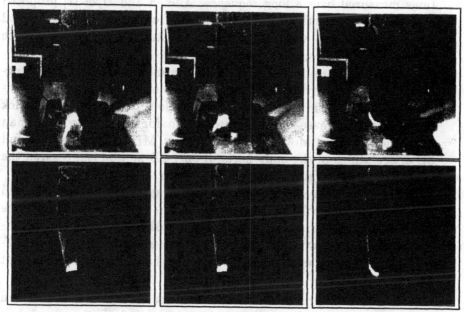

Figure 13. Walking perpendicular to the imaging plane.

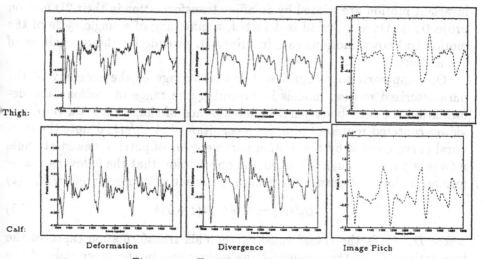

Thigh:

Calf:

| Deformation | Divergence | Image Pitch |

Figure 14. Tracking results for Figure 13

on 2D image motion and therefore change with viewing angle. This leads us to formulate a view-based recognition strategy.

There are general issues to consider for a recognition approach:

— Cyclic actions require that enough cycles have been observed before recognition becomes possible. This leads to a focus on the coarse motions measured over a long sequence. Allmen and Dyer [1] proposed a method for detection of cyclic motions from their spatio-temporal curves and surfaces while Rangarajan and Shah [29] used a scale-space approach to match motion trajectories.

- The view-point of observing the human motion may affect the recognition of the activity (for an extreme case, consider recognition of human "walking" from a top view of the person). Campbell and Bobick [9] proposed a phase-space representation for recognition of human body motion from Moving Light Displays (MLD) capturing the full 3D articulations of the body parts. If only 2D motion is measured the viewpoint plays a critical role in recognition.

- Self-occlusions are quite prevalent in human movement. Capturing and representing these self-occlusions is a complex task even in the presence of multiple cameras and availability of full 3D models. In our preliminary recognition work we do not capture and represent these self-occlusions, instead we focus on body parts that are visible throughout the activity.

Seitz and Dyer [34] proposed an approach for determining whether an observed motion is periodic and computing its period. Their approach is based on the observation that the 3D points of an object performing affine-invariant motion are related by an affine transformation in their 2D motion projections. Once a period is detected, a matching of a single cycle of the motion to known motions can, in principal, provide for the recognition of the activity.

Our approach to recognition takes advantage of the economy of the parameterized motion models in capturing the range of motions and deformations of each body part. In the absence of shape cues, we employ a viewer-centered representation for recognition. Let $C_v{}^{i_j}(t)$ denote the temporal curve created by the motion parameter a_j of patch i viewed at angle v (where $j \in a_0, ..., a_7$). We make the observation that the following transformation does not change the nature of the activity represented by $C_v{}^{i_j}(t)$

$$D_v{}^{i_j}(t) = S_i * C_v{}^{i_j}(t + T_i) \tag{15}$$

where $D_v{}^{i_j}(t)$ is the transformed curve. This transformation captures the translation, T_i, of the curve and the scaling, S_i, in the magnitude of the image-motion measured for parameter a_j. The scaling of the curve allows accounting for different distances between the human and the camera (while the viewing angle is kept constant) and accounts for the physiological variation across humans. Notice that this transformation does not scale the curve in the temporal dimension since the nature of the activity changes due to temporal scaling (e.g., different speeds of "walking" can be captured by this scaling). This temporal scaling can be expressed as an affine transformation

$$D_v^{i_j}(t) = S_i * C_v^{i_j}(\alpha_i t + T_i) \qquad (16)$$

where $\alpha_i > 1.0$ leads to a linear speed up of the activity and $\alpha_i < 1.0$ leads to its slow down.

The recognition of an activity can be posed as a matching problem between the curve created by parameter a_j over time and a set of known curves (corresponding to known activities) that can be subject to the above transformation. Recognition of an activity for some viewpoint v requires that a single affine transformation should apply to all parameters a_j, this can be posed as a minimization of the error (under some error norm)

$$E(v) = \sum_{j \in 0..7} \rho[D_v^{i_j}(t) - (S_i * C_v^{i_j}(\alpha_i t + T_i)), \sigma] \qquad (17)$$

Recognition over different viewpoints requires finding the minimum error between all views v, which can be expressed as

$$\min_v \sum_{j \in 0..7} \rho[D_v^{i_j}(t) - (S_i * C_v^{i_j}(\alpha_i t + T_i)), \sigma] \qquad (18)$$

Recognition over multiple body parts uses the inter-part hierarchy relationships to progressively find the best match. As demonstrated and discussed in Section 6, the motion parameters are stable over a wide range of viewpoints of the activity, so that they could be represented by a few principal directions.

Our formulation requires computing a *characteristic* curve $C_v^{i_j}$ for each activity and body part viewed at angle v. Constructing this characteristic curve can be achieved by tracking the patch motions over several subjects and employing Principal Component Analysis (PCA) to capture the dominant curve components. Given an observed activity captured by $D^{i_j}(t)$ (notice that the v is dropped since it is unknown), our approach determines the characteristic curve that minimizes the error function given in Equation 18 by employing the recently proposed affine eigentracking approach [6] on the curves.

We are currently constructing these characteristic curves for several human activities. It is worth noting that, depending on the spatio-temporal complexity of the observed activity, simpler models could be used for recognition. For example in the case of facial expressions the activity can be simply captured by the model first proposed in [39] and used in [7]. Each expression was divided into three temporal segments: the *beginning*, *apex* and *ending*. Figure 15 illustrates qualitatively the different aspects of detecting and segmenting a "smile." In this Figure the horizontal axis represents the time dimension (i.e., the image sequence), the axis perpendicular to the

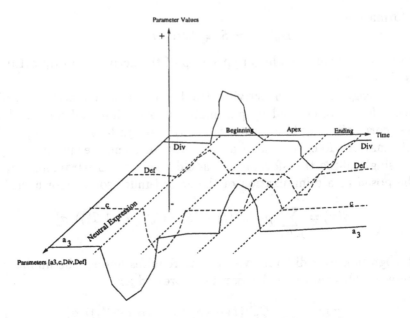

Figure 15.   The temporal model of the "smile" expression.

page represents each one of the parameters relevant to a "smile" (i.e., a_3, *Divergence, Deformation,* and *c*) and the vertical axis represents the values of these parameters. This diagram is an abstraction to the progression of a "smile," therefore the parameter values are not provided. Notice that Figure 15 indicates that the change in parameter values might not occur at the same frames at either the beginning or ending of actions, but it is required that a significant overlap be detectable to label a set of frames with a "beginning of a smile" label, while the motions must terminate before a frame is labeled as an "apex" or an "ending."

The detailed development of the "smile" model is as follows. The upward and outward motion of the mouth corners results in a negative curvature of the mouth (i.e., the curvature parameter *c* is negative). The horizontal and overall vertical stretching are manifested by positive divergence (Div) and deformation (Def). Finally, some overall upward translation is caused by the raising of the lower and upper lips due to the stretching of the mouth (a_3 is negative). Reversal of these motion parameters is observed during the ending of the expression.

The results of applying this recognition model to face expressions can be seen in Figure 16. Figure 16 shows the beginning of a "smile" expression while the head is rotating initially leftward and then rightward. The text that appears on the left side of each image represents a discrete interpretation of the underlying curves in terms of mid-level predicates which

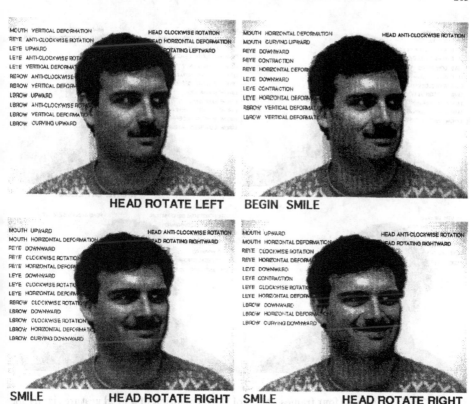

Figure 16. Four frames (four frames apart) of the beginning of a "smile" expression.

describe the facial motion [7]. Similarly, the text that appears on the right side represents the mid-level predicates of the head motion. The text below each image displays the recognized high-level description of the facial deformations and the head motions.

Figure 17 shows the recognition of head gestures based on the face motion recovery using a planar model. The gesture is recognized using the "curl" of the face. Other gestures were recognized in [25].

8. Discussion

We have demonstrated the use of parameterized optical flow methods for tracking and recognizing facial expressions and articulated motion. While the approach shows promise, there are a number of issues that still need to be addressed. First, the motion of human limbs in NTSC video (30 frames/sec) can be very large. For example, human limbs often move distances greater than their width between frames. This causes problems for a hierarchical gradient-based motion scheme such as the one presented here.

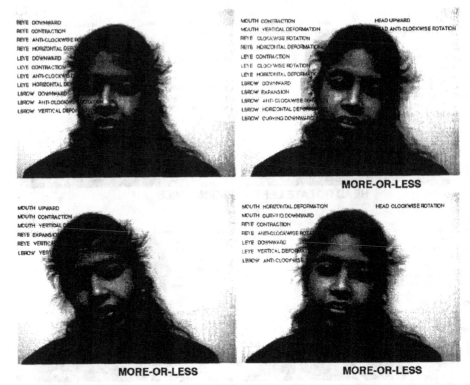

Figure 17. Four frames (four frames apart) of the recognition of a head gesture signifying the expression of "more-or-less".

To cope with large motions of small regions we will need to develop better methods for long-range motion estimation.

Unlike the human face, people wear clothing over their limbs which deforms as they move. The "motion" of the deforming clothing between frames is often significant and, where there is little texture on the clothing, may actually be the dominant motion within a region. A purely flow-based tracker such as the one here has no "memory" of what is being tracked. So if it is deceived by the motion of the clothing in some frame there is a risk that tracking will be lost. We are exploring ways of adding a template-style form of memory to improve the robustness of the tracking.

Self occlusion is another problem we have not addressed preferring to first explore the efficacy of the parameterized tracking and recognition scheme in the non-occlusion case. In extending this work to cope with occlusion, the template-style methods mentioned above may be applicable.

9. Conclusion

We have presented a method for tracking non-rigid and articulated human motion in an image sequence using parameterized models of optical flow and have shown how this representation of human motion can support the recognition of facial expressions and articulated motions. Unlike previous work on recovering human motion, this method assumes that the activity can be described by the motion of a set of parameterized patches (e.g. affine, planar, etc.). In the case of facial motion, the motion of facial features is estimated relative to the motion of the face. For the articulated motion of limbs we add an additional articulation constraint between neighboring patches. No 3D model of the person is required, features such as edges are not used, and the optical flow is estimated directly using the parameterized model. An advantage of the 2D parameterized flow models is that recovered flow parameters can be interpreted and used for recognition as described in [7]. Previous methods for recognition need to be extended to cope with the cyclical motion of human activities and we have proposed a method for performing view-based recognition of human activities from the optical flow parameters. Our current work is focused on the automatic segmentation of non-rigid and articulated human motion into parts and the development of robust view-based recognition schemes for articulate motion.

References

1. M. Allmen and C.R. Dyer. Cyclic motion detection using spatiotemporal surfaces and curves. In *ICPR*, pages 365–370, 1990.
2. A. Azarbayejani, T. Starner, B. Horowitz, and A. Pentland. Visually controlled graphics. *IEEE Transactions on Pattern Analysis and Machine Intelligence*, 15(6):602–604, June 1993.
3. A. Baumberg and D. Hogg. Learning flexible models from image sequences. In J. Eklundh, editor, *European Conf. on Computer Vision, ECCV-94*, volume 800 of *LNCS-Series*, pages 299–308, Stockholm, Sweden, 1994. Springer-Verlag.
4. A. G. Bharatkumar, K. E. Daigle, M. G. Pandy, and J. K. Aggarwal. Lower limb kinematics of human walking with the medial axis tranfromation. In *Proceedings of the Workshop on Motion of Non-rigid and Articulated Objects*, pages 70–76, Austin, Texas, November 1994.
5. M. J. Black and P. Anandan. The robust estimation of multiple motions: Parametric and piecewise-smooth flow fields. *Computer Vision and Image Understanding*, 63(1):75–104, January 1996.
6. M. J. Black and A. D. Jepson. Eigentracking: Robust matching and tracking of articulated objects using a view-based representation. In B. Buxton and R. Cipolla, editors, *European Conf. on Computer Vision, ECCV-96*, volume 1064 of *LNCS-Series*, pages 329–342, Cambridge, UK, 1996. Springer-Verlag.
7. M. J. Black and Y. Yacoob. Tracking and recognizing rigid and non-rigid facial motions using local parametric models of image motions. In *Proceedings of the International Conference on Computer Vision*, pages 374–381, Boston, Mass., June 1995.
8. A. Blake and M. Isard. 3D position, attitude and shape input using video tracking

of hands and lips. In *Proceedings of SIGGRAPH 94*, pages 185–192, 1994.

9. L.W. Campbell and A.F. Bobick. Recognition of human body motion using phase space constraints. In *Proceedings of the International Conference on Computer Vision*, pages 624–630, Boston, Mass., June 1995.

10. C. Cédras and M. Shah. Motion-based recognition: A survey. *Image and Vision Computing*, 13(2):129–155, March 1995.

11. R. Cipolla and A. Blake. Surface orientation and time to contact from image divergence and deformation. In G. Sandini, editor, *Proc. of Second European Conference on Computer Vision, ECCV-92*, volume 588 of *LNCS-Series*, pages 187–202. Springer-Verlag, May 1992.

12. François Dagognet. *Etienne-Jules Marey: A Passion for the Trace*. Zone Books, New York, 1992.

13. I. Essa, T. Darrell, and A. Pentland. Tracking facial motion. In *Proceedings of the Workshop on Motion of Non-rigid and Articulated Objects*, pages 36–42, Austin, Texas, November 1994.

14. I. A. Essa and A. Pentland. A vision system for observing and extracting facial action parameters. In *Proc. Computer Vision and Pattern Recognition, CVPR-94*, pages 76–83, Seattle, WA, June 1994.

15. D. Gavrila and L.S. Davis. 3d model-based tracking of humans in action: A multiview approach. In *Proc. Computer Vision and Pattern Recognition, CVPR-96*, San Francisco, CA, June 1996.

16. L. Goncalves, E. Di Bernardo, E. Ursella, and P. Perona. Monocular tracking of the human arm in 3D. In *Proceedings of the International Conference on Computer Vision*, pages 744–770, Boston, Mass., June 1995.

17. F. R. Hampel, E. M. Ronchetti, P. J. Rousseeuw, and W. A. Stahel. *Robust Statistics: The Approach Based on Influence Functions*. John Wiley and Sons, New York, NY, 1986.

18. S. X. Ju, M. J. Black, and Y. Yacoob. Cardboard people: A parameterized model of articulated motion. In *International Conference on Automatic Face and Gesture Recognition*, pages 38–44, Killington, Vermont, 1996.

19. M. Kass, A. Witkin, and D. Terzopoulos. Snakes: Active contour models. In *Proc. First International Conference on Computer Vision*, pages 259–268, June 1987.

20. J. J. Koenderink and A. J. van Doorn. Invariant properties of the motion parallax field due to the movement of rigid bodies relative to an observer. *Optica Acta*, 22(9):773–791, 1975.

21. H. Li, P. Roivainen, and R. Forcheimer. 3-D motion estimation in model-based facial image coding. *IEEE Transactions on Pattern Analysis and Machine Intelligence*, 15(6):545–555, June 1993.

22. N. Li, S. Dettmer, and M. Shah. Lipreading using eigensequences. In *International Workshop on Automatic Face and Gesture Recognition*, pages 30–34, Zurich, 1995.

23. J. Little and J. Boyd. Describing motion for recognition. In *International Symposium on Computer Vision*, pages 235–240, Miami, FL, November 1995.

24. K. Mase. Recognition of facial expression from optical flow. *IEICE Transactions*, E 74:3474–3483, 1991.

25. C. Morimoto, Y. Yacoob, and L.S. Davis. Recognition of head gestures using Hidden Markov Models. In *Proceedings of the International Conference on Pattern Recognition*, Vienna, Austria, 1996.

26. Eadweard Muybridge. *The Human Figure in Motion*. Dover Publications, Inc., New York, 1955.

27. S. A. Niyogi and E. H. Adelson. Analyzing and recognizing walking figures in xyt. In *Proc. Computer Vision and Pattern Recognition, CVPR-94*, pages 469–474, Seattle, WA, June 1994.

28. A. Pentland and B. Horowitz. Recovery of nonrigid motion and structure. *IEEE Transactions on Pattern Analysis and Machine Intelligence*, 13(7):730–742, July 1991.

29. K. Rangraajan, W. Allen, and M.A. Shah. Matching motion trajectories using scale-space. *Pattern Recognition*, 26(4):595–609, 1993.
30. J. Rehg and T. Kanade. Model-based tracking of self-occluding articulated objects. In *Proceedings of the International Conference on Computer Vision*, pages 612–617, Boston, Mass., June 1995.
31. K. Rohr. Towards model-based recognition of human movements in image sequences. *CVGIP: Image Processing*, 59:94–115, 1994.
32. M. Rosenblum, Y. Yacoob, and L.S. Davis. Human emotion recognition from motion using a radial basis function network architecture. In *Proceedings of the Workshop on Motion of Non-rigid and Articulated Objects*, Austin, Texas, November 1994.
33. H. S. Sawhney. 3D geometry from planar parallax. In *Computer Vision and Pattern Recognition, CVPR-94*, pages 929–934, Seattle, WA, 1994.
34. S.M. Seitz and C.R. Dyer. Affine invariant detection of periodic motion. In *Proc. Computer Vision and Pattern Recognition, CVPR-94*, pages 970–975, Seattle, WA, June 1994.
35. T. Starner and A. Pentland. Visual recognition of American Sign Language using Hidden Markov Models. In *International Workshop on Automatic Face and Gesture Recognition*, pages 189–194, Zurich, 1995.
36. D. Terzopoulos and K. Waters. Analysis and synthesis of facial image sequences using physical and anatomical models. *IEEE Transactions on Pattern Analysis and Machine Intelligence*, 15(6):569–579, June 1993.
37. S. Toelg and T. Pogio. Towards an example-based image compression architecture for video-conferencing. Technical Report CAR-TR-723, Center for Automation Research, U. of Maryland, 1994.
38. J. Wang, G. Lorette, and P. Bouthemy. Analysis of human motion: A model-based approach. In *7th Scandinavian Conf. Image Analysis*, Aalborg, Denmark, 1991.
39. Y. Yacoob and L.S. Davis. Computing spatio-temporal representations of human faces. In *Proc. Computer Vision and Pattern Recognition, CVPR-94*, pages 70–75, Seattle, WA, June 1994.
40. A. L. Yuille, D. S. Cohen, and P. W. Hallinan. Feature extraction from faces using deformable templates. In *Proc. Computer Vision and Pattern Recognition, CVPR-89*, pages 104–109, June 1989.

FACIAL EXPRESSION RECOGNITION USING IMAGE MOTION

IRFAN ESSA
College of Computing
Georgia Institute of Technology
Atlanta, GA 30332-0280, USA

AND

ALEX PENTLAND
Media Laboratory
Massachusetts Institute of Technology
Cambridge, MA 02141, USA

1. Introduction

The communicative power of the face makes machine understanding and recognition of human expression an important problem in computer vision. There is a significant amount of research on facial expressions in computer vision and computer graphics [11, 23]. Perhaps the most fundamental problem in this area is how to categorize active and spontaneous facial expressions to extract information about the underlying emotional states [6, 26]. Ekman and Friesen [10] have produced the most widely used system for describing visually distinguishable facial movements. This system, called the *Facial Action Coding System* or *FACS*, is based on the enumeration of all "action units" of a face which cause facial movements. As some muscles give rise to more than one action unit, the correspondence between action units and muscle units is approximate.

However, it is widely recognized that the lack of temporal and detailed spatial information (both local and global) is a significant limitation of the FACS model [11, 23]. Additionally, the heuristic "dictionary" of facial actions originally developed for FACS-based coding of emotion has proven difficult to adapt to machine recognition of facial expression.

We would like to *objectively* quantify facial movement during various facial expressions using computer vision techniques. Consequently, the goal of this chapter is to provide a method for extracting an extended FACS

271

M. Shah and R. Jain (eds.), Motion-Based Recognition, 271–298.
© *1997 Kluwer Academic Publishers.*

model (what we call FACS+), by coupling optical flow techniques with a physics-based model of both skin and muscle.

We will show that our method is capable of detailed, repeatable facial motion estimation in both time and space, with sufficient accuracy to measure previously-unquantified muscle coarticulations in facial expressions. We will further demonstrate that the parameters extracted using this method provide improved accuracy for recognition of facial expression.

Additionally, an interesting aspect of this work is that it has lead to the development of extremely simple, biologically-plausible motion energy detectors that accurately represent human expressions for motion-based recognition of facial motion.

The structure of this chapter is as follows: In the next section, a brief overview of work on facial expressions, spanning the areas of psychology, computer vision and computer graphics is presented. Section 3 presents our method for extraction of facial motion parameters from video. Section 4 discusses the the special features of the extracted parameters for representing facial expressions, which is then followed by a discussion of facial expression recognition in Section 5.

2. Background

2.1. REPRESENTATIONS OF FACIAL MOTION

Ekman and Friesen [10] have produced a system for describing "all visually distinguishable facial movements", called the *Facial Action Coding System or FACS*. It is based on the enumeration of all "action units" (*AU*s) of a face that cause facial movements. There are 46 *AU*s in FACS that account for changes in facial expression. The combination of these action units results in a large set of possible facial expressions. For example happiness expression is considered to be a combination of "pulling lip corners (*AU*12+13) and/or mouth opening (*AU*25+27) with upper lip raiser (*AU*10) and bit of furrow deepening (*AU*11)." However this is only one type of a smile; there are many variations of the above motions, each having a different intensity of actuation.

This representation and many variations of this representation have been excessively used for coding of human expressions by people, especially for emotion research [9, 16].

2.2. TRACKING FACIAL MOTION

There have been several attempts to track facial expressions over time. Mase and Pentland [20] were perhaps the first to track action units using optical flow. Although their method was simple, without a physical model and formulated statically rather than within a dynamic optimal estimation framework, the results were sufficiently good to show the usefulness of optical flow for observing facial motion.

Terzopoulos and Waters [32] developed a much more sophisticated method that tracked linear facial features to estimate corresponding parameters of a three dimensional wireframe face model, allowing them to reproduce facial expressions. One main limitation of this system is that it requires that facial features be highlighted with make-up for successful tracking. Although active contour models *i.e.*, *snakes* [17] are used, the system is still passive; the facial structure is passively shaped by the tracked contour features without any active control enforced on the model based on observations and vice versa.

Haibo Li, Pertti Roivainen and Robert Forchheimer [18] describe an approach in which a control feedback loop between computer graphics and computer vision processes is used for a facial image coding system. Their work is the most similar to ours, but both our goals and implementations and the details inherent in them differ. The main limitation of their work is the lack of detail in motion estimation as only large, predefined areas were observed, and only affine motion computed within each area. These limits may be an acceptable loss of quality for image coding applications. However, for our purposes this limitation is severe; it means we cannot observe the "true" pattern of muscle actuation because the method assumes the FACS model as the underlying representation.

2.3. RECOGNITION OF FACIAL MOTION

Recognition of facial expressions can be achieved by categorizing a set of such predetermined facial motions as in FACS, rather than determining the motion of each facial point independently. This is the approach taken by several researchers [19, 20, 36, 29, 4] for their recognition systems. Yacoob and Davis, who extend the work of Mase, detect motion (only in eight directions) in six predefined and hand initialized rectangular regions on a face and then use simplifications of the FACS rules for the six universal expressions for recognition. The motion in these rectangular regions, from the last several frames, is correlated to the FACS rules for recognition. Black and Yacoob extend this method, using local parameterized models of image motion to deal with large-scale head motions. These methods show about 90% in correctly recognizing expressions in their database of over 105 expressions [4, 36]. Mase on a smaller set of data (30 test cases) obtained an accuracy of 80% [19]. In many ways these are impressive results, considering the complexity of the FACS model and the difficulty in measuring facial motion within small windowed regions of the face.

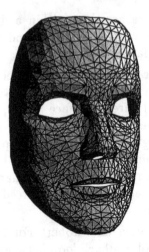

Figure 1. Geometric Model of a Face (Polygons/Vertices).

In our view perhaps the principle difficulty these researchers have encountered is the sheer complexity of describing human facial movement using FACS. Using the FACS representation, there are a very large number of AUs, which combine in extremely complex ways to give rise to expressions. Moreover, there is now a growing body of psychological research that argues that it is the dynamics of the expression, rather than detailed spatial deformations, that is important in expression recognition. Several researchers [1, 2, 6, 7, 9, 16] have claimed that the timing of expressions, something that is completely missing from FACS, is a critical parameter in recognizing emotions. This issue was also addressed in the NSF workshops and reports on facial expressions [11, 23]. To us this strongly suggests moving away from a static, "dissect-every-change" analysis of expression (which is how the FACS model was developed), towards a whole-face analysis of facial dynamics in motion sequences.

3. Visual Coding of Facial Motion

3.1. VISION-BASED SENSING: VISUAL MOTION

We use optical flow processing as the basis for perception and measurement of facial motion. We use Simoncelli's [31] method for optical flow computation, which uses a multi-scale, coarse-to-fine, Kalman filtering-based algorithm that provides good motion estimates and error-covariance information. Using this method we compute the estimated mean velocity vector $\hat{\mathbf{v}}_i(t)$, which is the estimated flow from time t to $t+1$. We also store the

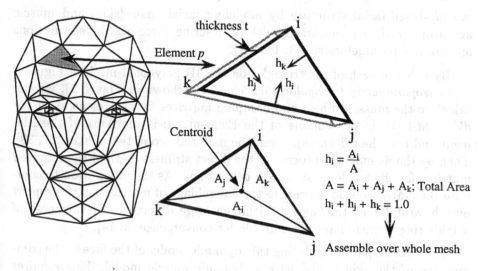

Figure 2. Using the FACS mesh to determine the continuum mechanics parameters of the skin using FEM.

flow covariances Λ_v between different frames for determining confidence measures and for error corrections in observations for the dynamic model (see Section 3.3).

3.2. FACIAL MODELING

A priori information about facial structure is an important parameter for our framework. Our face model is shown in Figure 1. This is an elaboration of the mesh developed by Platt and Badler [28]. We extend this into a topologically invariant physics-based model by adding anatomically-based muscles to it [12].

In order to conduct analysis of facial expressions and to define a new suitable set of control parameters (FACS+) using vision-based observations, we require a model with time dependent *states* and *state evolution* relationships. FACS and the related AU descriptions are purely static and passive, and therefore the association of a FACS descriptor with a dynamic muscle is inherently inconsistent. This problems motivated Waters [35] to develop a muscle model in a dynamic framework. By modeling the elastic nature of facial skin and the anatomical nature of facial muscles he developed a dynamic model of the face, including FACS-like control parameters. By implementing a procedure similar to that of Waters', we also built a dynamic muscle-based model of a face.

A physically-based dynamic model of a face was constructed by use of Finite Element methods. These methods give our facial model an *anato-*

mically-based facial structure by modeling facial tissue/skin, and muscle actuators, with a geometric model to describe force-based deformations and control parameters(see [3, 14, 21]).

By defining each of the triangles on the 3D polygonal mesh in Figure 1 as an *isoparametric triangular shell element*, (shown in Figure 2), we can calculate the mass, stiffness and damping matrices for each element (using $dV = tdA$, V: is the Volume of the Element, A: is the Area of the Element and t: is the thickness), given the material properties of skin [27, 33]. Then by the assemblage process of the direct stiffness method the required matrices for the whole mesh can be determined. As the integration to compute the matrices is done prior to the assemblage of matrices, each element may have different thickness t, although large differences in thickness of neighboring elements are not suitable for convergence [3, 14].

The next step in formulating this dynamic model of the face is the combination of the skin model with a dynamic muscle model. This requires information about the attachment points of the muscles to the face, or in our geometric case the attachment to the vertices of the geometric surface/mesh. The work of Pieper [27] and Waters [35] provides us with the required detailed information about muscles and muscle attachments.

3.3. DYNAMIC MODELING AND ESTIMATION

Initialization of Model on an image

In developing a representation of facial motion and then using it to compare to new data we need to locate a face and the facial features in the image followed by a registration of these features for all faces in the database. Previously we initialized our estimation process by manually translating, rotating and deforming our 3-D facial model to fit a face in an image. To automate this process we are now using the View-based and Modular Eigenspace methods of Pentland and Moghaddam [22, 24].

Using this method we can automatically extract the positions of the eyes, nose and lips in an image as shown in Figure 3(b). These feature positions are used to warp the face image to match the canonical face mesh (Figure 3(c) and (d)). This allows us to extract the additional "canonical feature points" on the image that correspond to the fixed (nonrigid) nodes on our mesh (Figure 3(f)). After the initial registering of the model to the image the coarse-to-fine flow computation methods presented by Simoncelli [31] and Wang [34] are used to compute the flow. The model on the face image tracks the motion of the head and the face correctly as long as there is not an excessive amount of rigid motion of the face during an expression.

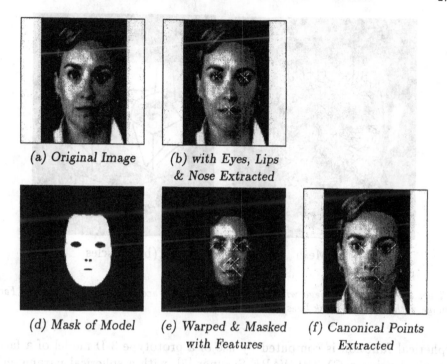

(a) Original Image

(b) with Eyes, Lips & Nose Extracted

(d) Mask of Model

(e) Warped & Masked with Features

(f) Canonical Points Extracted

Figure 3. Initialization on a face image using methods described by Pentland et al. [24, 22], using a canonical model of a face.

Images to face model

Simoncelli's [31] coarse-to-fine algorithm for optical flow computations provides us with an estimated flow vector, $\hat{\mathbf{v}}_i$. Now using a mapping function, $\mathcal{M}$, we would like to compute velocities for the vertices of the face model $\mathbf{v}_g$. Then, using the physically-based modeling techniques of Section 3.2 (described in more detail in [12]) and the relevant geometric and physical models, we can calculate the forces that caused the motion. Since we are mapping global information from an image (over the whole image) to a geometric model, we have to concern ourselves with translations (vector $\mathcal{T}$), and rotations (matrix $\mathcal{R}$). The Galerkin polynomial interpolation function $\mathbf{H}$ and the strain-displacement function $\mathcal{B}$, used to define the mass, stiffness and damping matrices on the basis of the finite element method are applied to describe the deformable behavior of the model [14, 25, 3].

We would like to use only a frontal view to determine facial motion and model expressions, and this is only possible if we are prepared to estimate the velocities and motions in the third axis (going into the image, the z-axis). To accomplish this, we define a function that does a spherical mapping, $\mathcal{S}(u, v)$, where are u and v are the spherical coordinates. The

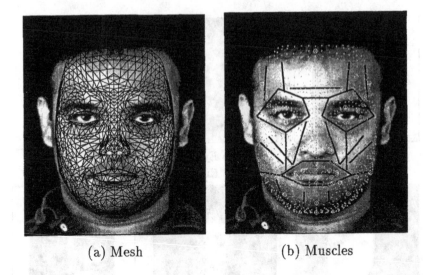

(a) Mesh   (b) Muscles

Figure 4. (a) Face image with a FEM mesh placed accurately over it and (b) Face
image with muscles (black lines), and nodes (dots).

spherical function is computed by use of a prototype 3-D model of a face,
acquired using a CYBERWARE Scanner [8], with a spherical parameterization; this canonical face model is then used to wrap the image onto the
shape. In this manner, we determine the mapping equation:

$$\mathbf{v}_g(x, y, z) = \mathcal{M}(x, y, z)\hat{\mathbf{v}}_i(x, y \mid z, y) \approx \mathbf{HSR}\,(\hat{\mathbf{v}}_i(x, y) + \mathbf{T})\,. \qquad (1)$$

For the rest of the chapter, unless otherwise specified, whenever we talk
about velocities we will assume that the above mapping has already been
applied. We are presently pursuing more accurate methods for observation
to model mapping for image to model correspondence.

Estimation and Control
Driving a physical system with the inputs from noisy motion estimates
can result in divergence or a chaotic physical response. This is why an
estimation and control framework needs to be incorporated to obtain stable
and well-proportioned results. Similar considerations motivated the control
framework used in [18]. Figure 6 shows the whole framework of estimation
and control of our active facial expression modeling system. The next few
sections discuss these formulations.

The continuous time Kalman filter (CTKF) allows us to estimate the
uncorrupted state vector, and produces an *optimal least-squares estimate*
under quite general conditions [5, 15]. The Kalman filter is particularly

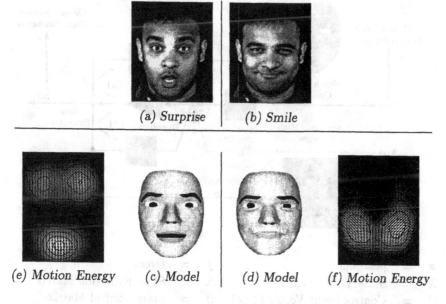

(a) Surprise | (b) Smile

(e) Motion Energy   (c) Model | (d) Model   (f) Motion Energy

Figure 5. Determining of expressions from video sequences. (a) and (b) show expressions of smile and surprise, (c) and (d) show a 3D model with surprise and smile expressions, and (e) and (f) show the spatio-temporal motion energy representation of facial motion for these expressions.

well-suited to this application because it is a recursive estimation technique, and so does not introduce any delays into the system (keeping the system active). The CTKF for the above system is:

$$\dot{\hat{X}} = A\hat{X} + BU + L\left(Y - C\hat{X}\right), \quad \text{where: } L = \Lambda_e C^T \Lambda_m^{-1}, \quad (2)$$

where $\hat{X}$ is the linear least squares estimate of X based on $Y(\tau)$ for $\tau < t$ and Λ_e the error covariance matrix for $\hat{X}$. The Kalman gain matrix L is obtained by solving the following Riccati equation to obtain the optimal error covariance matrix Λ_e:

$$\frac{d}{dt}\Lambda_e = A\Lambda_e + \Lambda_e A^T + G\Lambda_p G^T - \Lambda_e C^T \Lambda_m^{-1} C\Lambda_e. \quad (3)$$

The Kalman filter, Equation (2), mimics the noise free dynamics and corrects its estimate with a term proportional to the difference $(Y - C\hat{X})$, which is the innovations process. This correction is between the observation and our best prediction based on previous data. Figure 6 shows the estimation loop (the bottom loop) which is used to correct the dynamics based on the error predictions.

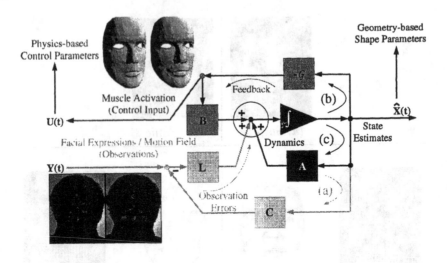

x	$=$	positions	f	$=$	forces
$\mathbf{X}$	$=$	State Vector $(x, \dot{x}, \ddot{x})$	$\mathbf{A}$	$=$	State Evolution Matrix
$\mathbf{U}$	$=$	Control Input Vector (f, x)	$\mathbf{B}$	$=$	State-Control Matrix
$\mathbf{Y}$	$=$	Measurement Vector $(x, \dot{x})$	$\mathbf{C}$	$=$	State Observation Matrix
$\mathbf{L}$	$=$	Kalman Gain Matrix	$\mathcal{G}$	$=$	Control Feedback Gain Matrix

Figure 6. Block diagram of the control-theoretic approach. Showing the estimation and correction loop (a), the dynamics loop (b), and the feedback loop (c).

The optical flow computation method has already established a probability distribution $(\Lambda_{\mathbf{v}}(t))$ with respect to the observations. We can simply use this distribution in our dynamic observations relationships. Hence we obtain:

$$\Lambda_m(t) = \mathcal{M}(x, y, z)\Lambda_{\mathbf{v}}(t), \quad \mathbf{Y}(t) = \mathcal{M}(x, y, z)\hat{\mathbf{v}}_i(t). \qquad (4)$$

Control, Measurement and Correction of Dynamic Motion
Now using a control theory approach we will obtain the muscle actuations. These actuations are derived from the observed image velocities. The control input vector $\mathbf{U}$ is therefore provided by the control feedback law: $\mathbf{U} = -\mathcal{G}\mathbf{X}$, where $\mathcal{G}$ is the *control feedback gain matrix*. We assume the instance of control under study falls into the category of an *optimal regulator* [15]. Hence, the optimal control law $\mathbf{U}^*$ is given by:

$$\mathbf{U}^* = -\mathbf{R}^{-1}\mathbf{B}^T\mathbf{P}_c\mathbf{X}^* \qquad (5)$$

where $\mathbf{X}^*$ is the optimal state trajectory and $\mathbf{P}_c$ is given by solving yet another *matrix Riccati equation* [15]. Here $\mathbf{Q}$ is a real, symmetric, positive

Expression	Magnitude of Control Point Deformation
AU 2 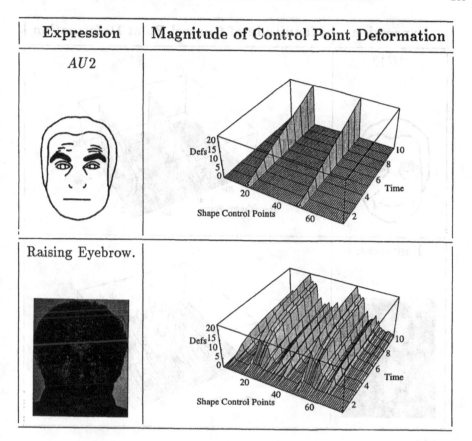	
Raising Eyebrow.	

Figure 7. FACS/CANDIDE *deformation vs. Observed deformation for the Raising Eyebrow expression. Surface plots (top) show deformation over time for FACS actions AU 2, and (bottom) for an actual video sequence of raising eyebrows.*

semi-definite *state weighting* matrix and **R** is a real, symmetric, positive definite *control weighting* matrix. Comparing with the control feedback law we obtain $\mathcal{G} = \mathbf{R}^{-1}\mathbf{B}^T\mathbf{P}_c$ This control loop is also shown in the block diagram in Figure 6 (upper loop (c)).

4. Analysis and Representations

The goal of this work is to develop a new representation of facial action that more accurately captures the characteristics of facial motion, so that we can employ them in recognition, coding and interpretation of facial motion. The current state-of-the-art for facial descriptions (either FACS itself or muscle-control versions of FACS) has two major weaknesses:

Expression	Magnitude of Control Point Deformation

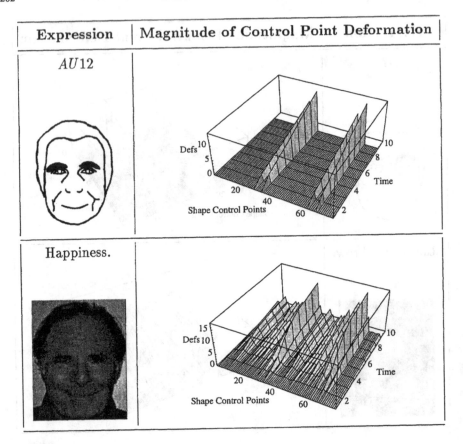

Figure 8.   FACS/CANDIDE deformation vs. Observed deformation for the Happiness expression. Surface plots (top) show deformation over time for FACS action AU 12, and (bottom) for an actual video sequence of happiness.

— The action units are purely local spatial patterns. Real facial motion is almost never completely localized; Ekman himself has described some of these action units as an "unnatural" type of facial movement. Detecting a unique set of action units for a specific facial expression is not guaranteed.

— There is no time component of the description, or only a heuristic one. From EMG studies it is known that most facial actions occur in three distinct phases: *application*, *release* and *relaxation*. In contrast, current systems typically use simple linear ramps to approximate the actuation profile. Coarticulation effects are not accounted for in any facial movement.

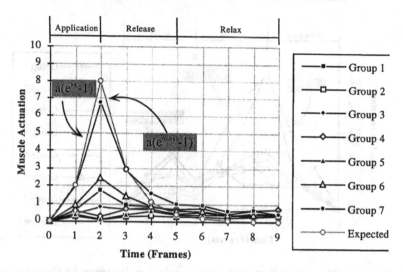

Figure 9.   *Actuations over time of the seven main muscle groups for the expressions of raising brow. The plots shows actuations over time for the seven muscle groups and the expected profile of application, release and relax phases of muscle activation.*

Other limitations of FACS include the inability to describe fine eye and lip motions, and the inability to describe the coarticulation effects found most commonly in speech. Although the muscle-based models used in computer graphics have alleviated some of these problems [35], they are still too simple to accurately describe real facial motion.

Our method lets us characterize the functional form of the actuation profile, and lets us determine a basis set of "action units" that better describes the spatial properties of real facial motion. A similar form of analysis has recently been undertaken by Terzopoulos and Waters [32], where they extract muscle actuations using snakes. However their method, because it is based on sparse features, does not extract the kind of detail that is essential for the coding, interpretation and recognition tasks we are interested in.

In the next few paragraphs, we will illustrate the resolution of our representation using the smile and eyebrow raising expressions. Questions of repeatability and accuracy will be addressed while discussing the data obtained during our expression recognition experiments.

4.1. SPATIAL PATTERNING

To illustrate that our new parameters for facial expressions are more spatially detailed than FACS, comparisons of the expressions of *raising eyebrow* and *smile* produced by standard FACS-like muscle activations and our visually extracted muscle activations are shown in Figure 7 and Figure 8.

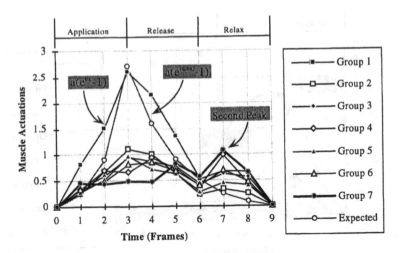

Figure 10. *Actuations over time of the seven main muscle groups for the expressions of*
smiling – lip motion. The plots shows actuations over time for the seven muscle groups
and the expected profile of application, release and relax phases of muscle activation.

The top row of Figure 7 shows $AU2$ ("Raising Eyebrow") from the
FACS model and the linear actuation profile of the corresponding geometric
control points. This is the type of spatio-temporal patterning commonly
used in computer graphics animation. The bottom row of Figure 7 shows
the observed motion of these control points for the expression of *raising*
eyebrow by Paul Ekman. This plot was achieved by mapping the motion
onto the FACS model and the actuations of the control points measured.
As can be seen, the observed pattern of deformation is very different than
that assumed in the standard implementation of FACS. There is a wide
distribution of motion through all the control points, not just around the
largest activation points.

Similar plots for happiness expression are shown in Figure 8. These
observed distributed patterns of motion provide a detailed representation
of facial motion that we will show is sufficient for accurate characterization
of facial expressions.

4.2. TEMPORAL PATTERNING

Another important observation about facial motion that is apparent in
Figure 7 and Figure 8 is that the facial motion is far from linear in time. This
observation becomes much more important when facial motion is studied
with reference to muscles, which is in fact the effector of facial motion and
the underlying parameter for differentiating facial movements using FACS.

The top rows of Figure 7 and Figure 8, that show the development of

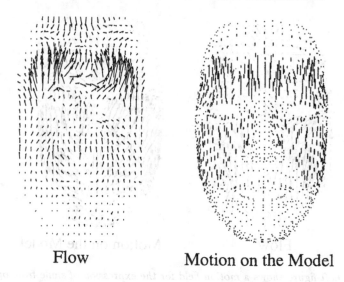

Flow   Motion on the Model

Figure 11. Left figure shows a motion field for the expression of raise eye brow from optical flow computation and the right figure shows the motion field after it has been mapped to a face model using the control-theoretic approach of Figure 6.

FACS expressions can only be represented by a muscle actuation that has a step-function profile. Figure 9 and Figure 10 show plots of facial muscle actuations for the observed smile and eyebrow raising expressions. For the purpose of illustration, in this figure the 36 face muscles were combined into seven local groups on the basis of their proximity to each other and to the regions they effected. As can be seen, even the simplest expressions require multiple muscle actuations.

Of particular interest is the temporal patterning of the muscle actuations. We have fit exponential curves to the activation and release portions of the muscle actuation profile to suggest the type of rise and decay seen in EMG studies of muscles. From this data we suggest that the relaxation phase of muscle actuation is mostly due to passive stretching of the muscles by residual stress in the skin.

Note that Figure 10 for the smile expression also shows a second, delayed actuation of muscle group 7, about 3 frames after the peak of muscle group 1. Muscle group 7 includes all the muscles around the eyes and as can be seen in Figure 9 is the primary muscle group for the raising eye brow expression. This example illustrates that coarticulation effects can be observed by our system, and that they occur even in quite simple expressions. By using these observed temporal patterns of muscle activation, rather than simple linear ramps, or heuristic approaches of the representing temporal changes, we can more accurately characterize facial expressions.

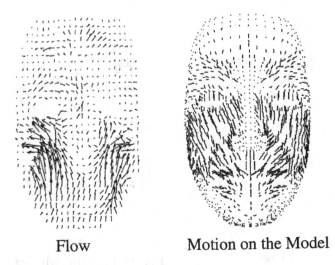

Flow   Motion on the Model

Figure 12. Left figure shows a motion field for the expression of smile from optical flow computation and the right figure shows the motion field after it has been mapped to a face model using the control-theoretic approach of Figure 6.

4.3. MOTION TEMPLATES FROM THE FACIAL MODEL

So far we have discussed how we can extract the muscle actuations of an observed expression. However our control-theoretic approach can also be used to extract the "corrected" or "noise-free" 2-D motion field that is associated with each facial expression.

The system shown in Figure 6 employs optimal estimation, within an optimal control and feedback framework. It maps 2-D motion observations from images onto a physics-based dynamic model, and then the estimates of corrected 2-D motions (based on the optimal dynamic model) are used to correct the observations model. Figure 11 and Figure 12 show the corrected flow for the expressions of raise eyebrow and smile, and also show the corrected flow after it has been applied to the face model as deformation of the skin.

By using this methodology to *back-project* the facial motion estimates into the image we can remove the complexity of physics-based modeling from our representation of facial motion, and instead use only the 2-D observations (e.g., motion energy) to describe the facial motion that is characteristic of each facial expression.

Note that this corrected motion representation is better than what could be obtained by measuring optical flow using standard optical flow techniques, because it incorporates the regularization constraints from our physical model of the face. Figure 5 (e) and (f) shows examples of this representation of facial motion energy. It is this representation of facial motion that we will use for generating spatio-temporal "templates" for recognition of facial expressions. It is important to note here that as much as the first model-based method is specific to faces, the second, motion-energy based method is quite general and not at all specific to faces. It uses a simple model for regularizing the flow and therefore a similar recognition technique can be utilized for many other motion-based recognition tasks.

5. Characterization of Facial Expressions

One of the main advantages of the methods presented here is the ability to use real imagery to define representations for different expressions. As we discussed in the last section, we do not want to rely on pre-existing models of facial expression as they are generally not well suited to our interests and needs. We would rather observe subjects making expressions and use the measured motion, either muscle actuations or 2-D motion energy, to accurately characterize each expression.

For this purpose we have developed a video database of people making expressions. Currently these subjects are video-taped while making an expression on demand. These "on demand" expressions have the limitation that the subjects' emotion generally does not relate to his/her expression. However we are for the moment more interested in characterizing facial motion and not human emotion. Categorization of human emotion on the basis of facial expression is an important topic of research in psychology and we believe that our methods can be useful in this area. We are at present collaborating with several psychologists on this problem.

At present we have a database of 20 people making expressions of *smile, surprise, anger, disgust, raise brow*, and *sad*. Some of our subjects had problems making the expression of sad, therefore we have decided to exclude that expression from our present study. We are working on expanding our database to cover many other expressions and also expressions with speech. The last frames of some of the expressions in our database are shown in Figure 13. All of these expressions are digitized as sequences at 30 frames per second and stored at the resolution of 450x380. All the results discussed in this chapter are based on expressions performed by 7 people with a total of 52 expressions. This database is substantially larger than that used by Mase [19] in his pioneering work on recognizing facial expressions. Yacoob and Davis [36] currently have the largest database (30 subjects and 105 expressions). Although our database is smaller than that of Yacoob and Davis, we believe that it is sufficiently large to demonstrate that we have achieved improved accuracy at facial expression recognition.

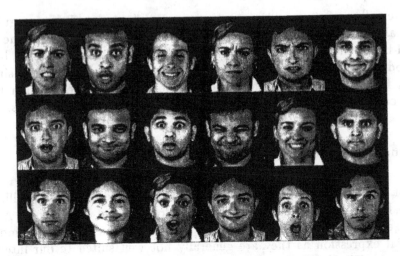

Figure 13. Expressions from video sequences for various people in our database. These expressions are captured at 30 frames per second at NTSC resolution (and cropped appropriately). We have by now developed a video database of over 30 people under different lighting conditions and backgrounds. We are also incorporating head movements into our database.

5.1. MODEL-BASED RECOGNITION

Recognition requires a unique "feature vector" to define each expression and a similarity metric to measure the differences between expressions. Since both temporal and spatial characteristics are important we require a feature vector that can account for both of these characteristics. We must also, however, account for the speed at which the expressions are performed. Since facial expressions occur in three distinct phases: *application, release* and *relaxation* (see Figure 9 and Figure 10), by dividing the data into these phases and by time warping it (using Dynamic Time Warping) for all expressions into a fixed time period of ten discrete samples, we can normalize the temporal time course of the expression. This normalization allows us to use the muscle actuation profiles to define a unique feature vector for a each facial motion.[1] Consequently, we have defined the peak actuation of each muscle between the application and release phases as the feature vector for each expression.

We defined an *standard* muscle activation feature vector for each of the expressions *smile, surprise, anger, disgust,* and *raise eyebrow* by choosing at random two subjects from our database of facial expressions. These standard peak muscle actuation features, which we call *muscle templates*

---

[1]It may be that a better way to account for temporal aspects of facial motion would be to use a phase-portrait approach [30]. Although we are investigating such methods, at present our results suggest little need for incorporating such detail.

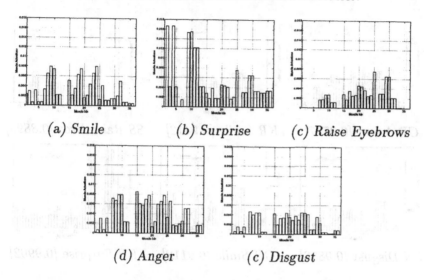

(a) Smile   (b) Surprise   (c) Raise Eyebrows

(d) Anger   (c) Disgust

Figure 14. Feature vectors of muscle templates for different expressions

are shown in Figure 14. As can be seen, the muscle templates for each expression are unique, indicating that they are good features for recognition. These feature vectors are then used for recognition of facial expression by comparison using a normalized dot product similarity metric.

Experiments: Physical Model Method By computing the muscle activation feature vectors for each of 7 subjects making about 5 expressions each (not in the training set), we can assess the recognition accuracy of this physical-model based expression recognition method.

Figure 15 shows the peak muscle actuations for 6 people making different expressions. The muscle template used for recognition is also shown for comparison. The dot products of the feature vectors with the corresponding expression's muscle template are shown below each figure.

This figure is important in that it illustrates the repeatability and accuracy with which we can estimate muscle actuations. The largest differences in peak muscle actuations are due to facial asymmetry and intensity of the actuation. The intensity difference is especially apparent in the case of the surprise expression where some people open their mouth less then others. Our analysis does not enforce any symmetry constraints and none of our data, including the muscle templates shown in Figure 14, portray exactly symmetric expressions.

Table 1 shows the results of dot products between peak muscle actuations of five randomly chosen expressions with each expression's muscle

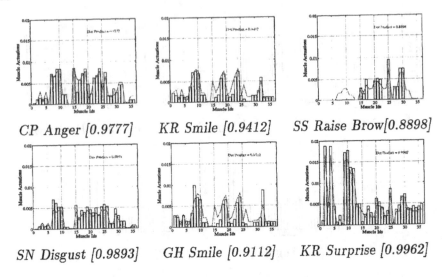

CP Anger [0.9777]　　　KR Smile [0.9412]　　　SS Raise Brow[0.8898]

SN Disgust [0.9893]　　　GH Smile [0.9112]　　　KR Surprise [0.9962]

Figure 15. Peak muscle actuations for several different expressions by different people. The dotted line shows the muscle template used for recognition. The normalized dot product of the feature vector with the muscle template is shown below each figure.

template. It can be seen that for the five instances shown in this table, each expression is correctly identified.

	Smile	Surprise	Anger	Disgust	Raise Brow
Smile	**0.91**	0.36	0.91	0.75	0.17
Surprise	0.32	**0.99**	0.34	0.28	0.20
Anger	0.62	0.28	**0.99**	0.47	0.81
Disgust	0.32	0.22	0.66	**0.88**	0.43
Raise Brow	0.75	0.27	0.84	0.24	**0.98**

TABLE 1. Some examples of recognition of facial expressions, using peak muscles actuations. A score of 1.0 indicates complete similarity.

	SM	SP	AN	DI	RB
SM	**0.97 ± 0.03**	0.63 ± 0.04	0.95 ± 0.01	0.86 ± 0.04	0.59 ± 0.16
SP	0.58 ± 0.03	**0.99 ± 0.01**	0.59 ± 0.04	0.57 ± 0.05	0.56 ± 0.09
AN	0.90 ± 0.05	0.55 ± 0.05	**0.97 ± 0.02**	0.91 ± 0.01	0.65 ± 0.14
Di	0.82 ± 0.06	0.57 ± 0.05	0.92 ± 0.03	**0.95 ± 0.03**	0.78 ± 0.10
RB	0.58 ± 0.05	0.57 ± 0.07	0.70 ± 0.05	0.78 ± 0.06	**0.96 ± 0.04**

SM: Smile, SP: Surprise, AN: Anger, DI: Disgust, RB: Raise Brows.

TABLE 2. *Mean and Standard Deviations of Similarity scores of all expressions in the database. Similarity metric is normalized dot products*

We have used this recognition method with all of our test data, consisting of 7 different people making the 52 expressions (including: *smile, surprise, raise eye brow, anger,* and *disgust*) Since some people did not make all expressions, we have 10 samples each for surprise, anger, disgust, and raise eyebrow expressions, and 12 for the smile expression. Expressions of sadness were excluded as we had only a few samples of the expression of sadness leading to inconclusive results. Table 2 shows the average and the standard deviation of all the expressions compared to each expression's muscle template. Of particular note is the excellent repeatability and consistency of the muscle activation estimates.

Table 3 shows the results of the overall recognition results in the form of a *classification matrix.* In our tests there was only one recognition failure, for the expression of anger. Our overall accuracy was 98.0%.

5.2. SPATIO-TEMPORAL MOTION ENERGY TEMPLATES FOR RECOGNITION

The previous section used estimated peak muscle actuations as a feature to detect similarity/dissimilarity of facial expressions. Now we consider a second, much simpler representation: the templates of facial motion energy.

Figure 16 shows the pattern of motion generated by averaging two randomly chosen subjects per expression from our facial expression image sequence database. Notice that each of these motion templates is unique and therefore can serve as an sufficient feature for categorization of facial expression. Note also that these motion-energy templates are sufficiently smooth that they can be subsampled at one-tenth the raw image resolution, greatly reducing the computational load.

Expressions	Smile	Surprise	Anger	Disgust	Raise Brow
Template					
Smile	**12**	0	1	0	0
Surprise	0	10	0	0	0
Anger	0	0	9	0	0
Disgust	0	0	0	10	0
Raise Brow	0	0	0	0	10
Success	100%	100%	90%	100%	100%

TABLE 3. *Results of Facial Expression Recognition using peak-muscle actuations. This result is based on 12 image sequences of smile, 10 image sequences of surprise, anger, disgust, and raise eyebrow. Success rate for each expression is shown in the bottom row. The overall recognition rate is 98.0%.*

We use the Euclidean norm of the difference of between the motion energy template and standard 2-D optical flow motion energy as a metric for measuring the similarity/dissimilarity of expressions. Note that metric works oppositely from the dot-product metric: the lower the value of this metric, more similar the expressions.

Experiments: Spatio-temporal motion energy templates Using the average of two people making an expression, we generate motion-energy template images for each of the five expressions. Using these templates (shown in Figure 16), we conducted recognition tests for our whole database of 52

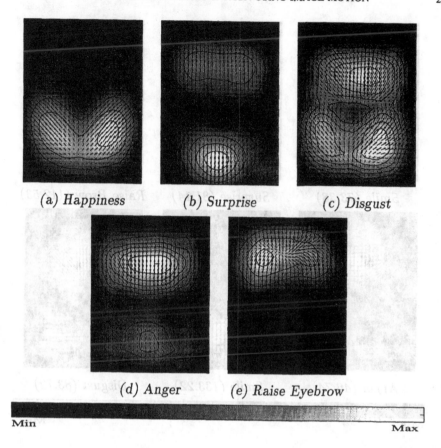

| (a) Happiness | (b) Surprise | (c) Disgust |

| (d) Anger | (e) Raise Eyebrow |

Min   Max

Figure 16. Spatio-temporal motion-energy templates for the five expressions, averaged using the data from two people.

image sequences. Figure 17 shows six examples of the motion energy images generated by different people. The similarity scores of these to the corresponding expression templates are shown below each figure.

Table 4 shows the results of this recognition test for five different expressions by different subjects. The scores show that all the five expressions were correctly identified. Conducting this analysis for the whole database of 52 expressions, we can generate a table which shows the mean and variance of the similarity scores across the whole database. These scores are shown in Table 5. The classification results of this methods over the whole database, displayed as a *confusion/classification matrix*, are shown in Table 6. This table shows that again we have just one incorrect classification of the anger expression. The overall recognition rate with this method is also 98.0%.

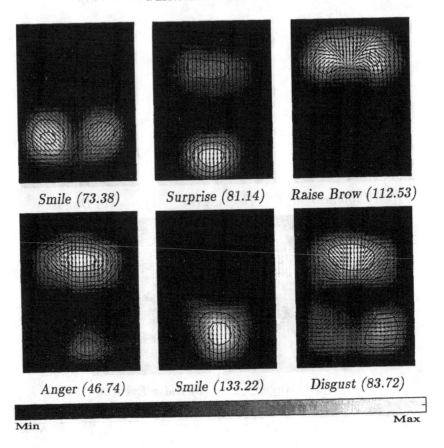

Figure 17. Motion templates for four expressions from 4 people. Their similarity scores are also shown.

6. Discussion and Conclusions

In this chapter we have presented two methods for representation of facial motion. Unlike previous efforts at facial expression characterization, coding, interpretation, or recognition that have been based on the Facial Action Coding System (FACS), we have developed new, more accurate representations of facial motion and then used these new representations for the coding, and recognition/identification task.

We accomplished this by analyzing image sequences of facial expressions and then probabilistically characterizing the facial muscle activation associated with each expression. This is achieved using a detailed physics-based dynamic model of the skin and muscles coupled with optimal estimates of optical flow in a feedback controlled framework. This analysis produces a muscle-based representations of facial motion, which is then used to recog-

Expressions	Smile	Surprise	Anger	Disgust	Raise Brow
Template					
Smile	**73.4**	255.0	230.3	209.4	294.6
Surprise	233.1	**81.1**	143.7	141.4	243.4
Anger	213.8	187.0	**46.7**	95.2	152.3
Disgust	154.0	178.3	126.4	**83.7**	227.3
Raise Brow	288.9	322.2	147.7	240.1	**46.8**

TABLE 4. *Example scores for recognition of facial expressions using spatio-temporal templates. Low Scores show more similarity to the template.*

	SM	SP	AN	DI	RB
SM	**94.1±34.7**	266.2 ± 52.3	234.5 ± 62.7	153.7 ± 59.7	306.6 ± 15.3
SP	230.9 ± 8.7	**123.6±70.7**	160.5 ± 38.3	173.5 ± 14.2	233.4 ± 14.1
AN	225.7 ± 16.5	199.2 ± 76.0	**98.3±46.3**	160.1 ± 29.1	147.0 ± 15.5
DI	149.0 ± 22.7	198.1 ± 54.0	140.3 ± 43.7	**99.3±23.4**	224.3 ± 16.2
RB	339.9 ± 32.9	321.6 ± 96.4	208.9 ± 33.0	293.2 ± 26.8	**106.8±27.0**

SM: Smile, SP: Surprise, AN: Anger, DI: Disgust, RB: Raise Brows.

TABLE 5. *Mean ± Standard Deviation of scores for recognition of facial expressions using the spatio-temporal templates over the whole database. Low Scores show more similarity to the template.*

nize facial expressions in two different ways.

The first recognition method we described uses the physics-based model directly, by recognizing expressions through comparison of estimated muscle activations. This method yields a recognition rate of 98% over our database of 52 sequences.

The second method uses the physics-based model to generate spatio-temporal motion-energy templates of the whole face for each different expression. These simple, biologically-plausible motion energy "templates" are then used for recognition. This method also yields a recognition rate of 98%. This level of accuracy at expression recognition is substantially better than has been previously achieved. However, it is important to point out that comparisons between different methods are quite subjective until an attempt is made to test the different methods on a similar and much larger data set.

We have also used this representation in real-time tracking and synthesis of facial expressions [13]. We are at the moment working on increasing the size of our database to also include other expressions and speech motions. We are also looking into model-based coding applications, biomedical applications and controlled experiments with psychologists.

Expressions	Smile	Surprise	Anger	Disgust	Raise Brow
Template					
Smile	12	0	0	0	0
Surprise	0	10	0	0	0
Anger	0	0	9	0	0
Disgust	0	0	1	10	0
Raise Brow	0	0	0	0	8
Success	100%	100%	90%	100%	100%

TABLE 6. *Results of Facial Expression Recognition using spatio-temporal motion energy templates. This result is on based on 12 image sequences of smile, 10 image sequences of surprise, anger, disgust, and raise eyebrow. Success rate for each expression is shown in the bottom row. The overall recognition rate is 98.0%.*

References

1. J. N. Bassili. Facial motion in the perception of faces and of emotional expression. *Journal of Experimental Psychology*, 4:373–379, 1978.

2. J. N. Bassili. Emotion recognition: The role of facial motion and the relative importance of upper and lower areas of the face. *Journal of Personality and Social Psychology*, 37:2049–2059, 1979.

3. Klaus-Jürgen Bathe. *Finite Element Procedures in Engineering Analysis*. Prentice-Hall, 1982.

4. M. J. Black and Y. Yacoob. Tracking and recognizing rigid and non-rigid facial motions using local parametric model of image motion. In *Proceedings of the International Conference on Computer Vision*, pages 374–381. IEEE Computer Society, Cambridge, MA, 1995.

5. R. G. Brown. *Introduction to Random Signal Analysis and Kalman Filtering*. John Wiley & Sons Inc., 1983.

6. V. Bruce. *Recognizing Faces*. Lawrence Erlbaum Associates, 1988.

7. J. S. Bruner and R. Taguiri. The perception of people. In *Handbook of Social Psychology*. Addison-Wesley, 1954.

8. Cyberware, Inc. 2110 Del Monte Avenue., Monterey, California 93940, USA, URL. http://www.cyberware.com/.

9. P. Ekman. The argument and evidence about universals in facial expressions of emotion. In H. Wagner and A. Manstead, editors, *Handbook of Social Psychophysiology*. Lawrence Erlbaum, 1989.

10. P. Ekman and W. V. Friesen. *Facial Action Coding System*. Consulting Psychologists Press Inc., 577 College Avenue, Palo Alto, California 94306, 1978.

11. P. Ekman, T. Huang, T. Sejnowski, and J. Hager (Editors). Final Report to NSF of the Planning Workshop on Facial Expression Understanding. Technical report, National Science Foundation, Human Interaction Lab., UCSF, CA 94143, 1993.

12. I. Essa. *Analysis, Interpretation, and Synthesis of Facial Expressions*. PhD thesis, Massachusetts Institute of Technology, MIT Media Laboratory, Cambridge, MA 02139, USA, 1994.

13. I. Essa, T. Darrell, and A. Pentland. Tracking facial motion. In *Proceedings of the Workshop on Motion of Nonrigid and Articulated Objects*, pages 36–42. IEEE Computer Society, 1994.

14. I. A. Essa, S. Sclaroff, and A. Pentland. Physically-based modeling for graphics and vision. In Ralph Martin, editor, *Directions in Geometric Computing*, pages 160–196. Information Geometers, U.K., 1993.

15. B. Friedland. *Control System Design: An Introduction to State-Space Methods*. McGraw-Hill, 1986.

16. C. E. Izard. Facial expressions and the regulation of emotions. *Journal of Personality and Social Psychology*, 58(3):487–498, 1990.

17. M. Kass, A. Witkin, and D. Terzopoulos. Snakes: Active contour models. *International Journal of Computer Vision*, 1(4):321–331, 1987.

18. H. Li, P. Roivainen, and R. Forchheimer. 3-d motion estimation in model-based facial image coding. *IEEE Trans. Pattern Analysis and Machine Intelligence*, 15(6):545–555, June 1993.

19. K. Mase. Recognition of facial expressions for optical flow. *IEICE Transactions, Special Issue on Computer Vision and its Applications*, E 74(10), 1991.

20. K. Mase and A. Pentland. Lipreading by optical flow. *Systems and Computers*, 22(6):67–76, 1991.

21. D. Metaxas and D. Terzopoulos. Shape and nonrigid motion estimation through physics-based synthesis. *IEEE Trans. Pattern Analysis and Machine Intelligence*, 15(6):581–591, 1993.

22. B. Moghaddam and A. Pentland. Face recognition using view-based and modular eigenspaces. In *Automatic Systems for the Identification and Inspection of Humans*,

volume 2277. SPIE, 1994.

23. C. Pelachaud, N. Badler, and M. Viaud. Final Report to NSF of the Standards for Facial Animation Workshop. Technical report, National Science Foundation, University of Pennsylvania, Philadelphia, PA 19104-6389, 1994.

24. A. Pentland, B. Moghaddam, and T. Starner. View-based and modular eigenspaces for face recognition. In *Computer Vision and Pattern Recognition Conference*, pages 84–91. IEEE Computer Society, 1994.

25. A. Pentland and S. Sclaroff. Closed form solutions for physically based shape modeling and recovery. *IEEE Trans. Pattern Analysis and Machine Intelligence*, 13(7):715–729, July 1991.

26. R. Picard. Affective computing. Technical Report 321, MIT Media Laboratory, Perceptual Computing Section, November 1995. Available as MIT Media Lab Perceptual Computing Techreport # 362 from http://www-white.media.mit.edu/vismod/.

27. S. Pieper, J. Rosen, and D. Zeltzer. Interactive graphics for plastic surgery: A task level analysis and implementation. *Computer Graphics, Special Issue: ACM Siggraph, 1992 Symposium on Interactive 3D Graphics*, pages 127–134, 1992.

28. S. M. Platt and N. I. Badler. Animating facial expression. *ACM SIGGRAPH Conference Proceedings*, 15(3):245–252, 1981.

29. M. Rosenblum, Y. Yacoob, and L. Davis. Human emotion recognition from motion using a radial basis function network architecture. In *The Workshop on Motion of Nonrigid and Articulated Objects*, pages 43–49. IEEE Computer Society, 1994.

30. E. Shavit and A. Jepson. Motion understanding using phase portraits. In *Looking at People Workshop*. IJCAI, 1993.

31. E. P. Simoncelli. *Distributed Representation and Analysis of Visual Motion*. PhD thesis, Massachusetts Institute of Technology, 1993.

32. D. Terzopoulus and K. Waters. Analysis and synthesis of facial image sequences using physical and anatomical models. *IEEE Trans. Pattern Analysis and Machine Intelligence*, 15(6):569–579, June 1993.

33. S. A. Wainwright, W. D. Biggs, J. D. Curry, and J. M. Gosline. *Mechanical Design in Organisms*. Princeton University Press, 1976.

34. J. Y. A. Wang and E. Adelson. Layered representation for motion analysis. In *Proceedings of the Computer Vision and Pattern Recognition Conference*, 1993.

35. K. Waters and D. Terzopoulos. Modeling and animating faces using scanned data. *The Journal of Visualization and Computer Animation*, 2:123–128, 1991.

36. Y. Yacoob and L. Davis. Computing spatio-temporal representations of human faces. In *Proceedings of the Computer Vision and Pattern Recognition Conference*, pages 70–75. IEEE Computer Society, 1994.

Part III

Lipreading

LEARNING VISUAL MODELS FOR LIPREADING

CHRISTOPH BREGLER
Computer Science Division
U.C. Berkeley
Berkeley, CA 94720
bregler@cs.berkeley.edu

AND

STEPHEN M. OMOHUNDRO
NEC Research Institute, Inc.
4 Independence Way
Princeton, NJ 08540
om@research.nj.nec.com

1. Introduction

This chapter describes learning techniques that are the basis of a "visual speech recognition" or "lipreading" system[1] . Model-based vision systems currently have the best performance for many visual recognition tasks. For geometrically simple domains, models can sometimes be constructed by hand using CAD-like tools. Such models are difficult and expensive to construct, however, and are inadequate for more complex domains. To do model-based lipreading, we would like a parameterized model of the complex "space of lip configurations". Rather than building such a model by hand, our approach is to have the system itself build it using machine learning. The system is given a collection of training images which it uses to automatically construct the models that are later used in recognition.

There are several phases of processing involved in our system. Ultimately, the recognition of the time sequence of images is performed using Hidden Markov Model technology similar to that used in speech recognition. Unlike speech recognition, however, there are extra phases to find,

[1]This is an extended version of [5].

M. Shah and R. Jain (eds.), Motion-Based Recognition, 301–320.
© 1997 Kluwer Academic Publishers.

track, and extract the lips from a sequence of images of a speaker. We will describe how the learned models are used to facilitate these tasks.

Some versions of our system do recognition based only on the visual input, while others use both visual and acoustic information. When visual and acoustic information is combined, it is necessary to deal with the fact that the acoustic sampling rate is higher than the visual image rate. We will describe how the learned models are used to interpolate between frames of video.

There is a common abstract learning task that underlies these different tasks. We use the expression "nonlinear manifold learning" for the task of inducing a smooth nonlinear surface in a high-dimensional space from a set of points drawn from that surface. This task is important throughout vision because it is often the case that the parameters of visual models are related by smooth nonlinear constraints. Learning such constraints and manipulating them in a computationally tractable way is therefore central to building learning-based visual recognition systems.

The first section of this chapter describes our representation for manifolds and the algorithm for learning it from data. The next two sections present experiments on learning synthetic models where performance can be directly evaluated. We then describe how the learned manifolds may be used for interpolation. We describe the "lip manifold" our system learns for visual speech recognition. We show how we use this for improving the performance of a snake-based lip tracker and for the task of interpolating between lip images. We present performance results for a single speaker using just visual information. Finally, we describe more complex experiments with multiple speakers and combined visual and acoustic information.

2. Smooth nonlinear manifold representation and induction

2.1. MOTIVATION

Human lips are geometrically complex objects whose shape varies with several distinct degrees of freedom that are controlled by the facial musculature of a speaker. For recognition, we would like to extract these degrees of freedom by using a computational representation of certain aspects of lip shape. If we represent a single configuration of the lips as a point in a feature space, then the set of all lip configurations achievable by a speaker will describe a smooth surface in the feature space. In differential geometry, such smooth surfaces are called "manifolds". For example, as a speaker smoothly opens her lips, the corresponding point in the lip feature space will move along a smooth curve. Changing the orientation of the lips would move the configuration point along a different curve in the feature space. Allowing both the lips to open and the orientation to vary would give rise to configurations that describe a two-dimensional surface in the feature space. The dimension of the "lip" surface is the same as the number of degrees of freedom of the lips including both intrinsic degrees of freedom due to the musculature and external degrees of freedom representing properties of the viewing conditions.

We would like to learn such manifolds from examples and to perform the computations on them required by recognition. We may abstract the problem as the "manifold learning problem": given a set of points drawn from a smooth manifold in a space, induce the dimension and structure of the manifold.

There are a variety of tasks which are important to perform on such learned surfaces. Perhaps the most important such task for recognition is the "nearest point" query. The system must return the point on the surface which is closest to a specified query point (Fig. 1a). This task arises in any recognition context where the entities to be recognized are smoothly parameterized (eg. recognition of objects which may be rotated, scaled, etc.) There would be one surface for each class representing its feature values as the various parameters are varied [17]. Under simple noise models, the best classification choice for recognition will be to choose the class of the surface whose closest point is nearest the query point. The choice of surface gives the class of the recognized entity and the closest point itself provides the best estimate for values of the parameters of that entity. We will see that this query also arises in several other situations in a recognition system. We would therefore like the surface representation to support it efficiently in order to perform fast indexing.

Other important classes of queries are "interpolation queries" and "prediction queries". For these, two or more points on a curve are specified and the goal is to interpolate between them or extrapolate beyond them. Knowledge of the constraint surface can dramatically improve performance over "knowledge-free" approaches like linear or spline interpolation. (Fig. 1b)

Another important set of queries are "completion queries". In these queries, the values of certain features are unknown and have to be determined by the remaining features. The explicit manifold representation restricts the range of the unknown features based on the remaining features. (Fig. 1c). This task generalizes the usual task of regression. If the variables of the feature space are split into an "input" set and an "output" set, then the graph of a mapping between these sets is a surface. Evaluation of the mapping is equivalent to specifying the "input" set and letting the surface determine the "output" set. Unlike traditional regression methods, however, if the surface is represented explicitly we can also specify the "output" variables and ask for constraints on the "input" variables. This

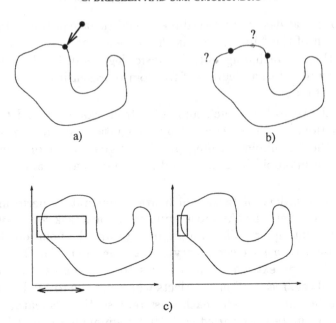

Figure 1.   Surface tasks a) Closest point query, b) interpolation and prediction, c) generalized regression

allows inversion of learned mappings. Even more importantly, we can use this type of query to combine multiple constraints allowing each to further constrain the variables of the other.

2.2. MANIFOLD REPRESENTATION BASED ON THE CLOSEST POINT QUERY

In this chapter we describe a manifold representation based on the closest point query. Our approach starts from the observation that if the data points were drawn from a *linear* manifold, then we could represent it computationally by a point on the surface together with a projection matrix that projects arbitrary vectors orthogonally to the surface such that the resulting vector is parallel to the surface. Given a set of points drawn from such a surface, principal components analysis can be used to discover the dimension of the linear space and to find the best fitting projection matrix in the least squares sense. The largest principal vectors span the space and there is a precipitous drop in the principle values at the dimension of the surface (This is similar to approaches described [12, 22, 21]). A principal components analysis will no longer suffice, however, when the manifold is nonlinear because even a 1-dimensional curve could be embedded so as to span all the dimensions of the space.

If a nonlinear manifold is smooth, however, then each local piece appears more and more linear under magnification. If we consider only those data points which lie within a local region, then to a good approximation we may model them with a linear patch. The principal values can be used to determine the most likely dimension of the patch and that number of the largest principal components span its tangent space. The idea behind our representations is to "glue" such local patches together using a partition of unity (ie. a set of smooth non-negative functions which sum to one everywhere).

The manifold is represented as a mapping from the embedding space to itself which takes each point to the nearest point on the manifold. K-means clustering is used to determine an initial set of "prototype centers" from the data points. A principal components analysis is performed on a specified number of the nearest neighbors of each prototype point. These "local PCA" results are used to estimate the dimension of the manifold and to find the best linear projection in the neighborhood of prototype i. The influence of these local models is determined by Gaussians centered on the prototype location with a variance determined by the local sample density. The projection onto the manifold is determined by forming a partition of unity from these Gaussians and using it to form a convex linear combination of the local linear projections:

$$P(x) = \frac{\sum_i G_i(x) P_i(x)}{\sum_i G_i(x)} \tag{1}$$

This initial model is then refined to minimize the mean squared error between the training samples and the nearest surface point using EM optimization [6] and gradient descent.

A related mixture model approach applied to input-output mappings was presented by [9].

2.3. SYNTHETIC EXPERIMENTS

To test this approach, we generated sample sets from artificial manifolds and applied the learning technique to them. (Section 3 describes the application of the technique along with the closest point query to tracking and interpolation of real lip images.)

Figure 2a shows 200 sample points drawn from a one-dimensional curve in a two-dimensional space. 16 prototype centers are chosen by k-means clustering. At each center, a local principal components analysis is performed on the closest 20 training samples. Figure 2b shows the prototype centers and the two local principal components as straight lines. In this case, the larger principal value is several times larger than the smaller one. The system therefore attempts to construct a one-dimensional learned man-

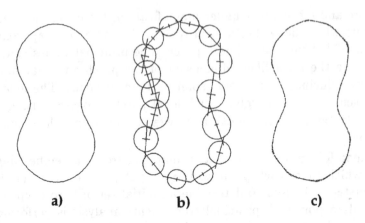

Figure 2.   Learning a 1-dimensional surface. a) The surface to learn b) The local patches and the range of their influence functions, c) The learned surface.

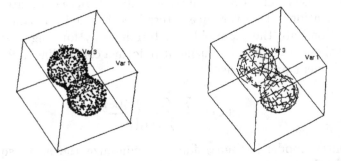

Figure 3.   Learning a two-dimensional surface in the three dimensions a) 1000 random samples on the surface b) The two largest local principle components at each of 100 prototype centers based on 25 nearest neighbors.

ifold. The circles in Figure 2b show the extent of the Gaussian influence functions for each prototype. Figure 2c shows the resulting learned surface. It was generated by randomly selecting 2000 points in the neighborhood of the surface and projecting them according to the learned model.

Figure 3 shows the same process applied to learning a two-dimensional surface embedded in three dimensions.

2.4. CLOSEST POINT QUERY COMPARED TO NEAREST NEIGHBOR

To quantify the performance of this learning algorithm, we studied it on the task of learning a two-dimensional sphere in three dimensions. It is easy to compare the learned results with the correct ones in this case. Figure 4a shows how the empirical error in the nearest point query decreases

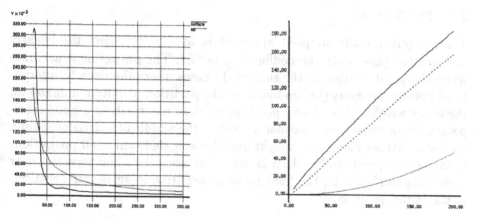

Figure 4. Quantitative performance on learning a two-dimensional sphere in three dimensions. a) Mean squared error of closest point queries as function of the number of samples for the learned surface vs. nearest training point b) The mean square root of the three principle values as a function of number of neighbors included in each local PCA.

as the number of training samples increases. It also shows the error made by a nearest-neighbor algorithm. With 50 training samples our approach produces an error which is one-fourth as large. Figure 4b shows how the average size of the local principal values depends on the number of nearest neighbors included. Because this is a two-dimensional surface, the two largest values are well-separated from the third largest. The rate of growth of the principal values is useful for determining the dimension of the surface in the presence of noise.

3. Using manifold representation for interpolation

So far we have described how to perform the closest point query on our learned manifold representation. We are also interested in interpolating between two given points using the manifold representation and the closest point query. Geometrically, linear interpolation between two points may be thought of as moving along the straight line joining the two points. This might cause the interpolated points to lie outside the space of reachable configurations. In our non-linear approach to interpolation, the point moves along a curve in the learned manifold that joins the two points. This constrains the interpolated point to only "meaningful" values. We have studied several algorithms for estimating the shortest manifold trajectory connecting two given points. For our performance results, we are interested in the point which is halfway along the shortest trajectory. We have studied three algorithms for finding a point on the surface which approximates this point.

3.1. "FREE-FALL"

The computationally simplest approach is to simply project the linearly interpolated point onto the nonlinear manifold. The projection is accurate when the point is close to the surface. In cases where the linearly interpolated point is far away (i.e. no weight of the partition of unity dominates all the other weights) the closest-point-query does not result in a good interpolant. For a worst case, consider a point in the middle of a circle or sphere. All local patches have same weight and the weighted sum of all projections is the center point itself, which is not a surface point. Furthermore, near such "singular" points, the final result is sensitive to small perturbations in the initial position.

3.2. "SURFACE-WALK"

A better approach is to "walk" along the surface itself rather than relying on the linear interpolant. Each step of the walk is initially taken to be linear and in the direction of the target point. The result of a step is immediately projected onto the manifold, however. The next step is then taken from this new point. When the target is finally reached, the arc length of the curve is approximated by the accumulated lengths of the individual steps. The point half way along the curve as measured by this arc length is then chosen. This algorithm is far more robust than the first one, because it only uses very local projections, even when the two input points are far away from each other. Figure 5b illustrates this algorithm.

3.3. "SURFACE-SNAKE"

In some ways this approach is a combination of the first two algorithms. It begins with linear interpolated points and iteratively moves the points toward the surface. The *Surface-Snake* is a sequence of n points preferentially distributed along a smooth curve with equal distances between them. An energy function is defined on such sequences of points so that the energy minimum tries to satisfy the three constraints of smoothness, equidistance, and nearness to the surface:

$$E = \sum_i \alpha ||v_{i-1} - 2v_i + v_{i+1}||^2 + \beta ||v_i - proj(v_i)||^2 \qquad (2)$$

E has value 0 if all v_i are equally distributed on a straight line and also lie on the surface. In general E can never be 0, if the surface is nonlinear, but a minimum for E represents an optimizing solution.

We begin with a straight line between the two input points and perform gradient descent in E to find this optimizing solution.

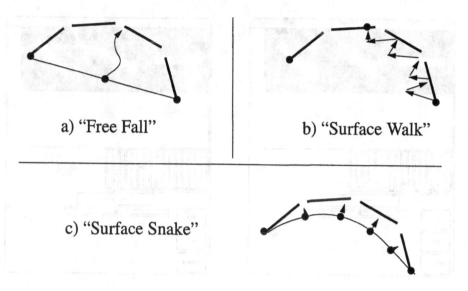

a) "Free Fall"   b) "Surface Walk"

c) "Surface Snake"

Figure 5.   Proposed interpolation algorithms.

For another approach to nonlinear interpolation using a different architecture see [2].

3.4. SYNTHETIC EXPERIMENTS

To quantify the performance of these approaches to interpolation, we generated a database of $16 * 16$ pixel images consisting of rotated white bars on
black background. The bars were rotated for each image by a specific angle. This represents a one-dimensional nonlinear surface embedded in a 256
dimensional image space. A nonlinear surface represented by 16 local linear
patches was induced from the 256 images. Figure 6a shows two bars and
their linear interpolation. Figure 6b shows the nonlinear interpolation using
the *Surface-Walk* algorithm. The slider bars below the image represent the
current weights for the linear patches which are necessary to produce the
interpolated image.

Figure 7 shows the average pixel mean squared error of linear and nonlinear interpolated bars. The x-axis represents the relative angle between
the two input points.

Figure 8 shows some iteration of a *Surface-Snake* interpolating 7 points
along a 1 dimensional surfaces embedded in a 2 dimensional space.

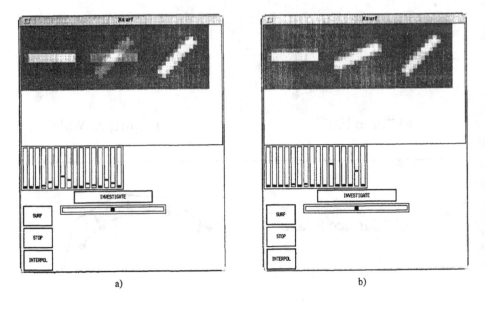

a)   b)

Figure 6.   a) Linear interpolation, b) nonlinear interpolation.

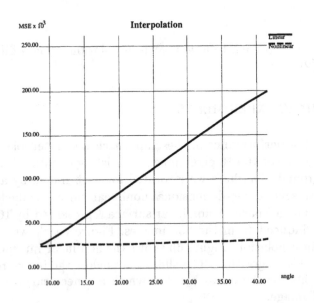

Figure 7.   Average pixel mean squared error of linear and nonlinear interpolated bars.

4. Application to visual speech recognition

We are using the manifold techniques described above in a system for visual speech recognition. We view certain feature vectors of human lips as

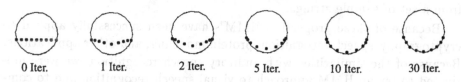

| 0 Iter. | 1 Iter. | 2 Iter. | 5 Iter. | 10 Iter. | 30 Iter. |

Figure 8. Surface-Snake iterations on an induced 1 dimensional surface embedded in 2 dimensions.

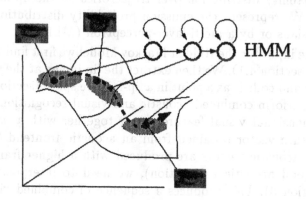

Figure 9. Lip trajectory approximated as HMM embedded in manifold representation

points which are constrained to lie on a low-dimensional nonlinear manifold embedded in the lip feature space. This manifold represents all possible lip configurations. While uttering a word or a sentence, the "lip feature point" moves along a trajectory on this manifold. (Fig. 9).

We model these trajectories using Hidden Markov Models (HMMs). An HMM represents a probability distribution over the space of strings over a specified set of output symbols. It consists of a set of hidden states and for each hidden state an emission distribution describing the probability of emitting each output symbol and a transition distribution describing the probability of making a transition to each hidden state. The hidden states and their transition distributions form an ordinary Markov model. An HMM generates a sequence of hidden states by starting in a start state and probabilistically transitioning to other states until an output state is reached. The corresponding string of symbols is generated by choosing a symbol according to the emission distribution of each hidden state in the sequence. The "forward-backward" algorithm is a dynamic programming algorithm for determining the probability of being in each hidden state at each location in an input string. It also produces the total probability that a given string was generated from a given HMM. The "Baum-Welch" algorithm is a technique based on EM for inducing the HMM parameters from a set of sample strings.

Because of these properties, HMM's have been successfully applied to cryptography, speech recognition, protein modeling, and other applications. Because of the similarities with auditory speech recognition, we were motivated to try an HMM approach to visual speech recognition and to combined visual-acoustic speech recognition. The domain of the HMM emission vectors is defined by the lip-manifold. A specific HMM word model represents a probability distribution over trajectories on the lip-manifold for a given word. We represent the emission probability distributions by a mixture of gaussians or by a multi-layer-perceptron (MLP).

To get the input for the Hidden Markov Model we first find and track the lip position (section 4.1). We then extract the lip image at the selected location and size and code it as a point in a lip-feature space (section 4.2). When we want to perform combined acoustic and visual recognition, we fuse the visual n-dimensional visual feature vector together with a m-dimensional acoustic feature vector obtained from an acoustic frontend (section 4.4). Because the acoustic vectors are produced with a higher frame rate (necessary for good acoustic recognition), we need to interpolate the visual vectors (section 3). This produces a sequence of combined visual-acoustic $n + m$-dimensional vectors as input for the HMM.

The parameters of the HMM are set by the Baum-Welch procedure from a set of example utterances. We train a separate HMM for each word that is to be recognized. Once learned, the HMM's may be presented with a sequence of pure visual feature vectors or a sequence of bimodal visual-acoustic vectors. Each HMM estimates the likelihood that it generated the sequence and the most likely HMM is selected as the recognized utterance. In the pure visual domain we are interested in the recognition performance on the word level (section 4.3). In the visual-acoustic domain we are interested in the improvement that visual information can make to purely acoustic recognition in continuous speech recognition (section 4.5).

4.1. CONSTRAINT BOUNDARY TRACKING

To track the position of the lips we are using an "active vision" technique related to "snakes" [10] and "deformable templates" [25]. In each image, a contour shape is matched to the boundary of the lips. The space of contours that represent lips is represented by a learned lip-contour-manifold. During tracking we try to find the contour (manifold-point) which maximizes the graylevel gradients along the contour in the image.

The boundary shape is parameterized by the x and y coordinates of 40 evenly spaced points along the contour. The left corner of the lip boundary is anchored at $(0,0)$ and all values are normalized to give a lip width

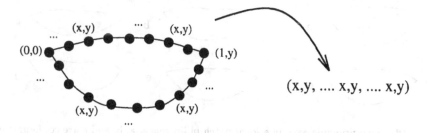

Figure 10. Lip contour coding

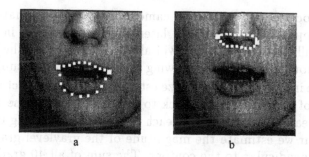

Figure 11. Snakes for finding the lip contours a) A correctly placed snake b) A snake which has gotten stuck in a local minimum of the simple energy function.

of 1 (Fig 10). Each lip contour is therefore a point in an 80-dimensional "contour-space" (because of the anchoring and scaling it is really a 77-dimensional space).

The training set consists of 4500 images of 6 speakers uttering random words. The training images are initially "labeled" with a conventional *snake* algorithm. The standard *snake* approach chooses a curve by trying to maximizing its smoothness while also adapting to certain image features along its length. These criteria are encoded in an energy function and the snake is selected by gradient descent. Unfortunately, this approach sometimes causes the selection of the boundary of incorrect neighboring objects (Fig. 11). To get a clean training sample, we cull the incorrectly aligned *snakes* from the database by hand.

We then apply the manifold learning technique described above to the database of correctly aligned lip snakes. The algorithm learns a 5-dimensional manifold embedded in the 80-dimensional contour space. 5 dimensions were sufficient to describe the contours with single pixel accuracy in the image. Figure 12 shows some of the lip models along two of the principal axes in the local neighborhood of one of the patches.

The tracking algorithm starts with a crude initial estimate of the lip position and size. In our training database all subjects positioned themselves

Figure 12. Two principle axes in a local patch in lip space. a, b, and c are configurations along the first principle axis, while d, e, and f are along the third axis.

at a similar location in front of the camera. The initial estimate is not crucial to our approach as we explain later. Currently work is in progress to integrate a full face finder, which will allow us to estimate the lip location and size without even roughly knowing the position of the subject.

Given the initial location and size estimate, we backproject an initial lip contour out of the lip-manifold back to the image (we choose the mean of one of the linear local patches). At each of the 40 points along the backprojected contour we estimate the magnitude of the graylevel gradient in the direction perpendicular to the contour. The sum of all 40 gradients would be maximal if the contour were perfectly aligned with the lip boundary. We iteratively maximize this term by performing a gradient ascent search over the 40 local coordinates. After each step, we anchor and normalize the new coordinates to the 80-dimensional shape space and project it back into the lip-manifold. This constraints the gradient ascent search to only to consider legal lip-shapes. Thus the search moves the lip-manifold point in the direction which maximally increases the sum of directed graylevel gradients. The initial guess only has to be roughly right because in the first few iterations we use large enough image filters that the shape is attracted even far from the correct boundary.

The lip contour searches in successive images in the video sequence are initialized with the contour found from the previous image. Additionally we add a temporal term to the gradient ascent energy function which forces the temporal second derivatives of the contour coordinates to be small. Figure 13 shows an example gradient ascent for a starting image and the contours found in successive images.

4.2. LIP IMAGE CODING AND INTERPOLATION

In initial experiments we directly used the contour coding as the input to the Hidden Markov Models, but found that the outer boundary of the lips is not distinctive enough to give reasonable recognition performance. The inner lip-contour and the appearance of teeth and tongue are impor-

Figure 13. A typical relaxation and tracking sequence of our lip tracker

tant for recognition. These features are not very robust for tracking the lips, however, because they disappear frequently when the lips close. For this reason the recognition features we use consist of the components of a graylevel matrix positioned and sized at the location found by the contour based lip-tracker. Empirically we found that a matrix of 24x16 pixel is enough to distinguish all possible lip configurations. Each pixel of the 24x16 matrix is assigned the average graylevel of a corresponding small window in the image. The size of the window is determined by the size of the extracted contour. Because a 24x16 graylevel matrix corresponds to a 384-dimensional vector, we also reduce the dimension of the recognition feature space by projecting the vectors to a linear subspace determined by a principal components analysis.

To interpolate missing lip-images, we induce a nonlinear manifold embedded in this lower dimensional subspace. The interpolation is done in the lower dimensional linear space and is also constrained by the learned manifold. Figure 14 shows an example interpolation of lip images in a 32-dimensional linear subspace. Figure 14a shows the linear interpolation, and Figure 14b shows the nonlinear interpolation constrained by an 8-dimensional manifold, using the manifold-snake interpolation technique.

4.3. ONE SPEAKER, PURE VISUAL RECOGNITION

The simplest of our experiments is based on a small speaker dependent task, the "bartender" problem. The speaker may choose between 4 different cocktails names[2], but the bartender cannot hear due to background noise. The cocktail must be chosen purely by lipreading. A subject uttered each of the 4 words 23 times. An HMM was trained for each of the 4 words using a mixture of Gaussians to represent the emission probabilities. With a test set of 22 utterances, the system got only a single word wrong (4.5% error).

This task is artificially simple, because the vocabulary is very small, the system is speaker dependent, and it does not deal with continuous or spontaneous speech. These are all current state-of-the-art problems in the

---

[2]We use the words: "anchorsteam", "bacardi", "coffee", and "tequilla". Each word takes about 1 second to utter on average.

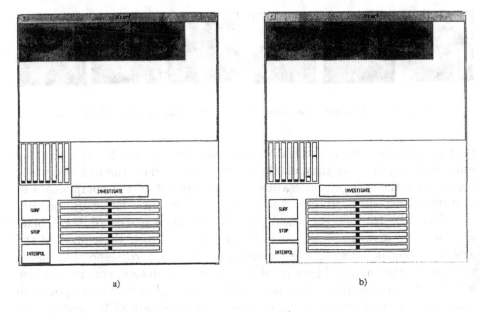

Figure 14. 24x24 images projected into a 32 dimensional subspace: a) linear interpolation b) nonlinear interpolation.

speech recognition community. For pure lip reading, however, the performance of this system is sufficiently high to warrant reporting. Subsequently we deal with the more state-of-the-art tasks using a system based on combined acoustic and visual modalities.

4.4. ADDITIONAL ACOUSTIC PROCESSING AND SENSOR FUSION

For the additional acoustic preprocessing we use an off-the-shelf acoustic front-end system, called RASTA-PLP [8] which extracts feature vectors from the digitized acoustic data with a constant rate of 100 frames per second.

Psychological studies have shown that human subjects combine acoustic and visual information at a rather high level. This supports a preceptual model that posits conditional independence between the two speech modalities [15]. We believe, however, that such conditional independence cannot be applied to a speech recognition system that combines modalities on the phoneme/viseme level. Visual and acoustic speech vectors are conditionally independent given the vocal tract position, but not given the phoneme class. Our experiments have shown that combining modalities at the input level of the speech recognizer produces much higher performance than combining them on higher levels.

4.5. MULTI-SPEAKER VISUAL-ACOUSTIC RECOGNITION

In this experiment we would like the visual lipreading system to improve the performance of acoustic speech recognition. We focus on scenarios where the acoustic signal is distorted by background noise or crosstalk from another speaker. Current state-of-the-art speech recognition systems perform poorly in such environments. Given the additional visual lip-information, we would like to determine how much error reduction can be achieved using the visual lip-manifold techniques.

We collected a database of 6 speakers spelling names or saying random sequences of letters. Letters can be thought of as small words, which makes this task a connected word recognition problem. Each utterance was a sequence of 3-8 letter names. The spelling task is notoriously difficult, because the words (letter names) are very short and highly ambiguous. For example the letters "n" and "m" sound very similar, especially in acoustically distorted signals. Visually they are more distinguishable (it is often the case that visual and acoustic ambiguities are complementary presumably for good evolutionary reasons). In contrast "b" and "p" are visually similar but acoustically different (voiced plosive vs. unvoiced plosive). With acoustic crosstalk from another speaker, the recognition and segmentation (i.e. when does one letter end and another begin) have additional difficulties. Information about which speaker's lips the acoustic signal is correlated to should make the recognizer more robust against crosstalk signals from other speakers.

Our training set consists of 2955 connected letters (uttered by the 6 speakers). We used an additional cross-validation set of 364 letters to avoid overfitting. In this set of experiments the HMM emission probabilities where estimated by a multi-layer-perceptron (MLP) [3]. The same MLP/HMM architecture has achieved state-of-the-art recognition performance on standard acoustic databases like the ARPA resource management task.

We have trained 3 different versions of the system: one based purely on acoustic signals using 9-dimensional RASTA-PLP features, and two that combine visual and acoustic features. The first bimodal system (AV) is based on the acoustic features and 10 additional coordinates obtained from the visual lip-feature space as described in section 4.2. The second bimodal system (Delta-AV) uses the same features as the AV-system plus an additional 10 visual "Delta-features" which estimate temporal differences in the visual features. The intuition behind these features is that the primary information in lip reading lies in the temporal change.

We generated several test sets covering the 346 letters: one set with clean speech, two with 10db and 20db SNR additive noise (recorded inside a moving car), and one set with 15db SNR crosstalk from another speaker

Task	Acoustic	AV	Delta-AV	rel. err.red.
clean	11.0 %	10.1 %	11.3 %	-
20db SNR	33.5 %	28.9 %	26.0 %	22.4 %
10db SNR	56.1 %	51.7 %	48.0 %	14.4 %
15db SNR crosstalk	67.3 %	51.7 %	46.0 %	31.6 %

TABLE 1. Results in word error (wrong words plus insertion and deletion errors caused by wrong segmentation)

uttering letters as well.

Table 1 summarizes our simulation results. For clean speech we did not observe a significant improvement in recognition performance. For noise-degraded speech the improvement was significant at the 0.05 level. This was also true of the crosstalk experiment which showed the largest improvement.

4.6. RELATED COMPUTER LIPREADING APPROACHES

One of the earliest successful attempts to improve speech recognition by combining acoustic recognition and lipreading was done by Petajan in 1984 [19]. More recent experiments include [14, 24, 4, 23, 7, 20, 16, 18, 13, 1, 11]. All approaches attempt to show that computer lip reading is able to improve speech recognition, especially in noisy environments. The systems were applied to phoneme classification, isolated words, or to small continuous word recognition problems. Reported recognition improvements are difficult to interpret and compare, because they are highly dependent on the complexity of the selected task (speaker dependent/independent, vocabulary, phoneme/word/sentence recognition), how advanced the underlying acoustic system is, and how simplified the visual task was made (eg. use of reflective lipmarkers, special lipstick, or special lighting conditions). We believe that our system based on learned manifold techniques and Hidden Markov Models is the most complete system applied to a complex speech recognition task to date but it is clear that many further improvements are possible.

5. Conclusion and Discussion

This chapter can only begin to describe the many applications of manifold learning in vision. We have also not described certain hierarchical geometric data structures that can dramatically improve the computational performance of these techniques. We have shown how we are using them in the domain of lip reading and that they give significantly improved recognition performance. It would be difficult to build traditional computer vision models of human lips and so the fact that our system builds these models by learning is significant. Many lip reading research groups mark a subject's lips with special reflective tape, paint, or lipstick or wire the subject with strain gauges. The techniques described in this chapter show that such artifices are unnecessary and that video images may be directly used for visual speech recognition.

References

1. A. Adjoudani and C. Benoit. On the integration of auditory and visual parameters in an hmm-based asr. In *NATO Advanced Study Institute on Speechreading by Man and Machine*, 1995.
2. D. Beymer, A. Shashua, and T. Poggio. Example based image analysis and synthesis. *M.I.T. A.I. Memo No. 1431*, Nov 1993.
3. H.A. Bourlard and Morgan N. *Connectionist Speech Recognition, A Hybrid Approach*. Kluwer Academic Publishers, 1993.
4. C. Bregler, H. Hild, S. Manke, and A. Waibel. Improving connected letter recognition by lipreading. In *Int. Conf. Acoustics, Speech, and Signal Processing*, volume 1, pages 557–560, Minneapolis, 1993. IEEE.
5. C. Bregler and S.M. Omohundro. Nonlinear manifold learning for visual speech recognition. In W. Eric L. Grimson, editor, *Proceedings of the Fifth International Conference on Computer Vision*, pages 494–499, 10662 Los Vaqueros Circle, P.O. Box 3014, Los Alamitos, CA 90720-1264, June 1995. IEEE Computer Society Press.
6. A.P. Dempster, N.M. Laird, and D.B. Rubin. Maximum likelihood from incomplete data via the EM algorithm. *Journal of the Royal Statistical Society B*, 39, 1977.
7. Alan J. Goldschen. *Continuous Automatic Speech Recognition by Lipreading*. PhD thesis, Dept. of Electrical Engineering and Computer Science, George Washington University, 1993.
8. H. Hermansky, N. Morgan, A. Bayya, and P. Kohn. Rasta-plp speech analysis technique. In *Proc. ICASSP*, San Francisco, 1992.
9. M.I. Jordan and R. A. Jacobs. Hierarchical mixtures of experts and the em algorithm. *Neural Computation*, 6(2), March 1994.
10. Michael Kass, Andrew Witkin, and Demetri Terzopoulus. Snakes: Active contour models. *International Journal of Computer Vision*, 1(4):321–331, 1987.
11. R. Kaucic, B. Dalton, and A. Blake. Real-time lip tracking for audio-visual speech recognition applications. In *4th European Conf. Computer Vision*, April 1996.
12. M. Kirby, F. Weisser, and A. Dangelmayr. A model problem in represention of digitial image sequences. *Pattern Recognition*, 26(1), 1993.
13. J. Luettin, N. A. Thacker, and S. W. Beet. Visual speech recognition using active shape models and hidden markov models. In *to appear in IEEE Int. Conf. on Acoustics, Speech, and Signal Processing*, 1996.
14. Kenji Mase and Alex Pentland. Lip reading: Automatic visual recognition of spoken words. *Opt. Soc. Am. Topical Meeting on Machine Vision*, pages 1565–1570, June 1989.
15. Dominic W. Massaro and Michael M. Cohen. Evaluation and integration of visual and auditory information in speech perception. *Journal of Experimental Psychology: Human Perception and Performance*, 9:753–771, 1983.
16. Javier R. Movellan. Visual speech recognition with stochastic networks. In G. Tesauro, D. Touretzky, and T. Leen, editors, *Advances in Neural Information Processing Systems*, volume 7. MIT press, Cambridge, 1995.

17. H. Murase and S.K. Nayar. Visual learning and recognition of 3-d objects from appearance. *Int. J. Computer Vision*, 14(1):5–24, January 1995.

18. L. Nan, S. Dettmer, and M. Shah. Visual lipreading using eigensequences. In *Proc. of the Int. Workshop on Automatic Face- and Gesture-Recognition, Zurich, 1995*, 1995.

19. Eric D. Petajan. *Automatic Lipreading to Enhance Speech Recognition*. PhD thesis, University of Illinois at Urbana-Champaign, 1984.

20. Peter L. Silsbee. Sensory integration in audiovisual automatic speech recognition. In *28th Annual Asilomar Conf. on Signals, Systems, and Computers*, pages 561–565, November 1994.

21. P. Simard, Y. LeCun, and J. Denker. Efficient pattern recognition using a new transformation distance. In *Advances in Neural Information Processing Systems*, 1993.

22. M. Turk and A. Pentland. Eigenfaces for recognition. *Journal of Cognitive Neuroscience*, 3(1):71–86, 1991.

23. Greg J. Wolff, K. Venkatesh Prasad, David G. Stork, and Marcus E. Hennecke. Lipreading by neural networks: Visual preprocessing, learning and sensory integration. In Jack D. Cowan, Gerald Tesauro, and Joshua Alspector, editors, *Advances in Neural Information Processing Systems*, volume 6, pages 1027–1034. Morgan Kaufmann, 1994.

24. Ben P. Yuhas, Moise H. Goldstein, Terence J. Sejnowski, and Robert E. Jenkins. Neural network models of sensory integration for improved vowel recognition. *Proc. IEEE*, 78(10):1658–1668, October 1990.

25. Alan L. Yuille, David S. Cohen, and Peter W. Hallinan. Facial feature extraction by deformable templates. Technical Report 88-2, Harvard Robotics Laboratory, 1988.

CONTINUOUS AUTOMATIC SPEECH RECOGNITION BY LIPREADING

ALAN J. GOLDSCHEN
The Mitre Corporation[†]
McLean, VA 22101

OSCAR N. GARCIA
Wright State University
Dayton, OH 45435

AND

ERIC D. PETAJAN
Bell Laboratories - Lucent Technologies
Murray Hill, NJ 07974

1. Introduction

An automatic speechreading recognizer uses information about motions produced by the oral-cavity region[1] of a speaker uttering a sentence. The ability to automatically 'lipread' a speaker using a sequence of image frames is an example of motion-based recognition.

We assert that such a machine is capable of performing automatic speech recognition through the use of several sources of information. This process is analogous to those sources of information that humans use (Erman et al. [10], Cohen and Massaro [8]). Current speech recognizers use only acoustic information from the speaker, and in noisy environments often use secondary sources of information such as a grammar. One important secondary source of information is observable articulatory information. This information originates primarily from the speaker's oral-cavity region and secondarily from the speaker's facial region (e.g, gestures, expressions, head-position, eyebrows, eyes, ears, mouth, teeth, tongue, cheeks, jaw, neck, and hair (Pelachaud et al. [37])) and may be acquired automatically through

[†]This work has no reference to Mitre past or present.
[1]The *oral-cavity region* corresponds to the mouth region which includes the lips, teeth, tongue, and outer skin area.

M. Shah and R. Jain (eds.), Motion-Based Recognition, 321–343.

machine vision techniques. The oral-cavity articulatory information complements the acoustic information, and is not corrupted by the processes causing acoustical noise (Silsbee [45]). Automatic machine recognition improves when acoustical information combines with observed articulatory information.

The application of motion-based recognition techniques for human-to-machine speech communication focuses on the recognition of a spoken sentence using information embedded in the sequence of image frames. Our speaker-dependent, visual (optical or video-based) automatic speech recognizer is an example of motion-based recognition. This recognizer uses visual information from the oral-cavity shadow of a speaker uttering a spoken sentence (Goldschen [14], Goldschen et al. [15]). The system performs recognition of a spoken sentence without using any syntactic, semantic, acoustic, or contextual guides.

We introduce 13 (mostly dynamic) oral-cavity features for visual recognition, present phones that appear visually similar (visemes[2]) for our speaker, and present the recognition results for our Hidden Markov Models (HMMs) using visemes, trisemes[3], and generalized trisemes[4]. The system achieves a 25 percent sentence recognition rate for sentences having a perplexity of 150 without using any syntactic, acoustic, or contextual guides. We begin by reviewing human perception as well as speech perception, discussing phone-to-viseme mapping results, and describing other systems that use visual information for automatic speech recognition.

1.1. HUMAN AND SPEECH PERCEPTION

In human speech perception experiments, visual information is complementary to acoustic information because many phones that are acoustically close are visually distant (Summerfield [48], Cohen and Massaro [8]). Visually similar phones such as /p/, /b/, /m/ form a *viseme*. Cohen and Massaro [8] report that visual information from the speaker's face influences human perception and understanding. They conclude that the speaker's face is particularly helpful when the acoustic information is ambiguous or is degraded, especially by noise or hearing impairment. For example, the phone /p/ appears visually similar to the phones /b/ and /m/; at a signal-to-noise ratio of zero /p/ is acoustically similar to the phones /t/, /k/, /f/, /th/, and /s/ (Summerfield [48]). Using both sources of information, humans (or machines) can determine the phone /p/. The *McGurk effect*

---

[2] A viseme is specific oral-cavity movements that corresponds to a phone as defined by Fisher [12]

[3] A triseme is a triplet of visemes.

[4] A generalized triseme is clustered (similar) trisemes.

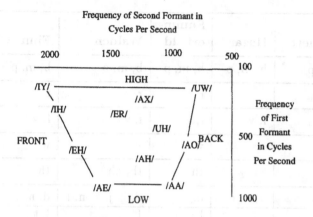

Figure 1. Vowel Triangle reproduced (from Peterson and Barney 1952). The Vowel Triangle illustrates that vowels tend to follow the place of articulation within the oral-cavity.

occurs when the fusion of acoustical and visual sources causes humans to perceive a phone different from either the auditorally or visually sensed phone (McGurk [32]). The perception of speech in noise improves greatly when presented with both acoustical and visual information because the information sources are complementary. Silsbee [45] demonstrates that this complementary concept also holds for *bimodal speech recognizers*, which are speech recognizers that use both acoustical and visual information.

1.2. PHONE-TO-VISEME MAPPINGS

Phones are typically classified into two groups: the vowel and diphthong group (vowels) and the non-vowel and non-diphthong group (consonants). Table 1 depicts a consonant phone-to-viseme mapping, primarily from the published works of human speechreading teachers (Burchett [5], Hazard [17], speech perception (Summerfield [48]), and hard-of-hearing lipreaders (Walden [51]). Although other researchers have built visual automatic speech recognizers, Finn [11] and Goldschen [14] use an algorithmic approach to determine consonant phone-to-viseme mapping.

Figure 1 illustrates the vowel triangle reproduced from Peterson and Barney [42]. The vowel triangle indicates that vowel phone-to-viseme mapping is nearly one-to-one. That is, each vowel viseme contains one phone. The placement of vowels along the first and second formant frequencies tends to follow tongue movements: high, front, low, and back. The first formant decreases with tongue height and the second formant decreases as the tongue shifts backward.

Successful visual automatic speech recognition of a speaker uttering a sentence requires accurate viseme models, especially when using image

Burchett	Hazard	Summ-erfield	Walden	Finn	Golds-chen
b, m, p	b, m, p	b, m, p	b, m, p	b, m, p	b, m, p
r		r	r	r	r
f, v	f, v	f, v	f, v	f, v	f, v, w
w	w	w	w	w	
th		th	th, dh	th	th
d, n, t	d, n, t	d, t	d, g, j, k, n, t	d, n,	d, dh, epi
l		l	l	l, t	g, k, l
k, g	k, g	n, k, g		k, g	n, t
s, z	s, z	s, z	s, z	s, z	s, sh, z
ch, j, sh, zh	ch, j, sh	sh, zh	sh, zh	ch, jh, sh, zh	zh
		y		y	y
h				hh	hh hv ch dx, nx, q en jh ng h#

TABLE 1. Mapping of consonant phones-to-visemes.

frames of the oral-cavity (Goldschen [14]). Similarly, Lee [24] states that accurate acoustic phone models are necessary for continuous speech using acoustic information. Our research started with the development of an algorithm to automatically determine visemes from a speaker (Goldschen [14]). Garcia et al. [13] provides preliminary conclusions of this research effort and Goldschen et al. [16] summarizes the phone-to-viseme mapping process.

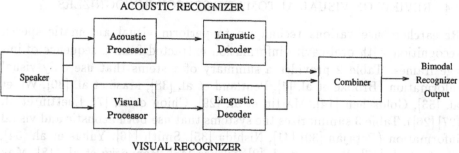

ACOUSTIC RECOGNIZER

VISUAL RECOGNIZER

(a) BIMODAL RECOGNIZER THAT COMBINES OUTPUT FROM
THE ACOUSTIC AND VISUAL RECOGNIZERS

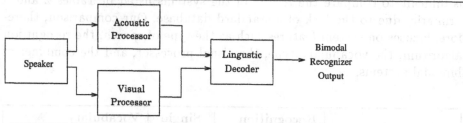

(b) BIMODAL RECOGNIZER FROM A SINGLE LINGUISTIC DECODER

Figure 2. Two general architectures for bimodal speech communication using acoustic
and visual information. The acoustic recognizer consists of an acoustic processor and
linguistic decoder, while a visual recognizer consists of a visual processor and linguistic
decoder. Part (a) illustrates a bimodal recognizer with an acoustic recognizer and a
visual recognizer. The combiner merges these two separately recognized outputs. Part
(b) illustrates a bimodal recognizer with a single linguistic decoder.

1.3. GENERAL BIMODAL RECOGNITION ARCHITECTURES

Figure 2 depicts two architectures for bimodal speech recognition and was
extended from Jelinek [19] [20]. The acoustic processor converts the acoustic
signal into a sequence of acoustic feature vectors, where each feature vector
corresponds to one frame (or sample) of speech. Analogously, the visual
processor converts the visual signal into a sequence of visual feature vectors,
where each visual feature vector corresponds to the oral-cavity of an image
frame. The linguistic decoder converts a sequence of feature vectors into
sequence of words using a knowledge of the spoken language. Jelinek [19]
defines an *acoustic recognizer* as the combination of an acoustic processor
and a linguistic decoder. A *visual recognizer* is the combination of a visual
processor and linguistic decoder (Goldschen [14]). A *bimodal recognizer*, as
depicted in Figure 2, performs recognition using both visual and acoustical
information.

1.4. REVIEW OF VISUAL AUTOMATIC SPEECH RECOGNIZERS

Researchers use various techniques to perform visual automatic speech recognition with oral-cavity information extracted from a sequence of image frames. Table 2 provides a summary of systems that use only visual information (Brooke et al. [4], Pentland et al. [38], Mase et al. [30], Wu et al. [53], Goldschen [14], Martin et al. [29], Chiou et al. [7], Luettin et al. [27] [26]). Table 3 summarizes the systems that use both acoustic and visual information (Petajan [39] [41], Nishida [35], Smith [46], Yuhas et al. [54], Stork et al. [47], Bregler et al. [2], Silsbee [45], Hennecke et al. [18], Mak et al. [28], Matthews et al. [31], Meier et al. [33], Tomlinson et al. [50]). It is difficult to compare the results of the systems listed in Tables 2 and 3 primarily due to the lack of a standard database. Our comparison, therefore, focuses on system features such as the type of speech, the recognition algorithm, the vocabulary type, the visual processor, and the combiner of bimodal systems.

Author	Recognition Method	Single Speaker	Vocabulary Type (Perplexity)
Brooke/Petajan 1986	Radial Motion	Yes	Phones (3)
Pentland/Mase 1989	Optical Flow	No	Digits (10)
Wu et al. 1991	Neural Networks	No	Phones (5)
Goldschen 1993	HMMs	Yes	Continuous Sent. (150)
Martin et al. 1995	Optical Flow	No	Phones (3)
Chiou et al. 1996	HMMs	Yes	Words (10)
Luettin et al. 1996	HMMs	No	Digits (4)

TABLE 2. Comparison of visual automatic speech recognition system that use only visual information. (References are included in the text.)

Figure 3 illustrates a state space of visual automatic speech recognition systems along three axes: the type of speech, the connectedness of speech, and the size of the vocabulary (Petajan [39], Brooke et al. [4], Pentland et al. [38], Smith [46], Yuhas et al. [54], Mase et al. [30], Wu et al. [53], Stork et al. [47], Bregler et al. [2], Goldschen [14], Silsbee [45], Hennecke et al. [18], Mak et al. [28], Martin et al. [29], Chiou et al. [7], Luettin et al. [27] [26], Matthews et al. [31], Meier et al. [33], Tomlinson et al. [50]). The *type of speech* indicates whether the system uses a single speaker or multiple speakers for training and testing. We did not use the term *speaker-independent* since each system requires some initial setup such as the adjustment of thresholds or lighting parameters. Those systems that

Author	Recognition Method	Single Speaker	Vocabulary Type (Perplexity)
Petajan 1984	Linear Time Warping	Yes	Words,Letters Digits (100)
Petajan et al. 1988	Dynamic Time Warping	Yes	Letters, Digits (36)
Nishida 1986	Differences	No	Word Boundaries
Smith 1989	Derivatives	Yes	Words (4)
Yuhas et al. 1989	Neural Networks	Yes	Vowel Phones (9)
Stork et al. 1992	Neural Networks	No	Consonant Phones (10)
Bregler et al. 1993	Neural Networks	Yes	Connected Letters (26)
Silsbee 1993	HMMs	Yes	Words (500)
Hennecke et al.1994	Neural Networks	No	Phones (5)
Mak et al. 1994	Differences	No	Syllable Bound.(5 sent)
Matthews et al. 1996	HMMs	No	Letters
Meier et al. 1996	Neural Networks	Yes	Continuous Letters (26)
Tomlinson et al.1996	HMMs	Yes	Digit Triplets (10)

TABLE 3. Comparison of bimodal systems that use both acoustic and visual information. (References are given in the text.)

perform recognition with multiple speakers (Pentland et al. [38], Mase et al. [30], Stork et al. [47], Hennecke et al. [18], Martin et al. [29], Luettin et al. [27] [26] and Matthews et al. [31]) capture information about the oral-cavity from their model – not from the oral-cavity image itself. The *connectedness of speech* indicates whether the system performs recognition of isolated words, connected speech, or continuous speech. Most systems depicted in Figure 3 perform isolated word recognition. Bregler et al. [3] performs connected letter recognition, while Goldschen[14], Meier [33], and Tomlinson et al. [50] perform continuous speech recognition. The motion-based recognition discipline of visual automatic speech recognition is quite small (as Figure 3 illustrates).

The following sections summarize the visual processor (feature extraction), the visual recognition algorithm, and the combining of acoustic and visual information for bimodal recognizers.

1.4.1. *Visual Processor (Feature Extraction)*

The visual processor determines how the system calculates visual features for motion-based recognition. The literature describes two major approaches

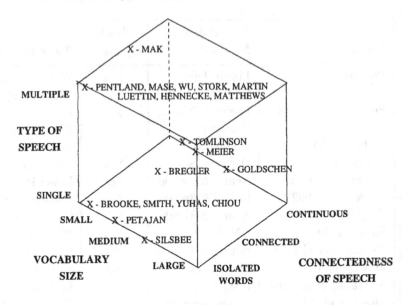

Figure 3. Description of visual automatic speech recognition systems along three axis: Type of Speech, Vocabulary Size, and Connectedness of Speech. (References are given in the text.)

for building a visual processor. The *first approach* finds the oral-cavity region in the image frame, and computes oral-cavity features directly from the image frame (e.g., area, width, and related derivative features). This process often requires pre-processing of the image to account for lighting variations, color distortion, rotation, centering, and scaling. Most of the systems in Figure 3 or in Tables 2 and 3 calculate visual features with this approach. These systems track various features such as the oral-cavity, teeth, and tongue, and do not depend solely on contours and shape of the oral-cavity.

The *second approach* does not calculate visual features directly from the image frame, but rather from a model of the oral-cavity. The system tracks certain 'control points' from each image so that changes of the oral-cavity in the image are reflected in the model. Although this approach overcomes difficulties associated with the first approach, it is assumed that the model produces all possible distinctly identifiable oral-cavity shapes and sizes. Furthermore, these models tend to focus on contours and lack critical information about the teeth and tongue. Despite these drawbacks, this approach facilitates the development of systems using multiple speakers. Pentland et al. [38], Mase et al. [30], Martin et al. [29] use *optical flow* algorithms to capture movements of the four muscle regions around the oral-cavity. Stork et al. [47] and Hennecke et al. [18] model the oral-cavity contours with *deformable templates*, a set of parabolic equations. Bregler et al. [2] and Chiou et al. [7] track oral-cavity movements across image frames with 'snakes' that model open or closed curve contours from a set of strategically placed control points (Yuille et al. [55]). Luettin et al. [27] tracks the inner and outer lip boundaries with *active shape models*, a set of statistically derived shapes from the training data. Matthews et al. [31] uses multi-scale nonlinear image decomposition with sieves and scale histograms to model the oral-cavity motion.

1.4.2. *Visual Recognizer*
The recognition algorithms listed in Tables 2 and 3 are similar to the acoustic speech recognition algorithms (Rabiner et al. [44], Deller et al. [21]). Petajan [39] [41] performs template matching with linear and dynamic time warping. Goldschen [14], Silsbee [45], Chiou et al. [7], Luettin et al. [27], and Tomlinson et al. [50] perform recognition with trained HMMs. Yuhas et al. [54], Wu et al. [53] Stork et al. [47], Bregler et al. [2], Hennecke et al. [18], and Meier et al. [33] use trained neural networks for recognition.

1.4.3. *Bimodal Recognition*
The bimodal recognition systems, listed in Table 3, use visual information to improve the robustness of their acoustic recognizers, especially in noisy environments. The first system (Petajan [39]) used visual information to significantly improve the recognition accuracy of the acoustic recognizer. The ability to design robust recognizers continues to be the primary focus of bimodal recognizers. These systems demonstrate that visual information offers greater improvements at low signal-to-noise ratios, than at high signal-to-noise ratios.

Some researchers use separate acoustical and visual recognizers, as depicted in Figure 2(a). Nishida [35] and Mak et al. [28] use visual information to find word or syllable boundaries to improve the performance of their acoustic recognizers. Petajan et al. [41] uses oral-cavity heuristic rules to integrate the output of the visual recognizer with the acoustic recognizer. Smith [46] uses the visual recognizer to distinguish among four words that his acoustic recognizer confuses. Silsbee [45] uses a weighted average of the two scores obtained from his HMM-based acoustic recognizer and his HMM-based visual recognizer.

Bimodal recognizers that use a feature vector containing both acoustic and visual information, as depicted in Figure 2(b), tend to use standard acoustic speech recognition algorithms. Bregler et al.[2] and Meier et al. [33] perform the integration at a neural network layer, while Tomlinson et al. [50] combines the information prior to the HMM recognizer.

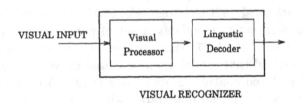

VISUAL RECOGNIZER

Figure 4. Description of the automatic visual speech recognition system.

2. An Automatic Speech Recognition System by Lipreading

2.1. INTRODUCTION

We describe a motion-based recognition method to perform continuous automatic speech recognition using only visual information obtained from the shadow of the oral-cavity region of a speaker (Garcia et al. [13], Goldschen [14], Goldschen et al. [15]). The system attempts to determine the correct spoken sentence using images of the oral-cavity captured by a camera. We identify the important features of the oral-cavity region of our speaker for visual recognition using a correlation matrix and a principal component analysis. For each image frame, the system computes a feature vector containing these important oral-cavity features. The feature vectors are vector quantized into a sequence of codebook entries, obtained from the codebook generated by a K-Means clustering algorithm. To model the sequence oral-cavity movements during a spoken utterance, the system uses trained HMMs to recognize the sequence of codebook entries. We use discrete density HMMs with the output probability density function associated with each state of the HMM (Goldschen [14]). These discrete density functions are tied together (some HMM states share the same density function). The tied state approach allows for better training of the discrete density functions.

2.2. DESCRIPTION OF THE SYSTEM

The visual recognizer consists of the visual processor and linguistic decoder as depicted in Figure 4 (Goldschen [14], Goldschen et al. [15]). The visual processor converts visual features of the oral-cavity region into a sequence of codewords. The (HMM) linguistic decoder converts the sequence of codewords into a recognized sentence.

2.2.1. Database description

The database consists of visual recordings captured to disk at 60 frames-per-second of one speaker with a beard and mustache reading the 450 TIMIT sentences [36]. The recording of the speaker's lower facial region was obtained with a head-mounted harness containing a camera, two side-mounted incandescent lights with a microphone. This very large database contains each sentence in ASCII, acoustical, and visual formats.

The database contains phone-segmented boundaries (hand-segmented timing marks) for 67 of the 450 TIMIT sentences spoken by our speaker. These 67 sentences contain a total of 11214 image frames. We break the 450 sentences in the database as follows. The phone-segmented sentences identify the phone-to-viseme mapping for our speaker and train the initial viseme HMMs. We then randomly select 300 training sentences, which include the 67 phone-segmented sentences, to build context-independent viseme HMMs, context-dependent triseme HMMs, and generalized triseme HMMs. We test the system with the remaining 150 sentences.

Petajan ([39] [40] [41]) and Goldschen [14] performed a significant amount of image processing to the visual portion of this database. Figure 5 illustrates a sequence of the image frames of the oral-cavity region. The image processing system first creates an image with the edges of the oral-cavity, nostrils, and facial regions at the video frame rate. After identifying the nostrils and oral-cavity regions in the first frame, the system then uses a nostril tracking algorithm to quickly identify the oral-cavity in the remaining frames (Petajan [39]). Using separate threshold values for the nostrils and the oral-cavity, the system creates a smoothed binary image by filling in closed oral-cavity regions, and ignores regions such as the nostrils. The oral-cavity region is then rotated in an image by the angle that the horizontal axis produces when connecting an imaginary line between the two nostrils. This angular rotation ensures that the oral-cavity region is horizontal. The oral-cavity is then centered within the image using the extreme abscissa and ordinate values for the oral-cavity region. Next, noise pixels are removed from the image using a median filter that iterates over an image until no further filtering occurs (Goldschen [14]).

2.2.2. Visual processor

The visual processor depicted in Figure 4 receives a sequence of visual images similar to the oral-cavity depicted in Figure 5, and computes the oral-cavity features for each image frame. Using the codebook, each oral-cavity feature is vector quantized into a codebook entry. This section describes the methodology to determine the appropriate oral-cavity features and to generate the codebook.

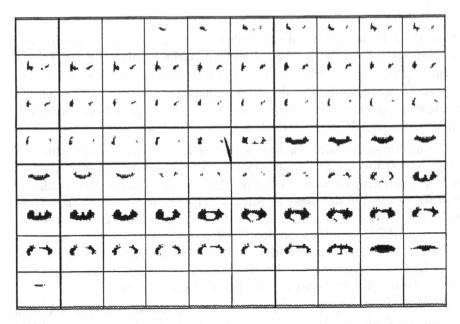

Figure 5. Sequence of Oral-Cavity images centered, rotated, and smoothed for our speaker uttering the sentence: *She is thinner than I am.*

Feature Selection   We initially hypothesize the following seven static oral-cavity features (Garcia et al. [13]):

1. **Area:** The number of black pixels in an oral-cavity binary image.
2. **Contour blob (CBlob):** The number of connected (closed) regions in the original, not smoothed, image.
3. **Height:** The vertical distance between the maximum y and minimum y ordinates of the shadow of the oral-cavity.
4. **Image Blob (IBlob):** The number of regions in a smoothed binary image (after filtering).
5. **Perimeter:** The number of pixels along the edge of the shadow of the oral-cavity.
6. **Rounding:** The ratio of the width to the height.
7. **Width:** The horizontal distance between the maximum x and minimum x abscissa of the shadow of the oral-cavity.

The area, height, perimeter, and width features are taken from Petajan [39] and Finn [11]. The rounding feature was based on the lip rounding and the spatial normalization concepts of Montgomery and Jackson [34]. Petajan, from his experience (Petajan [39] [41]), claims that visually different oral-cavity images contain similar values for the area and perimeter due to the presence of the teeth or tongue. A count of the number of connected regions gives an indication of their presence. In Goldschen [14], we hypothesize that the dynamic features would enable a system to achieve successful visual automatic speech recognition. For each of the 67 phone-segmented sentences, the system computes 35 oral-cavity features, as listed in Table 4.

Static Features	First Derivative	Second Derivative	Magnitude of First Derivative	Magnitude of Second Derivative
Area	Area'	Area"	|Area'|	|Area"|
CBlob	CBlob'	CBlob"	|CBlob'|	|CBlob"|
Height	Height'	Height"	|Height'|	|Height"|
IBlob	IBlob'	IBlob"	|IBlob'|	|IBlob"|
Perimeter	Perimeter'	Perimeter"	|Perimeter'|	|Perimeter"|
Rounding	Rounding'	Rounding"	|Rounding'|	|Rounding"|
Width	Width'	Width"	|Width'|	|Width"|

TABLE 4. Initial oral-cavity features calculated from each image.

We reduce the dimensionality of the feature space using a correlation matrix of the 35 features and a principal component analysis (Garcia et al. [13], Goldschen [14], Goldschen et al. [16]). From this analysis the following clusters emerge (assuming a minimum correlation of 0.6 between two features):

1. *Area, Height, Width, IBlob, Perimeter*
2. *CBlobɪ, CBlobɪ, |CBlobɪ|, |CBlobɪɪ|,*
3. *Heightɪ, Heightɪɪ, Widthɪ, Widthɪɪ*
4. *|Roundingɪ|, |Roundingɪɪ|, |Widthɪ!, ||Widthɪɪ|*
5. *Perimeterɪ, Perimeterɪɪ*
6. *|Areaɪ|, |Areaɪɪ|*
7. *|Heightɪ|, |Heightɪɪ|*
8. *|IBlobɪ|, |IBlobɪɪ|*
9. *IBlobɪ, IBlobɪɪ*
10. *|Perimeterɪ|, |Perimeterɪɪ|*
11. *Roundingɪ, Roundingɪɪ*
12. *CBlob*
13. *Area*
14. *Areaɪ*
15. *Areaɪɪ*

We select the features with two feature selection algorithms from Jolliffe [22], which involves obtaining the dominant eigenvalue from a principal component analysis. Garcia et al. [13] and Goldschen et al. [16] describe this feature selection process in greater detail. We chose to retain the following 13 features from our original 35 features: *IBlob, Rounding, Width, Area1, Height1, Rounding1, AreaII, HeightII, IBlobII, PerimeterII, |Rounding1|, |HeightII|, |PerimeterII|*. Ten of these features pertain to the derivatives, confirming our belief that the dynamics of the oral-cavity features are important for computerized lipreading analysis (Garcia et al. [13]).

Codebook Generation   The sequence of feature vectors, containing the 13 features kept, characterizes the oral-cavity region in the image frame. The system then vector quantizes each feature vector into a codeword. These codewords are the output symbols, or observation sequences, for the discrete density HMMs. A K-Means non-hierarchical clustering algorithm (from Andeberg [1]) generates a codebook of 64 codewords using the Euclidean distance as our metric. The size of the HMM density function associated with each HMM state is 64. The codebook was built from the 11214 images in the 67 phone-segmented sentences.

2.2.3. *The linguistic decoder*

The linguistic decoder, as depicted in Figure 4, accepts a sequence of codeword entries from the visual processor and uses a-priori knowledge about the grammar from the set of training sentences. We build context-independent viseme HMMs, context-dependent triseme HMMs, and generalized triseme HMMs to capture the co-articulatory effects during continuous speech production. As mentioned earlier, the first step prior to building any models for continuous speech is to identify the phone-to-viseme mapping for our speaker.

Phone-to-Viseme Mapping   Garcia et al. [13] and Goldschen et al. [14] describe the process of mapping the phones into visemes for our speaker. To determine the phones that correspond to a viseme for our speaker, we train HMMs using the Forward-Backward (F-B) algorithm (Rabiner [43], Levinson et al. [25]) on each phone of the 67 hand-segmented sentences from the visual database. After training, the system determines visemes by identifying those phone HMMs that recognize similar phone sequences. Although other lipreading systems have been built, this system (Goldschen [14] and Goldschen et al. [16]) is the first to empirically determine the phone-to-viseme mapping for all phones.

We obtain the phone-to-viseme mapping for our speaker by determining which of the 56 phones phone HMMs are similar (recognize similar observation sequences). Juang and Rabiner [23] propose the following HMM similarity metric D (Equation 1):

$$D(\Omega_1, \Omega_2) = \frac{1}{T} * [\log \Pr(O^2|\Omega_1) - \log \Pr(O^2|\Omega_2)] \qquad (1)$$

Equation 1 indicates how likely observation sequence O^2 of length T (used to train HMM Ω_2) compares to another HMM Ω_1, and is the ratio of $\Pr(O^2|\Omega_1)$ to $\Pr(O^2|\Omega_2)$. The conditional probabilities are computed using the HMM forward algorithm described in Rabiner [43] and Levinson et al. [25]. Because D of equation 1 is non-symmetric, we use Equation 2 to compute the average similarity D_s between HMMs.

$$D_s(\Omega_1, \Omega_2) = \frac{1}{2} * [D(\Omega_1, \Omega_2) + D(\Omega_2, \Omega_1)] \qquad (2)$$

The system calculates a similarity matrix using D_s of Equation 2 to indicate how similar one phone HMM is to another phone HMM. Using this similarity metric, we use the Average Linkage hierarchical clustering algorithm from Andeberg [1]:

Step 1. Begin with n clusters with each cluster consisting of one entity. Let the clusters be labeled 1 through n.

Step 2. Search the similarity matrix, D_s, for the most similar pair of clusters. This is the smallest value of D_s and label it d_{ij}.

Step 3. Reduce the number of clusters by 1 by merging together clusters i and j. Label the result of the merge r and update the other clusters with new cluster r as follows:

- For each cluster k, compute $D_s(k, r) = Avg(D_s(k, i), D_s(k, j))$

- Now delete all rows and columns associated with i and j.

Step 4. Perform steps 2 and 3 a total of n-1 times, at which point there will be only one cluster.

The Average Linkage method is similar to the Single Linkage method which at step 3 creates the new cluster by taking the minimum value or the Complete Linkage method which at step 3 takes the maximum value.

Before using the Average Linkage hierarchical clustering algorithm, we divide the phones into three groups to facilitate the clustering in a more homogeneous space. The first group contains consonant phones including their closures. That is, each closure is represented by the original consonant (e.g., /bcl/ observations and /b/ observations are used to train the /b/ phone HMM). This group enables us to compare our results with that of previous researchers that performed consonant phone-to-viseme mappings (as mentioned in Table 1). The second group contains the consonant phones plus closures, with the closures being treated as independent phones. The purpose of this group is to investigate whether the closures cause changes in the consonant phone-to-viseme mapping. The third group contains the vowel and diphthong phones.

The viseme groups that emerge when the closures are modeled with their respective consonants are depicted in the *Goldschen* column of Table 1. These viseme groups appear to be fairly consistent with the results of other researchers (Burchett [5], Hazard [17], Summerfield [48], Walden [51], Finn [11]). As Table 1 illustrates, the opinions of the researchers vary for the grouping of phones /n/ and /l/. Our result suggests that phones /n/ and /l/ are part of a viseme group that includes a large number of phones for our speaker.

Despite the previous results, our data includes the segmentation of closures that are separated from their consonant phones. Therefore, we use the clustering algorithm again, while modeling the closures separately from their corresponding phones. The same viseme groups emerge as before except that the system obtains the viseme group *(/bcl/,/m/,/pcl/)* instead of *(/b/,/m/,/p/)* and obtains *(/b/,/p/,/r/)* instead of *(/r/)*. These results clearly indicate (as we suspected, although other researchers had not considered them) that closure affects the grouping of visemes for our speaker.

The left column of Table 5 depicts the viseme grouping of our vowel phones. In most cases the solitary vowel or diphthong phone forms its own viseme group with the exception of the viseme groups *(/ax/, /ih/, /iy/)* and *(/ae/, /eh/)*. Recalling the vowel triangle from Peterson and Barney [42], the viseme group *(/ax/,/ih/,/iy/)* includes vowels associated with similar tongue height (high). It appears, at least for our speaker, that the high back vowel /ax/ is not distinguishable from the high front vowels /iy/ and /ih/. Waters [52] observes that all unstressed (unaccented) vowels often turn into the schwa /ax/. The vowel triangle also shows that /ae/ and /eh/ are front vowels with similar tongue height (low).

The phone-to-viseme mappings for our speaker are summarized in Table 5. We decided to separate the phones /r/ and /w/ into two groups based on our phonetic and speechreading knowledge as detailed in Garcia et al. [13]. We reduce the number of consonants from 37 phones to 19 visemes and also reduce the number of vowels and diphthongs from 19 to 16 visemes. Overall this yields a total mapping reduction from 56 phones to 35 visemes.

Context-Independent Viseme HMMs We create context-dependent viseme HMMs for our speaker using the HMM F-B algorithm. That is, we create one HMM for each of the 35 visemes listed in Table 5. We train each viseme HMM using the observations sequences from every phone in the viseme that occurs in the 67 phone-segmented sentences.

Vowel/Diphthongs		Consonants		
aa	ae, eh	b, p	bcl, m, pcl	ch
ah	ao	dh, epi	dx, nx, q	f, v
aw	ax,ih,iy	en	hh	hv
axr	ay	jh	ng	r
er	ey	s, sh, z	th	w
ix	ow	y	zh	h#
oy	uh	d, dcl, g, gcl, k, kcl, l, n, t, tcl		
uw	ux			

TABLE 5. Final phone-to-viseme mapping used by our visual automatic speech recognition system. This mapping provides a total reduction from 56 phones to 35 visemes.

Context-Dependent Triseme HMMs We built the triseme HMMs by creating a triseme HMM for every triplet of visemes that occurs in the 300 training sentences. The triseme HMMs attempt to capture the co-articulatory transitions between visemes. We had a total of 2702 trisemes from the 300 training sentences, which is far smaller than the number of possible trisemes ($35^3 = 42,875$). These 2702 trisemes have an insufficient number of samples to train each corresponding triseme HMM.

To train the context-dependent triseme HMMs, we initialize all triseme HMMs with the context-independent viseme HMM that corresponds to the middle viseme of the triseme. For each of the 300 training sentences, the system concatenates the corresponding triseme HMMs together and trains using the HMM F-B algorithm with embedded re-estimation.

Generalized Triseme HMMs We create generalized triseme HMMs by merging (clustering) similar context-dependent triseme HMMs together. We built the generalized triseme HMMs to reduce the number of context-dependent triseme HMMs. By having a smaller number of HMMs, more training samples exist for each HMM.

Since perfect phone boundaries did not exist for all of the trisemes, the trisemes could not be clustered as done previously during the phone-to-mapping process. In order to perform the clustering, the 2702 context-dependent triseme HMMs become generators to generate the observation sequences required for the HMM similarity metric of Equation 2. This is a novel approach to solving the problem. After generating the observation sequences for each triseme, we cluster only those trisemes that share the

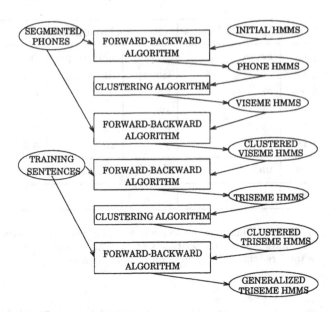

Figure 6. Illustrates the training process used to build viseme models for our continuous visual automatic speech recognition system.

same middle viseme and include the middle viseme in this group. (This additional step ensures that if a context-dependent triseme does not occur in our test data, it has the corresponding context-independent viseme.) The Average Linkage clustering algorithm, again, merges together trisemes with the same middle viseme. These clustered triseme HMMs becomes the generalized triseme HMMs. We then train the 934 generalized triseme HMMs with the 300 training sentences using the HMM F-B algorithm with embedded re-estimation.

2.2.4. *Summary of HMM training*

Figure 6 summarizes the overall training process. The 67 hand-segmented sentences are used to train the initial 56 phone HMMs using the F-B algorithm. Similar phone HMMs are clustered into 35 viseme HMMs using the Average Linkage clustering algorithm. The 67 phone segmented sentences are again used with the F-B algorithm to create the context-independent viseme HMMs. The 300 training sentences are used to generate the 2702 context-dependent triseme HMMs using the F-B algorithm with embedded re-estimation. The context-dependent triseme HMMs are clustered using the Average Linkage clustering algorithm to form the 934 generalized triseme HMMs.

2.2.5. *Testing the visual automatic speech recognition system*

After creating the visual recognizer as depicted in Figure 4, we test the performance using the 150 test sentences (Goldschen [14]).

From the set of 150 test sentences in the visual database, a sentence to be recognized is chosen, and the appropriate codebook entries are determined from the visual features. The recognized sentence is the sentence with the maximum Viterbi probability (Rabiner [43] Levinson et al. [25]) from all the 150 sentences. We repeat this for each of the 150 test sentences. The test process is done three times, first to test the context-independent viseme HMMs, second to test context-dependent triseme HMMs, and third to test generalized triseme HMMs. We determine the recognition rate as the number of sentences correctly recognized. We did not compute viseme or word error rates since the test sentences spoken by our speaker were not segmented (into phone-segmented boundaries).

The generalized triseme HMMs achieve a recognition rate of 25.3 percent, the context-dependent triseme HMMs achieve a recognition rate of 12.7 percent, and the context-independent viseme HMMs achieve a recognition rate of 2 percent [14].

We expect the context-dependent triseme HMMs to achieve a recognition rate higher than the context-independent viseme HMMs, since the trisemes contain information about the neighboring visemes. Because there are so many occurrences of a viseme as compared to a triseme, the viseme HMMs are trained for many possible interpretations (co-articulations) of the viseme.

Likewise, we expect the generalized triseme HMMs to achieve a higher recognition rate than the context-dependent triseme HMMs, since every generalized triseme occurs multiple times in the training sentences. Because we added the context-dependent viseme in the appropriate cluster when creating the generalized trisemes, every generalized triseme in the test sentences is trained.

It is difficult to judge the quality of our results because no other system performs continuous, automatic speech recognition using only visual information. We believe that the recognition results would improve by using a grammar, just as an acoustic recognizer (Lee [24]) improves from a word recognizer rate of 31 percent to 76 percent. Cerio [6], Cornett [9], and Tobin [49] discuss audiologists performing isolated-word speech recognition studies in an attempt to determine the percentage of correctly recognized words. When using only acoustic information, individuals who possess normal hearing will correctly recognize about 40 percent of the isolated nonsense words. Expert lipreaders, using only visual information, will correctly recognize about 30 percent of these nonsense words. In contrast to the 30 percent recognition rate for nonsense words by expert lipreaders, our system achieved a 25 percent recognition rate for a group of sentences without the aid of acoustic and grammatical information.

3. Conclusions

We describe the motion-based recognition application of visual automatic speech recognition. We discuss human perception issues for bimodal recognition and that acoustic and visual sources of information are complementary to each other. We then review visual automatic speech recognition systems, indicating that comparisons among the systems are difficult due to the lack of a standard database. Those systems that support multiple speakers capture their oral-cavity features from models about the oral-cavity – not from the sequence of image frames. These systems, however, require modeling assumptions about the shape and contour of the oral-cavity and do not model the teeth and tongue. Although no such system currently exists, we believe that a large vocabulary, speaker-independent, visual automatic speech recognition system will emerge as progress continues in the fields of signal processing, pattern recognition, and motion-based recognition.

Motion-based recognition uses information from a sequence of image frames, similar to our visual automatic speech recognition system. This system performs continuous speech recognition using only features captured from the oral-cavity and does not use acoustic, syntactical, or grammatical information. We present dynamic features, some never introduced before, to describe the actions of the oral-cavity region during speech. We use a correlation matrix and a principal component analysis to reduce the oral-cavity feature space from 35 to 13 features. After finding the oral-cavity region from an image frame, most of the 13 features can easily be calculated with modern machines. We believe, additionally, that these dynamic features are very important for lipreading research, especially for human lipreading. Most students are taught to lipread by the placement of (static) oral-cavity features. Our research indicates, however, that the movement of (dynamic) oral-cavity features allow for successful lipreading by machine.

Using HMM and hierarchical clustering algorithms, our system identifies viseme groups that are consistent with the consonant viseme groups selected by expert human lipreaders and the vowel triangle (from Peterson and Barney [42]). Future researchers in bimodal automatic speech recognition may want to use vowel visemes to augment the acoustic information because vowels appear to be uniquely characterized by oral-cavity features. It appears that the consonant phone-to-viseme mapping is many-to-one (Finn [11] Goldschen [14]) and the vowel phone-to-viseme mapping is nearly one-one (Goldschen [14]).

We expect future work to examine whether additional oral-cavity features could be used, and focus on utilizing a trigram, bigram or word-pair grammar. We recommend that speaker-independent visual automatic speech recognition be investigated since most successful continuous acoustic automatic speech recognition systems using HMMs work quite well with multiple speakers. An investigation with multiple speakers would greatly contribute to this research by confirming the consistency of the viseme groups obtained herein.

References

1. Michael Andeberg. *Cluster Analysis for Applications*. Academic Press, New York, NY, 1973.

2. Christoph Bregler, Hermann Hild, Stefan Manke, and Alex Waibel. Improving connected letter recognition by lipreading. In *International Joint Conference on Speech and Signal Processing*, volume 1, pages 557–560. IEEE, April 1993.

3. Christoph Bregler, Stephen Omohundro, and Yochai Konig. A hybrid approach to bimodal speech recognition. In *28th Annual Asilomar Conference on Signals, Systems, and Computers*. IEEE, October 1994.

4. N. Michael Brooke and Eric D. Petajan. Seeing speech: Investigations into the synthesis and recognition of visible speech movements using automatic image processing and computer graphics. In *Proceedings of the International Conference on Speech Input/Output: Techniques and Applications*, pages 104–109, London, 1986.

5. J. Burchett. *Lipreading: A Handbook of Visible Speech*. The Royal National Institute for the Deaf, London, England, 1965.

6. Roberta Cerio. Personal communications, February 1989.

7. Greg Chiou and Jenq-Neng Hwang. Lipreading from color motion video. In *International Conference on Acoustics, Speech, and Signal Processing*, pages 2156–2159. IEEE, May 1996.

8. Michael Cohen and Dominic Massaro. What can visual speech synthesis tell visible speech recognition. In *Proceedings of the 28th Asilomar Conference on Signals, Systems, and Computers*, pages 566–571. IEEE, October 1994.

9. Orin Cornett. Personal communications, February 1989.

10. L. Erman and V. Lesser. The Hearsay-II speech understanding system: A tutorial. In A. Waibel and K. Lee, editors, *Readings in Speech Recognition*, pages 235–245. Morgan Kaufmann Publishers, 1990.

11. Kathleen Finn. *An Investigation of Visible Lip Information to be used in Automatic Speech Recognition*. PhD thesis, Georgetown University, Washington, DC, 1986.

12. C. G. Fisher. Confusions among visually perceived consonants. *Journal of Speech and Hearing Research*, 11:796–804, 1968.

13. Oscar Garcia, Alan Goldschen, and Eric Petajan. Feature extraction for optical automatic speech recognition or automatic lipreading. Technical Report GWU-IIST-92-32, The George Washington University, November 1992. Department of Electrical Engineering and Computer Science.

14. Alan Goldschen. *Continuous Automatic Speech Recognition by Lipreading*. PhD thesis, The George Washington University, Washington, DC, 1993.

15. Alan Goldschen, Oscar Garcia, and Eric Petajan. Continuous optical automatic speech recognition. In *Proceedings of the 28th Asilomar Conference on Signals, Systems, and Computers*, pages 572–577. IEEE, October 1994.

16. Alan Goldschen, Oscar Garcia, and Eric Petajan. Rationale for phoneme-viseme mapping and feature selection in visual speech recognition. In David Stork, editor, *Speechreading by Man and Machine: Models, Systems, and Applications*, NATO

Advanced Study Institute. Springer-Verlag, (in press).

17. Elizabeth Hazard. *Lipreading: For the Oral Deaf and Hard-of-Hearing Person.* Charles C. Thomas, Springfield, Illinois, 1971.

18. Marcus Hennecke, K. Prasad, and David Stork. Using deformable templates to infer visual speech dynamics. In *28th Annual Asilomar Conference on Signals, Systems, and Computers.* IEEE, October 1994.

19. Frederick Jelinek. Continuous speech recognition by statistical methods. *Proceedings of the IEEE*, 64:532–556, 1976.

20. Frederick Jelinek. Self-organized continuous speech recognition. In Jean-Paul Haton, editor, *Automatic Speech and Analysis Recognition*, pages 231–238. Reidel Publishing Company, 1982.

21. Jr. John Deller, John Proakis, , and John Hansen. *Discrete-Time Processing of Speech Signals.* Macmillan Publishing Company, New York, NY, 1993.

22. I.T. Jolliffe. *Principal Component Analysis.* Springer-Verlag, New York, NY, 1986.

23. Biing Juang and Lawrence Rabiner. A probabilistic distance measure for hidden markov models. *ATT Technical Journal*, 64(2):391–408, February 1985.

24. Kai Fu Lee. *Automatic Speech Recognition: The Development of the Sphinx System.* PhD thesis, Carnegie-Mellon University, Pittsburgh, PA 15213, 1989.

25. Stephen Levinson, Lawrence Rabiner, and Man Mohan Sondhi. An introduction to the application of theory of probabilistic function of a markov process to automatic speech recognition. *The Bell System Technical Journal*, 62(4):1035–1074, April 1983.

26. Juergen Luettin, Neil Thacker, and Steve Beet. Speech reading using shape and intensity information. In *International Conference on Spoken Language Processing*, pages 58–61. IEEE, October 1996.

27. Juergen Luettin, Neil Thacker, and Steve Beet. Visual speech recognition using active shape models and hidden markov models. In *International Conference on Acoustics, Speech, and Signal Processing*, pages 817–820. IEEE, May 1996.

28. M.W. Mak and W.G. Allen. Lip-motion analysis for speech segmentation in noise. *Speech Communication*, 14:279–296, 1994.

29. Glenn Martin and Mubarak Shah. Lipreading using optical flow. In *Proceedings National Conference on Undergraduate Research*, March 1995.

30. Kenji Mase and Alex Pentland. Automatic lipreading by optical flow analysis. *Systems and Computer in Japan*, 22(6):67–76, 1991.

31. Iain Matthews, J Bangham, and Stephen Cox. Audiovisual speech recognition using multiscale nonlinear image decomposition. In *International Conference on Spoken Language Processing*, pages 38–41. IEEE, October 1996.

32. Harry McGurk and John MacDonald. Hearing lips and seeing voices. *Nature*, 264:746–748, December 23/30 1976.

33. Uwe Meier, Wolfgang Hurst, and Paul Duchnowski. Adaptive bimodal sensor fusion for automatic lipreading. In *International Conference on Acoustics, Speech, and Signal Processing*, pages 833–836. IEEE, May 1996.

34. Allen Montgomery and Pamela Jackson. Physical characteristics of the lips underlying vowel lipreading performance. *Journal of Acoustical Society of America*, 73(6):2134–2144, June 1983.

35. Nishida. Speech recognition enhancement by lip information. *ACM SIGCHI Bulletin*, 17(4):198–204, April 1986.

36. NIST, Gaithersburg, MD 20899. *DARPA TIMIT CD-ROM*, November 1988.

37. Catherine Pelachaud, Norman Badler, and Marie-Luce Viaud. Final report to NSF of the standards for facial animation workshop. Technical report, University of Pennsylvania, Philadelphia, PA, October 1994.

38. Alex Pentland and Kenji Mase. Lip reading: Automatic visual recognition of spoken words. Technical Report MIT Media Lab Vision Science Technical Report117, Massachusetts Institute of Technology, January 15 1989.

39. Eric Petajan. *Automatic Lipreading to Enhance Speech Recognition.* PhD thesis, University of Illinois at Urbana-Champaign, 1984.

40. Eric Petajan. Automatic lipreading to enhance speech recognition. In *Proceeding of the IEEE Conference on Computer Vision and Pattern Recognition*, pages 40–47, San Francisco, CA, 1985. IEEE.

41. Eric Petajan, Bradford Bischoff, David Bodoff, and N. Michael Brooke. An improved automatic lipreading system to enhance speech recognition. In *CHI-88*, pages 19–25. ACM, 1988.

42. Gordan Peterson and Harold Barney. Control methods used in a study of the vowels. *Journal of Acoustical Society of American*, 24:175–184, March 1952.

43. Lawrence Rabiner. A tutorial on hidden markov models and selected applications in speech recognition. In Alex Waibel and Kai-Fu Lee, editors, *Readings in Speech Recognition*, pages 267–296. Morgan Kaufmann Publishers, Inc., 1990.

44. Lawrence Rabiner and Bing-Hwang Juang. *Fundamentals of Speech Recognition*. Prentice-Hall, 1993.

45. Peter Silsbee. *Computer Lipreading for Improved Accuracy in Automatic Speech Recognition*. PhD thesis, The University of Texas at Austin, 1993.

46. Steve Smith. Computer lip reading to augment automatic speech recognition. *Speech Tech*, pages 175–181, 1989.

47. David Stork, Greg Wolff, and Earl Levine. Neural network lipreading system for improved speech recognition. *International Joint Conference of Neural Networks*, 1992.

48. Quentin Summerfield. Some preliminaries to a comprehensive account of audio-visual speech perception. In Barbara Dodd and Ruth Campbell, editors, *Hearing by Eye: The Psychology of Lipreading*, pages 3–51. Lawrence Earlbaum Associated, 1987.

49. Henry Tobin. Personal communications, February 1989.

50. M. Tomlinson, M. Russell, and N. Brooke. Integrating audio and visual information to provide highly robust speech recognition. In *International Conference on Acoustics, Speech, and Signal Processing*, pages 821–824. IEEE, May 1996.

51. Brian Walden, Robert Prosek, Allen Montgomery, Charlene Scherr, and Carla Jones. Effects of training on the visual recognition of consonant. *Journal of Speech and Hearing Research*, 20:130–145, 1977.

52. Gill Waters. Speech production and perception. In Chris Rowden, editor, *Speech Processing*, pages 1–33. McGraw-Hill International, 1992.

53. Jian-Tong Wu, Shinichi Tamura, Hiroshi Mitsumoto, Hideo Kawai, Kenji Kurosu, and Kozo Okazaki. Neural network vowel-recognition jointly using voice features and mouth shape image. *Pattern Recognition*, 24(10):921–927, 1991.

54. Ben Yuhas, Moise Goldstein, and Terrence Sejnowski. Integration of acoustic and visual speech signals using neural networks. *IEEE Communications Magazine*, pages 65–71, 1989.

55. A. Yuille, P. Hallinan, and D. Cohen. Snakes: Active contour models. *International Journal on Computer Vision*, 8:99–112, 1992.

VISUALLY RECOGNIZING SPEECH USING EIGENSEQUENCES

NAN LI, SHAWN DETTMER AND MUBARAK SHAH
Computer Vision Lab
Computer Science Department
University of Central Florida
Orlando, FL 32816

1. Introduction

Humans have the very complex ability to interpret facial expressions, gestures, even the so called "body language". Hearing-impaired people further develop this ability since most of them can perform some lipreading and/or understand sign language. It is well known that the visual information about speech through lipreading is very useful for human speech recognition. Humans use visual information to enhance our speech recognition, even when the available visual signal is noisy, distant or incomplete.

It is essential for computer systems to possess the ability to recognize meaningful gestures and lip movement if computers are to interact naturally with people. Currently, human-computer interface is mainly through a keyboard and/or mouse. Physically challenged people may have difficulties with such input devices and may require a new means of entering commands or data into the computer. Gesture, speech, and touch inputs are a few possible means of addressing such users' needs to solve this problem. For example, using Computer Vision, a computer can recognize and perform the user's gesture and vocal commands thus alleviating the need for a keyboard. Applications for such a vision system are the remote control of a robotic arm, guiding a remote computer presentation system, and executing computer operational commands such as opening a window or executing a program.

Lipreading is a very difficult task, especially since certain phonemes can appear visually identical (phonemes are minimal meaningful units of sound from which two words can be distinguished). For instance, the phonemes "b", "p", and "m" sound different but look the same when spoken [2].

M. Shah and R. Jain (eds.), Motion-Based Recognition, 345–371.
© *1997 Kluwer Academic Publishers.*

Acoustically-based automatic speech recognition (ASR) is still not completely speaker independent, is limited in vocabulary and is sensitive to noise [2]. Combining acoustic and visual speech recognition is one possibility to better achieve lipreading capability.

The task of lipreading using computers is a complicated one. Many ideas and methods have been put forward. The general problem of lipreading remains unsolved. We present a method for lipreading which uses eigensequences. We consider the problem of recognizing the spoken English alphabet. In our approach, gray level values of all the pixels in all frames in a sequence representing a spoken letter are put in one large vector. Several such vectors corresponding to the training sequences are used to compute eigenvectors (eigensequence), for each spoken letter. The recognition of an unknown sequence representing a spoken letter is performed by computing the ratio of the energy of projection of the sequence on the model eigenspace and the energy of the sequence. For a perfect match, this ratio tends to 1.

The success of any pattern recognition method usually depends on two stages: i) feature selection and extraction which results in easily separable clusters of patterns with minimal influence from class independent factors. ii) correct and efficient partitioning of the pattern clusters. In the eigenvector based work of face recognition, image intensity is just taken as the feature and eigenvectors are derived for a compact representation of the face patterns. Classification is performed with ease in a much lower dimensional space.

In the case of lips sequences, the samples may contain global head movements and the change in the duration of the articulations. To reduce those class independent variations spatial registration of the mouth position and temporal warping of the sequence are introduced as the the first step during training and recognition. In our method, we choose for each letter a sequence sample as the reference. The training sequences of individual letters are registered and warped against the corresponding reference sample. This creates well matched groups of sequences for each letter and those in each group have the same length as the reference sample. For an efficient recognition, instead of representing all the sequences of all possible lengths by a common basis, we use the principal eigenvectors derived from each group of training sequences as the model for the corresponding letter. This can be equivalent to characterizing the letter by a number of major components common to those sequences. Recognition is based on to what extent those components are contained in a novel sequence. The use of major components for modeling is a non-linear procedure in which noise and minor difference between the samples of the same group can be discarded as minor components so that they can not influence the decision.

Our approach is based on the demonstrated success of the eigenvector approach using static images for face recognition problem [10, 7] and similar approach for illumination planning [6]. Turk and Pentland [10] decompose face images into a small set of characteristic feature images called "eigenfaces", which are essentially the principal components of the initial training set of face images. Recognition is performed by projecting a new image into the subspace spanned by the eigenfaces, then classifying face by comparing its position in face space with positions of known individuals. Murase and Nayar [6] address the problem of illumination planning for object recognition. For each object, they obtain a large number of images by varying pose and illumination. Images of all objects, together, constitute the planning set. The planning set is compressed using the K-L transform to obtain a low-dimensional subspace.

Instead of using the eigenvectors of a set of still images, we use *eigensequences* of a spatio-temporal sequence of images for the lipreading problem. We believe that lip movements for the same letter are expected to follow the similar spatio-temporal patterns. Therefore, eigensequences are suitable for the lipreading problem. Previously, separate spatial and temporal eigen decompositions have been used [4]. Since lip movement is essentially spatiotemporal in nature, to exploit this statistical redundancy in an integral way, in this paper we use the spatiotemporal eigen decomposition, in which the set of eigenvectors span the space of all possible sequences.

In order to recognize continuous utterances, we have developed a method for extracting letters from connected sequences. Our method uses the average frame difference function of a sequences and extract subsequences corresponding to individual letters by detecting the beginnings and endings of letters. This detection, in turn, is based on the peaks and valleys in the smoothed version of average frame difference function.

We have experimented with several sequences of English alphabet letters "A" to "J", and obtained very encouraging results. Since, in real life we speak the same letter with different speeds at different occasions, these sequences were variable in length. We use dynamic time warping to align sequences to a fixed length. Our eigensequence based approach for lipreading is very simple and straightforward; the major computation during recognition is simple dot product.

2. Related Work

In Petajan et al. [8], the lipreading task is performed by using the mouth opening area of a speaker to create a codebook. The mouth window is located by tracking the nostrils. The mouth image is then binarized, then a threshold is applied, so that only the mouth opening created a dark area. This large set of mouth images was reduced by clustering to about 255 clusters. A representative of each cluster is stored in a codebook of mouth images, which are ordered by increasing size of dark area, and identified with an index value. Once the mouth images codebook has been created, an inter-cluster distance table is computed, for faster computation during the matching process. The models of spoken words (the spoken letters and digits from zero to nine) are stored and vector quantized. Vector quantization replaces each mouth opening image of a sequence by the index of the closest image in the codebook, thus creating a vector of indices representing the sequence. Recognition is done by computing the distance between vector quantized word samples and every vector quantized word model. The model with smallest distance represents the sample.

In Finn and Montgomery's approach [2] twelve dots were placed around the mouth of a speaker and tracked during the experiments; a total of fourteen distances were measured, and used as a feature vector. The data were normalized relative to time and overall amplitude of distance measurements. The recognition consisted of computing a total root mean square value between two utterances: the model with smallest difference was considered the correct model.

A different scheme was developed by Mase and Pentland [5]. They observed that the most important features that affect mouth shape relate to the elongation of the mouth, and to the mouth opening, affecting upper and lower lips. Using optical flow, the authors expressed the two principal types of motions of the mouth as functions with respect to time: mouth opening $O(t)$ and elongation of the mouth $E(t)$. $O(t)$ and $E(t)$ are computed in each frame, then smoothed and normalized to a fixed variance. Word boundaries were taken to be times when $O(t) = 0$, i.e. when the mouth is closed, and can easily be located on the $O(t)$ plots. Templates were used for recognition, and matching was performed, after a resampling step that normalizes the time to speak one word (time warping).

Kirby et al. [4] used a linear combination of the fixed set of eigenvectors of the ensemble averaged covariance matrix to express mouth images. A spoken word made up of P images can then be expressed as a $Q \times P$ matrix of coefficients computed with respect to the set of Q eigen images. A template matching technique was then used for identification of particular words.

Goldschen [3] used visual information from the oral-cavity shadow of the speaker for continuous speech recognition. His system uses Hidden Markov Models for distinguishing optical information. The HMM's were trained to recognize a set of sentences using visemes, trisemes (triplets of visemes), and generalized trisemes (clustered trisemes).

Bregler and Konig [1] created a hybrid system that uses both acoustic and visual information. They use a procedure similar to "snakes" and "deformable templates" to locate and track the lip contours. Then either the principal components of the contours are used, or the principal components of a gray level matrix centered around the lips are used. The latter uses matrix coding in an approach they termed "Eigenlips". Their work showed improvement for the combined architecture over just acoustic information alone in the presence of noise.

3. Eigensequences

Eigen decomposition represents the signal by a linear combination of a set of statistically independent orthogonal bases. This representation is most compact in the sense that, taking few number of the bases, the proportion of the signal energy projected on them is statistically maximum among all possible linear decompositions.

Consider a sequence of mouth images, $I_1, I_2, \ldots, I_P$, where each image has M rows and N columns, and P is the number of frames in the sequence. The gray level value of all pixels are then collected throughout the sequence in a long vector (of size MNP) as follows:

$$u^j = (I_1(1,1), \ldots, I_1(M,N), I_2(1,1), \ldots, I_2(M,N), \ldots, I_P(1,1), \ldots, I_P(M,N)),$$

where $I_n(x, y)$ is the value of the pixel at location (x, y) in frame n. Matrix A is then made from these vectors, u^j as follows:

$$A = \left[u^1, u^2, \ldots, u^s \right]. \tag{1}$$

The eigenvectors of a matrix $L = AA^T$ are defined as

$$L\phi_i = \lambda_i \phi_i \quad 1 \leq i \leq n$$

where ϕ_i is the eigenvector and λ_i is the corresponding eigenvalue. The eigenvectors ϕ_i are called the *eigensequences*.

The matrix L is a $MNP \times MNP$ matrix, which is exceedingly large even for small M and N. However, eigenvectors, ϕ_i, can be computed from matrix $C = A^T A$, which is a smaller matrix, $s \times s$, where s is the number of sequences. Let α_i and λ_i respectively be the eigenvector and eigenvalue of matrix C, then

$$C\alpha_i = \lambda_i \alpha_i, \tag{2}$$
$$A^T A\alpha_i = \lambda_i \alpha_i. \tag{3}$$

Premultiplying the above by A, we get

$$AA^T(A\alpha_i) = \lambda_i(A\alpha_i). \tag{4}$$

Since $L = AA^T$, $A\alpha_i$ are the eigenvectors of L.

In equation 1 above we have assumed that all u^j vectors, hence sequences, are of the same length. Since this can not be guaranteed in the real world, we apply a warping algorithm [9] to obtain sequences of equal length.

Any unknown sequence, u^x, can be represented as a linear combination of eigensequences as follows:

$$u^x = \sum_{i=1}^{n} a_i \phi_i. \tag{5}$$

The linear coefficients, a_i, can be computed by finding the dot product of vector u^x with the eigensequences as:

$$a_i = u_x^T . \phi_i \quad 1 \le i \le n \tag{6}$$

4. Model Generation and Matching

We use eigensequences to solve the lipreading problem. First, several training sequences for each spoken letter are used to compute eigensequences, and the Q (typically 3 to 5) most significant eigensequences are selected and used as a model in recognition. Assume that we are given a novel sequence, representing an unknown spoken letter. In order to recognize this sequence, we first determine its projection on the eigenspace of model letters (by computing the linear coefficients, a_i's), then compute the energy (described below) for each possible match. The letter with the highest energy is selected as a possible match.

In our approach, each model is a set of eigensequences, e.g., the model for letter ω is a set $\left\{\phi_1^\omega, \phi_2^\omega, \ldots, \phi_Q^\omega\right\}$, where the superscript denotes the letter, and the subscript denotes the eigensequence number. To generate a model, one training sequences for each letter is selected as a reference sequence, and the remaining training sequences for that letter are warped to the reference sequence to obtain fixed length sequences.

The projection of a novel sequence, u^x, on the eigenspace of letter ω is given by:

$$a_i^\omega = u_x^T . \phi_i^\omega, \quad 1 \le i \le Q; \quad \omega \in \{A, \ldots, Z\}. \tag{7}$$

Note that before computing this projection, the novel sequence, u^x, is warped to the reference sequence of letter ω in order to make it of the same length. Due to slight head movement during the utterances, the frames in the novel sequence may be spatially misaligned with respect to the reference model sequence. We compensate for this by registering the frames using maximum correlation. Correlation of a window in the novel frame is computed in the reference frame for possible displacement of ± 20 pixels in the y direction, and the best displacement is selected to register the frames.

Let the energy of projection of u^x on the eigenspace of letter ω be $\sum_{i=1}^{Q} (a_i^\omega)^2$, and the energy of u^x is $||u^x||$. We will use the ratio of these two energies, E^ω, defined below for matching.

$$E^\omega = \frac{\sum_{i=1}^{Q} (a_i^\omega)^2}{||u^x||} \tag{8}$$

For a perfect match, this ratio will be equal to 1. E^ω is computed for all model letters, and the letter with the highest energy is selected as a match.

As stated earlier, we are using the Q most significant eigensequences in our method. If we use all the eigensequences, then the novel sequence u^x can be expressed as:

$$u^x = \sum_{i=1}^{n} b_i \phi_i \quad 1 \le i \le n. \tag{9}$$

where a_i's and b_i's are related as follows:

$$a_i = \begin{cases} b_i & \text{if } 1 \le i \le Q \\ 0 & \text{otherwise} \end{cases} \tag{10}$$

Now, consider the normalized distance between u^x and its projection

$$D = \frac{\sum_{i=1}^{n} (b_i - a_i)^2}{\sum_{i=1}^{n} b_i^2}, \tag{11}$$

which is equivalent to

$$D = 1 - \frac{\sum_{i=1}^{Q} a_i^2}{||u^x||} \tag{12}$$

where $||u^x|| = \sum_{i=1}^{n} b_i^2$. Consequently, minimizing the normalized distance D is equivalent to maximizing the energy ratio, E^ω, defined in equation 8.

It is important to use the *normalized* distance in matching. As noted earlier, before a novel sequence is projected onto each model eigenspace, it is spatially registered to the reference sequence of that model. Therefore,

the part of the (spatial) sequence taking part in the projection is not fixed
and may vary with each mapping. Therefore, the absolute distance is not
suitable.

The recognition process, in summary, requires that, first, the models be
generated. For model generation, we

- Warp all the training sequences for a spoken letter with respect to a
  reference sequence.
- Perform spatial registration using correlation.
- Represent each letter by its Q most significant eigensequences:
  $\left\{ \phi_1^\omega, \phi_2^\omega, \ldots, \phi_Q^\omega \right\}$.

Then, for matching, the following steps are taken:

- Warp the unknown sequence with respect to the reference sequence of
  each model.
- Perform spatial registration.
- For each model, compute the ratio of the energy of the projection of
  the unknown sequence into the model's eigenspace, and the energy of
  the unknown sequence.
- Determine best match by the maximum of all ratios computed.

5. Modular Vs. Global Eigenspaces

In the space of all possible sequences, the lip sequences map to the clusters
of individual letters. The task of lipreading then becomes that of determining which cluster an unknown sequence belongs to. We can use two
methods of eigen representation. One method is to compute the eigenvectors of the entire space (*global eigenspace*) and discriminate the lip patterns
by the distance to the respective cluster centers. The other is to use the
modular eigenspace, in which the principal eigenvectors which give the most
compact description of individual clusters are constructed, and the distance
from the input to the subspaces spanned by the principal eigensequences is
used.

We use modular eigenspaces in our approach, that is, separate eigensequences are computed for each spoken letter. The global approach would use
training sequences of all letters to compute global eigensequences. As noted
earlier, before computing eigensequences, we must convert all the sequences
to some fixed length. An important advantage of the modular eigenspace is
that sequences for construction of each model are only warped among that
group. Whereas in the global approach, it is difficult to select any reference
letter to which all other sequences can be warped, because the sequences
significantly differ from each other.

6. Extracting letters from connected sequences

The approach used in this chapter treats each spoken letter as a basic unit for recognition. It is assumed that the lip movements for a given letter can be expected to follow similar spatiotemporal patterns. Consequently, a good method for automatically isolating and extracting letters from a continuous sequence is needed for successful recognition.

For simplicity, we assume that our task is to recognize independent letters from lip sequences. The speaker is required to begin each letter with the mouth in the closed position, a constraint which was enforced with no difficulty during the experiments. The separation of the letters is based on the temporal variation, between successive frames. This is determined by computing the *average absolute intensity difference function*, $f(n)$, as defined below:

$$f(n) = \frac{1}{MN} \sum_{x=1}^{M} \sum_{y=1}^{N} ||I_n(x,y) - I_{n-1}(x,y)|| \tag{13}$$

Figure 1 shows the plot of the average frame difference function, f, for a connected sequence seq-a. From this plot, it is easy to see that the value of f during the articulation intervals is not necessarily greater than that during the non-articulation intervals, so separation of letters by using direct thresholding will not succeed. However, we note that the articulation intervals in this function correspond to clusters of big peaks and the non-articulation intervals correspond to the valleys between peaks, which may also have small local peaks.

Our approach begins with separating those clusters of peaks. First, the frame difference function, f, is smoothed to obtain function g. Then the global valleys are detected in g. These valleys occur between two consecutive letters. For each valley in g, starting from the frame number corresponding to the location of a valley in g, the hillside on the left and the hillside on the right in f, where f crosses a preset threshold, are identified. Next the first valley on left of right hillside, and the first valley on the right of left hillside in f are determined. The left valley is the end of a previous letter, and the right valley is the beginning of the next letter. The threshold, T, used for determining hillsides in f should satisfy the following constraint:

$$\max_i(p_I(i)) < T \leq \min_j(p_L(j)),$$

where, p_I is the value of a local peak in the non-articulation interval and p_L is value of a (left-most and right-most) outer-most peak during the articulation. Since the outermost peaks usually are large, a large margin can be allowed for the setting of T.

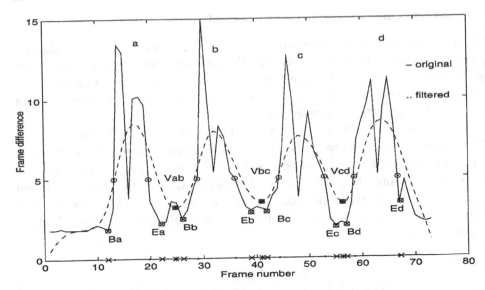

Figure 1. Plots of the frame difference function, f, and smoothed (filtered) version of f, g, for connected sequence seq-a. The valleys in g are shown as Vab, Vbc, Vcd. The beginning and endings of letters "A", "B", "C" and "D" respectively are shown as Ba, Ea, Bb, Eb, Bc, Ec, Bd and Ed.

The plots for f and g and detected beginning and end of letters in connected sequences a and b are shown in Figures 1 and 2.

7. Warping

Warping is used twice in our method for lipreading. First, during the generation of model eigensequences, second during the matching of a novel sequence with the model eigensequence. In this section, we briefly describe warping. Temporal warping of two sequences uses the Dynamic Programming Algorithm of Sakoe and Chiba [9]. The columns of each frame of a sequence are concatenated to form one vector, and a sequence of vectors is created. Thus for each pair of sequences we have:

$$A = [a_1, a_2, \ldots, a_i, \ldots, a_I]$$
$$B = [b_1, b_2, \ldots, b_j, \ldots, b_J]$$

where a_n is the nth vector of sequence A, and b_n is the nth vector of sequence B.

The algorithm employed uses the *DP-equation* in symmetric form with a slope constraint of 1. Therefore, g(i,j) is computed as follows:

Initial Condition:

$$g(1,1) = 2d(1,1)$$

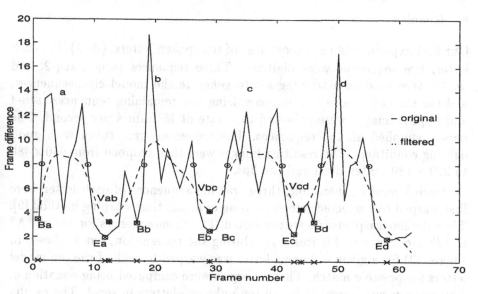

Figure 2. Plots of the frame difference function, f, and smoothed (filtered) version of f, g, for connected sequence seq-b. The valleys in g are shown as Vab, Vbc, Vcd. The beginning and ending of letters "A", "B", "C" and "D" respectively are shown as Ba, Ea, Bb, Eb, Bc, Ec, Bd and Ed.

where $d(i, j) = ||a_i - b_j||$.

The *DP-equation* we used:

$$g(i, j) = min \begin{bmatrix} g(i - 1, j - 2) + 2d(i, j - 1) + d(i, j) \\ g(i - 1, j - 1) + 2d(i, j) \\ g(i - 2, j - 1) + 2d(i - 1, j) + d(i, j) \end{bmatrix}$$

The minimum equation used for the calculation of g at point (i, j) gives the path from the previous point to the current point, thus creating a path from $(1, 1)$ to (I, J). Each point on this path indicates which frames from the input sequence match to frames in the reference sequence, which creates a warped sequence that uses the frames of the input sequence and is the same length as the reference sequence. Having a slope constraint of 1 allows for three possible cases for matching the lengths of the sequences. Two frames from the input sequence match to one frame in the reference sequence, in which case the two input frames are averaged to create one. There is a one-to-one correspondence, in which case the input frame is unchanged. Or, one frame from the input sequence matches to two frames in the reference sequence, in which case the input frame is just repeated.

8. Results

Our first experiments used sequences of ten spoken letters. (A–J). For each letter, five sequences were digitized. Three sequences (seq-1, seq-2, and seq-3) were used as a training set to generate the model eigensequences, and the method was tested on recognizing two remaining sequences (seq-4 and seq-5). Images were collected at a rate of 15 frames per second. One person supplied all the sequences. The sequences were taken with good lighting conditions. The resulting images were then cropped from 640 × 480 to 220 × 180 centered around the lips.

During model generation, three training sequences of each letter were first warped to a selected reference using dynamic time warping method [9]. Then the eigensequences were computed. The eigensequences for letters "A" to "J" are shown in Figures 3-12. During the recognition, seq-4 (shown in Figure 13) for a given unknown letter was warped to each of the ten model letters for possible match. Then, energies were computed using equation 8. This process was repeated for all ten unknown letters in seq-4. The results for matching are summarized in Figure 17. The matching of seq-5 (shown in Figure 14) was performed as for seq-4, and the results are summarized in Figure 18. The recognition rate is 90% for both sequences.

We also experimented with two connected sequences shown in Figures 15, and 16. First, the method discussed in section 6 was used to isolate spoken letters. Then extracted sequence corresponding to each letter was matched with the models as discussed above. The results are summarized in the tables shown in Figures 19-20. The recognition rate is 100% for both connected sequences.

Note that the values of energy ratios shown in the tables in Figures 17–20 are densely centered around 1, therefore high precision is needed to distinguish between them. This can be easily improved by subtracting the average image from each frame of the sequences, including frames in the models and the unknown sequences. The subtraction will not alter the relative distance between the sequences but reduce their energy on the whole, so the dynamic range of the representative energy ratio will be increased.

In our next set of extensive experiments, we studied the effect of reduced resolution on our method. In this case, we also used sequences of ten spoken letters (A–J). For these tests, we digitized twenty connected sequences, where each sequence was of the letters A through J. The average frame difference function was used to isolate the individual letters from these sequences. The images were cropped as stated above, but during the warping, every image was reduced in size to 29 × 19. Ten sequences for each letter were used for model generation, and the remaining ten were used as unknown sequences.

The results are shown in the table in Figure 21. Using only the highest ratios for matches, we had 10 matches out of a possible 10 for the letters "A", "B", "F", and "H". There were 8 of out 10 matches for letters "E", "G", and "J", while "I" matched 7 times, and "C" and "D" matched 6. Considering the second highest ratios, the letters "C", "D", and "I" had 3 additional matches, and "E", "G", and "J" had 1. And going to the third highest ratios, the letters "C" and "E" had 1 additional match. Thus, considering only the highest ratios as matches, we achieved an 83% correct recognition rate. Considering the two top ratios for matching, that rate jumps to 95%, and allowing for the three top ratios gives a recognition rate of 97%.

9. Conclusions

We presented a method for lipreading which uses eigensequences. In our approach, gray level values of all the pixels in all frames in a sequence representing a spoken letter are put in one large vector. Several such vectors corresponding to the training sequences are used to compute eigenvectors (eigensequence), for each spoken letter. The recognition of an unknown sequence representing a spoken letter is performed by computing the ratio of energy of projection of the sequence on the model eigenspace and the energy of the sequence.

Future work will include the experimentation of the proposed method with sequences of other letters "K" to "Z", and digits "0" to "10". Since the proposed spatiotemporal eigen decomposition results in a more compact representation, it can also be used to solve the image compression problem.

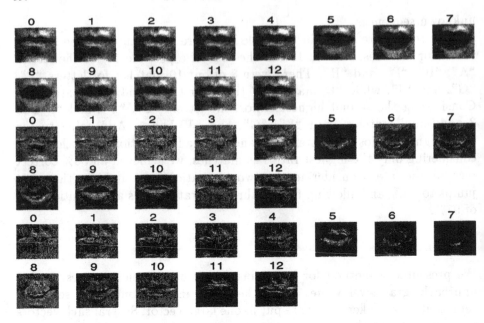

Figure 3. Three eigensequences of letter "A".

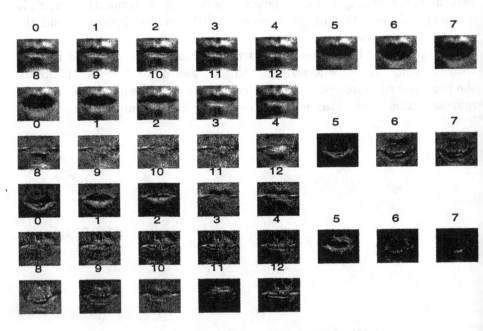

Figure 4. Three eigensequences of letter "B".

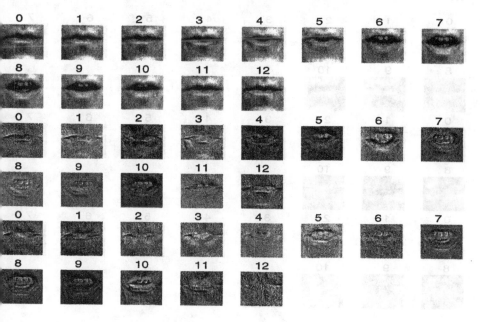

Figure 5. Three eigensequences of letter "C".

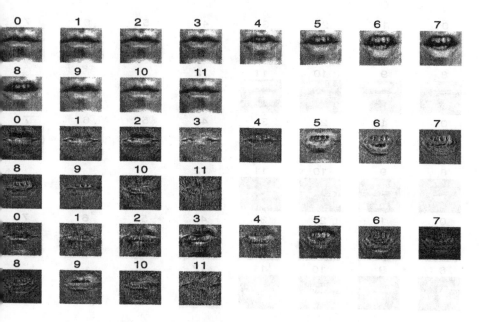

Figure 6. Three eigensequences of letter "D".

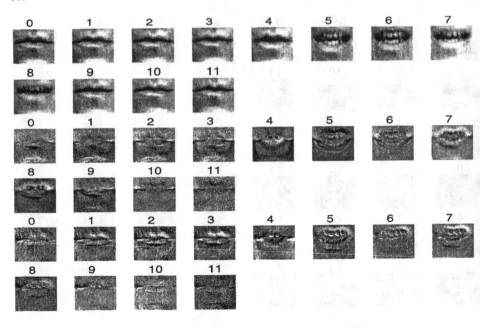

Figure 7. Three eigensequences of letter "E".

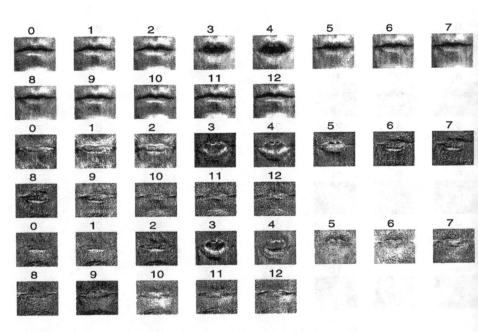

Figure 8. Three eigensequences of letter "F".

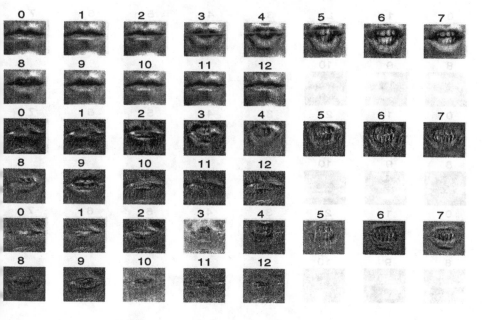

Figure 9. Three eigensequences of letter "G".

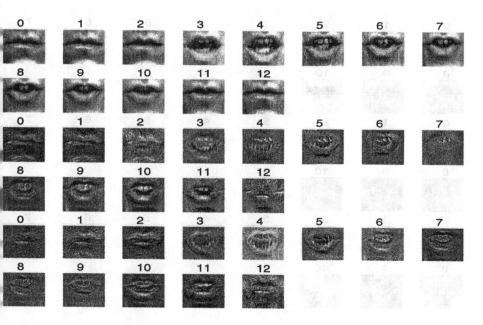

Figure 10. Three eigensequences of letter "H".

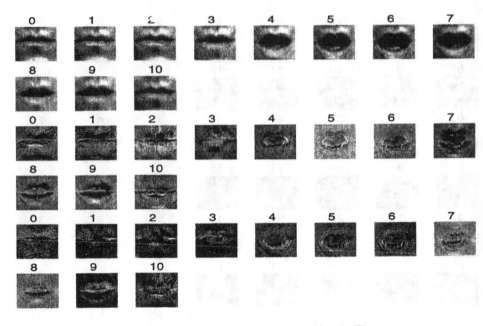

Figure 11. Three eigensequences of letter "I".

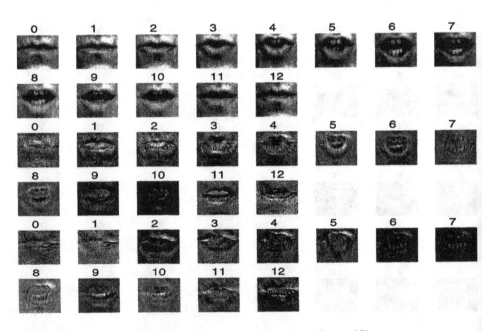

Figure 12. Three eigensequences of letter "J".

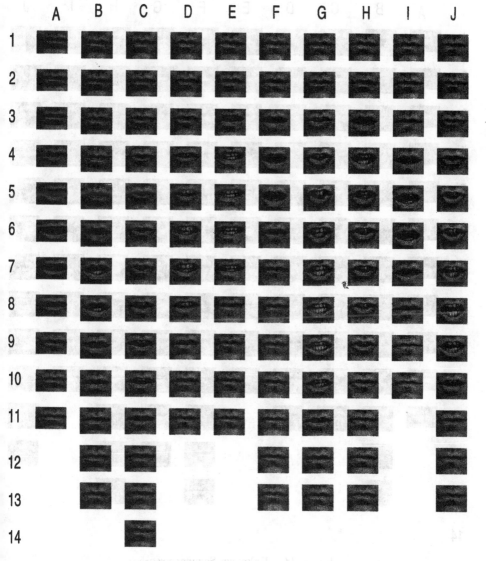

Figure 13. Sequence seq-4.

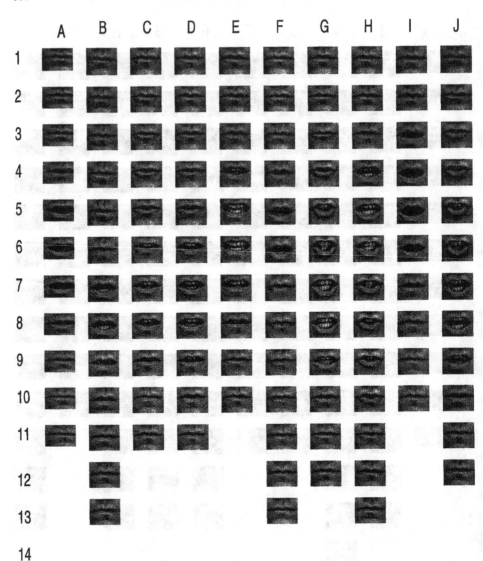

Figure 14. Unknown Sequence seq-5.

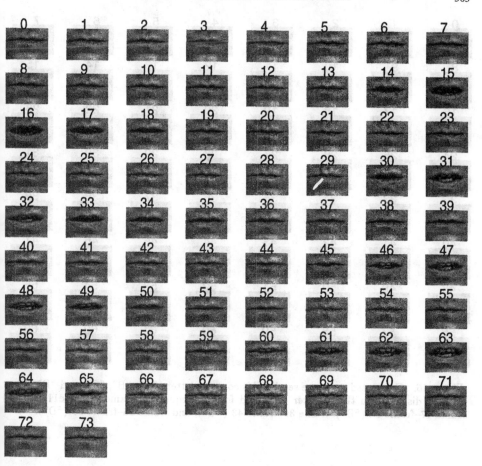

Figure 15. Connected sequence seq-a. This contains letters "A", "B", "C", and "D". The method discussed in this paper extracted four subsequences: frame 12–frame 22 ("A"), frame 26–frame 39 ("B"), frame 42–frame 55 ("C"), and frame 57–frame 67 ("D").

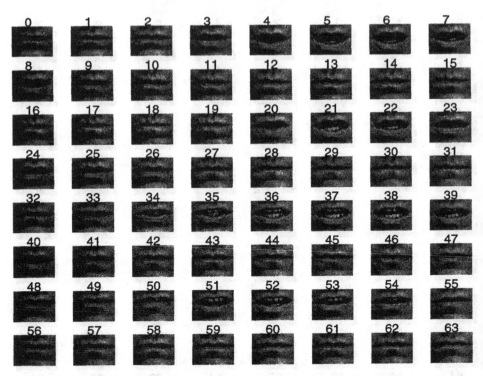

Figure 16.   Connected sequence seq-b. This contains letters "A", "B", "C", and "D". The method discussed in this chapter extracted four subsequences: frame 1–frame11 ("A"), frame 17–frame 29 ("B"), frame 30–frame 43 ("C"), and frame 46–frame 58 ("D").

References

1. C. Bregler and Y. Konig. Eigenlips for Robust Speech Recognition. In *Proc. IEEE Inter. Conf. on Acoustics, Speech, and Signal Processing*, 1994.
2. K. E. Finn and A. A. Montgomery. Automatic Optically-Based Recognition of Speech. *Pattern Recognition Letters*, 8:159–164, 1988.
3. Alan Jeffrey Goldschen. *Continuous Automatic Speech Recognition by Lipreading.* PhD thesis, George Washington University, School of Engineering and Applied Science, 1993.
4. M. Kirby, F. Weisser, and G. Dangelmayr. A Model Problem in the Representation of Digital Image Sequences. *Pattern Recognition*, 26(1):63–73, 1993.
5. K. Mase and A. Pentland. Lip Reading: Automatic Visual Recognition of Spoken Words. Technical Report 117, M.I.T. Media Lab Vision Science, 1989.
6. Murase, H. and Nayar, S. Illumination planning for object recognition in structured environment. In *IEEE CVPR-94*, pages 31–38, 1994.
7. Pentland, A., Moghaddam, B., Starner, T. View-based and modular eigenspaces for face recognition. In *IEEE CVPR-94*, pages 84–91, 1994.
8. E. D. Petajan, B. Bischoff, D. Bodoff, and N. M. Brooke. An Improved Automatic Lipreading System to Enhance Speech Recognition. In *SIGCHI '88: Human Factors in Computing Systems*, pages 19–25, October 1988.
9. H. Sakoe and S. Chiba. Dynamic Programming Algorithm Optimization for Spoken Word Recognition. *IEEE Trans. on Acoustics, Speech, and Signal Processing*, ASSP-26(1):43–49, February 1978.
10. Turk, M., and Pentland, A. Eigenfaces for recognition. *Journal of Cognitive Neuroscience*, pages 71–86, 1991.

Model	Unknown A	Unknown B	Unknown C	Unknown D	Unknown E
A	0.9975	0.9905	0.9859	0.9495	0.9748
B	0.9894	0.9926	0.9890	0.9579	0.9789
C	0.9887	0.9889	0.9971	0.9540	0.9815
D	0.9851	0.9817	0.9902	0.9966	0.9900
E	0.9840	0.9843	0.9858	0.9914	0.9903
F	0.9931	0.9927	0.9929	0.9873	0.9865
G	0.9813	0.9786	0.9884	0.9836	0.9833
H	0.9792	0.9801	0.9829	0.9843	0.9831
I	0.9948	0.9862	0.9799	0.9673	0.9582
J	0.9831	0.9825	0.9879	0.9827	0.9753
Best match	A	F	C	D	E

Model	Unknown F	Unknown G	Unknown H	Unknown I	Unknown J
A	0.9933	0.9827	0.9884	0.9936	0.9846
B	0.9892	0.9889	0.9880	0.9845	0.9878
C	0.9904	0.9912	0.9906	0.9893	0.9921
D	0.9880	0.9902	0.9906	0.9858	0.9887
E	0.9820	0.9876	0.9874	0.9838	0.9860
F	0.9981	0.9914	0.9936	0.9945	0.9895
G	0.9827	0.9968	0.9816	0.9797	0.9902
H	0.9875	0.9869	0.9976	0.9819	0.9861
I	0.9911	0.9759	0.9881	0.9952	0.9797
J	0.9825	0.9885	0.9819	0.9848	0.9937
Best match	F	G	H	I	J

Figure 17. Results for sequence seq-4. The entries in the table are the energy ratios, the perfect match has ratio equal to 1. Recognition is 90%. Every input letter, except for letter "B", was recognized correctly.

Model	Unknown A	Unknown B	Unknown C	Unknown D	Unknown E
A	0.9965	0.9910	0.9809	0.9882	0.9773
B	0.9893	0.9948	0.9882	0.9889	0.9882
C	0.9889	0.9864	0.9965	0.9942	0.9913
D	0.9852	0.9804	0.9941	0.9967	0.9927
E	0.9857	0.9863	0.9875	0.9893	0.9944
F	0.9939	0.9925	0.9932	0.9811	0.9919
G	0.9821	0.9761	0.9779	0.9883	0.9801
H	0.9778	0.9796	0.9895	0.9901	0.9859
I	0.9953	0.9873	0.9659	0.9774	0.9649
J	0.9842	0.9836	0.9879	0.9879	0.9846
Best match	A	B	C	D	E

Model	Unknown F	Unknown G	Unknown H	Unknown I	Unknown J
A	0.9914	0.9841	0.9870	0.9948	0.9877
B	0.9867	0.9859	0.9882	0.9814	0.9854
C	0.9872	0.9906	0.9885	0.9886	0.9921
D	0.9781	0.9875	0.9884	0.9788	0.9878
E	0.9839	0.9836	0.9821	0.9825	0.9847
F	0.9962	0.9914	0.9917	0.9920	0.9915
G	0.9772	0.9745	0.9751	0.9771	0.9904
H	0.9809	0.9873	0.9949	0.9804	0.9825
I	0.9920	0.9773	0.9868	0.9949	0.9806
J	0.9774	0.9879	0.9813	0.9794	0.9933
Best match	F	F	H	I	J

Figure 18. Results for sequence seq-5. The entries in the table are the energy ratios, the perfect match has ratio equal to 1. Recognition is 90%. Every input letter except letter "G", which was matched to letter "F", was recognized correctly.

Model	Unknown A	Unknown B	Unknown C	Unknown D
A	0.9949	0.9846	0.9770	0.9770
B	0.9920	0.9928	0.9822	0.9818
C	0.9921	0.9873	0.9930	0.9907
D	0.9904	0.9878	0.9910	0.9926
E	0.9846	0.9800	0.9846	0.9875
F	0.9925	0.9901	0.9904	0.9815
G	0.9826	0.9819	0.9836	0.9838
H	0.9860	0.9839	0.9860	0.9831
I	0.9903	0.9710	0.9715	0.9650
J	0.9856	0.9821	0.9848	0.9817
Best match	A	B	C	D

Figure 19. Results for connected sequence seq-a. This sequence contained letters "A", "B", "C" and "D". First the subsequence corresponding to these letters were extracted using the method described in the chapter, then the extracted subsequence were matched with ten model letters. The recognition is 100%.

Model	Unknown A	Unknown B	Unknown C	Unknown D
A	0.9953	0.9856	0.9871	0.9833
B	0.9895	0.9927	0.9900	0.9847
C	0.9914	0.9844	0.9911	0.9901
D	0.9864	0.9825	0.9894	0.9921
E	0.9824	0.9779	0.9876	0.9884
F	0.9911	0.9884	0.9879	0.9895
G	0.9797	0.9760	0.9869	0.9856
H	0.9843	0.9790	0.9826	0.9853
I	0.9935	0.9797	0.9819	0.9670
J	0.9872	0.9840	0.9874	0.9840
Best match	A	B	C	D

Figure 20. Results for connected sequence seq-b. This sequence contained letters "A", "B", "C" and "D". First the subsequence corresponding to these letters were extracted using the method described in the chapter, then the extracted subsequence were matched with ten model letters. The recognition rate is 100%

	A	B	C	D	E	F	G	H	I	J
1st	10	10	6	6	8	10	8	10	7	8
2nd			3	3	1		1		3	1
3rd			1		1					

Figure 21. Results for reduced resolution series of tests using 10 unknown sequences. The numbers indicate the number of matches out of 10 that were correct. The first row shows results for the highest ratios. The second row shows the number of additional matches from the second highest ratios. And the last row shows the same for the third highest ratios. The success rate is 83% from the highest ratios, 95% from top two highest ratios, and 97% from top three highest ratios.

Computational Imaging and Vision

1. B.M. ter Haar Romeny (ed.): *Geometry-Driven Diffusion in Computer Vision.* 1994   ISBN 0-7923-3087-0
2. J. Serra and P. Soille (eds.): *Mathematical Morphology and Its Applications to Image Processing.* 1994   ISBN 0-7923-3093-5
3. Y. Bizais, C. Barillot, and R. Di Paola (eds.): *Information Processing in Medical Imaging.* 1995   ISBN 0-7923-3593-7
4. P. Grangeat and J.-L. Amans (eds.): *Three-Dimensional Image Reconstruction in Radiology and Nuclear Medicine.* 1996   ISBN 0-7923-4129-5
5. P. Maragos, R.W. Schafer and M.A. Butt (eds.): *Mathematical Morphology and Its Applications to Image and Signal Processing.* 1996

                                                          ISBN 0-7923-9733-9
6. G. Xu and Z. Zhang: *Epipolar Geometry in Stereo, Motion and Object Recognition.* A Unified Approach. 1996   ISBN 0-7923-4199-6
7. D. Eberly: *Ridges in Image and Data Analysis.* 1996   ISBN 0-7923-4268-2
8. J. Sporring, M. Nielsen, L. Florack and P. Johansen (eds.): *Gaussian Scale-Space Theory.* 1997   ISBN 0-7923-4561-4
9. M. Shah and R. Jain (eds.): *Motion-Based Recognition.* 1997

                                                          ISBN 0-7923-4618-1

Kluwer Academic Publishers – Dordrecht / Boston / London